A Guide to the Breeding of Tropical Silk Moths (Lepidoptera: Saturniidae)

Die Zucht von tropischen Wilden Seidenspinnern (Lepidoptera: Saturniidae)

Frank Meister

English Translation by Frank Haase

In memory of

Rudolf E. J. Lampe

12.04.1933 – 12.07.2011,

one of the most outstanding saturniid breeders
of recent decades

In Erinnerung an

Rudolf E. J. Lampe

12.04.1933 – 12.07.2011,

einen der bedeutendsten Saturniiden-Züchter
der letzten Jahrzehnte

A Guide to the Breeding of Tropical Silk Moths

(Lepidoptera: Saturniidae)

Die Zucht von tropischen Wilden Seidenspinnern

(Lepidoptera: Saturniidae)

Frank Meister

English Translation by
Frank Haase

Verlag Dr. Friedrich Pfeil

Impressum

Bibliographic information published by the Deutsche Nationalbibliothek

The Deutsche Nationalbibliothek lists this publication in the Deutsche Nationalbibliografie; detailed bibliographic data are available in the Internet at http://dnb.d-nb.de.

Bibliografische Information der Deutschen Nationalbibliothek

Die Deutsche Nationalbibliothek verzeichnet diese Publikation in der Deutschen Nationalbibliografie; detaillierte bibliografische Daten sind im Internet über http://dnb.d-nb.de abrufbar.

Author's address

Frank Meister
Sternstraße 15
17291 Prenzlau, Germany
e-mail: fmeister66@gmx.de

Translator's address

Frank Haase
Biegerstraße 3
97478 Knetzgau, Germany

Anschrift des Autors

Frank Meister
Sternstraße 15
17291 Prenzlau
E-Mail: fmeister66@gmx.de

Anschrift des Übersetzers

Frank Haase
Biegerstraße 3
97478 Knetzgau

Cover photograph

Larva of *Gonimbrasia hecate* Rougeot, 1955. – Photograph courtesy of Andreas Riekert.

Titelbild

Raupe von *Gonimbrasia hecate* Rougeot, 1955. – Foto: Andreas Riekert, mit freundlicher Erlaubnis.

Pre-Press: Verlag Dr. Friedrich Pfeil, München
Press: Advantage Printpool, Gilching

Printed in the European Union

ISBN 978-3-89937-140-6

Printed on permanent and durable acid-free paper

Verlag Dr. Friedrich Pfeil
Wolfratshauser Straße 27
81379 München, Germany
phone: +49 89 742827-0
fax: +49 89 7242772
e-mail: info@pfeil-verlag.de
www.pfeil-verlag.de

Druckvorstufe: Verlag Dr. Friedrich Pfeil, München
Druck: Advantage Printpool, Gilching

Printed in the European Union

ISBN 978-3-89937-140-6

Gedruckt auf alterungsbeständigem und säurefreiem Papier

Verlag Dr. Friedrich Pfeil
Wolfratshauser Straße 27
81379 München
Tel. +49 89 742827-0
Fax +49 89 7242772
E-Mail: info@pfeil-verlag.de
www.pfeil-verlag.de

Contents / Inhalt

Foreword

Very little has appeared in print for decades about the methodology of breeding the Saturniidae. The more recent of the older works to appear in book form are those of Arnold Voelschow 1902, A. U. E. Aue 1933, Brian O. C. Gardiner 1982 and Stephen E. Stone 1991. There is no doubt that these standard works on the breeding of saturniids belong in the library of every breeder of this family. Nevertheless, some of the experiences described therein have since been superseded and are, to some extent, obsolete. Among other things, this is because earlier authors had less material available for breeding than we do now. A very helpful book is Ekkehard Friedrich's "Handbuch der Schmetterlingszucht", published in 1975. This book describes in detail, with many helpful illustrations, the raising of moths from the egg to the adult. It remained unsurpassed for many years until Rudolf E. J. Lampe's "Saturniidae of the World – Pfauenspinner der Welt" appeared in 2010. To be sure, this book makes only an initial attempt at describing the methodology of breeding, yet the almost three thousand photos of saturniids and their pre-imaginal stages are unique in their complexity and quality – the result of more than forty year's experience of working with this family. This is the first time an author has provided such a wealth of details about the course of development, the times of molting, and the plants suitable as food (c.f. Lampe 2010).

In just the last ten years, an enormous amount of breeding material has reached Europe – tropical saturniids that have never before been bred. The material derives from either the expeditions of entomological hobbyists around the world or from the captures of native hunters in their own countries. What in earlier times seemed impossible is today self-evident. Within only twenty-four hours, almost any place on Earth can be reached by airplane. In recent years, the infrastructure of many tropical countries has improved enormously. Traveling has become completely problem-free. World-wide, the medium of the internet has made breeding material available immediately after mating through a number of sources like www.actias.de or www.insectnet.com. International postal service and the enormous diminution of shipping time make

Vorwort

Seit einigen Jahrzehnten wurde nicht mehr sehr viel über die Methodik der Zucht von Saturniiden geschrieben. Die letzten älteren Werke in Buchform dazu stammen von Arnold Voelschow, 1902, A. U. E. Aue, 1933, Brian O. C. Gardiner, 1982 oder Stephen E. Stone, 1991. Diese Standardwerke der Saturniiden-Zucht gehören ohne Zweifel in jede Bibliothek eines Züchters dieser Familie. Die dort beschriebenen Erfahrungen sind heute dennoch teilweise überholt und manchmal nicht mehr sehr zeitgemäß. Das liegt unter anderen auch daran, dass den Autoren damals noch nicht so viel Zuchtmaterial zur Verfügung stand wie heute. Ein sehr hilfreiches Buch ist das 1975 erschienene Werk: »Handbuch der Schmetterlingszucht« von Ekkehard Friedrich. Es wird in diesem Buch, in sehr detaillierter Weise, ergänzt mit hilfreichen Abbildungen, über die Methodik der Zucht vom Ei bis zum Falter berichtet. Für viele Jahre unübertroffen dürfte das im Jahr 2010 erschienene Buch »Saturniidae of the World – Pfauenspinner der Welt« von Rudolf E. J. Lampe sein. Zwar wird im Buch nur ansatzweise über die Zuchtmethodik geschrieben, aber die fast 3000 Abbildungen von Saturniiden und ihren Präimaginalstadien aus über 40 Jahren Zuchterfahrung sind einzigartig in ihrer Komplexität und Qualität. Erstmalig liefert hier ein Autor auch detaillierte Angaben zum Zuchtverlauf, den Häutungsdaten und den Futterpflanzen (s. Lampe, 2010).

Gerade in den letzten 10 Jahren gelangte Zuchtmaterial von einer ungeheueren Vielzahl noch nie gezüchteter tropischer Saturniiden-Arten nach Europa. Das Material stammte entweder aus Expeditionen von Hobbyentomologen in alle Teile der Welt oder von einheimischen Fängern in den Ländern selbst. Was früher unmöglich schien, ist heute selbstverständlich. Mit dem Flugzeug erreicht man heute binnen 24 Stunden fast jeden Ort der Welt. Die Infrastruktur in vielen tropischen Ländern hat sich in den letzten Jahren enorm verbessert. Das Reisen ist unproblematisch geworden. Durch das Medium Internet kann Zuchtmaterial heute auf diversen Plattformen wie www.actias.de oder www.insectnet.com sofort nach der Kopula einer breiten Öffentlichkeit in der ganzen Welt angeboten werden. Die enorme Verkürzung von Transferzeiten im weltweiten

Fig. 1. Larva of *Automeris egeus* (Kramer, 1775). – Photograph courtesy of Andreas Riekert.
Abb. 1. Raupe von *Automeris egeus* (Kramer, 1775). – Foto: Andreas Riekert, mit freundlicher Erlaubnis.

Fig. 2. Larva of *Nudaurelia capdevillei* Rougeot, 1979. – Photograph courtesy of Andreas Riekert.
Abb. 2. Raupe von *Nudaurelia capdevillei* Rougeot, 1979. – Foto: Andreas Riekert, mit freundlicher Erlaubnis.

Fig. 3. Larva (L4) of *Caio harrietae* (Forbes, 1944), Peru. – Photograph by Frank Meister.
Abb. 3. Raupe (L4) von *Caio harrietae* (Forbes, 1944), Peru. – Foto: Frank Meister.

Fig. 4. Larva of *Bunaea alcinoe alcinoe* (Cramer, 1780). – Photograph courtesy of Carsten Anderssohn.
Abb. 4. Raupe von *Bunaea alcinoe alcinoe* (Cramer, 1780). – Foto: Carsten Anderssohn, mit freundlicher Erlaubnis.

it possible to send eggs almost anywhere in the world and have them arrive in a week or less. So today it is quite easy for any breeder to obtain materials like eggs, pupae and cocoons. The breeding of tropical saturniids has boomed in the last ten years. A huge number of totally different species can be successfully bred. Unfortunately, most of these efforts in recent years have not been documented, or they have not been published. Either a scientifically significant source of photos of pre-imaginal stages and lists of edible plants for several hundred saturniid species slumber in secrecy or they are lost. Using the internet, sporadic attempts are made to document breeding. So numerous are the sources that a great deal of data in the form of text or in pictures have become available, but this material has very little organization. So it's almost impossible to quickly find, for example, the particular plant on which a certain species feeds. As a result many breeders prefer to search in standard works or in periodicals. Very helpful in this regard are the extensive descriptions of Paukstadt and Paukstadt throughout the 1990's in "Galathea" and "Entomologische Zeitschrift". The first, fully successful breedings of Asian saturniids were published here. Also noteworthy is the CD available from Kirby Wolfe of the USA, a world-famous breeder of tropical saturniids. Also from the USA, the internet site of Bill Oehlke (http://www3.islandtelecom.com/~oehlkew/) offers a great many illustrations of currently known species with their larval stages but little worthwhile advice about breeding. At least, details are given of the proper plants to feed the depicted species. Furthermore, the most important German sites, www.actias.de, and www.saturniidae-breeder.de provide information about the plants eaten by particular species. Additionally, there are a few internet sites not worth mentioning, although they would seem to be devoted to the theme of breeding. Sadly, what one all too often finds is only photos of pre-imaginal stages. Plants for fodder are only occasionally mentioned. One finds too often that compilations of information about breeding have only marginal value. Collections of tips on breeding are found only through searching. Quite often such themes are actively discussed in forums. Unfortunately, these discussions too often lack details or consist of clearly subjective opinions without evidence of definite value for the ultimate goal of breeding the Saturniidae. Consequently, beginners frequently have diffi-

Postverkehr ermöglicht es zudem, Eier aus fast allen Teilen der Welt innerhalb einer Woche an den Empfänger zu versenden. Somit ist es heute für jeden Züchter sehr leicht, Zuchtmaterial wie Eier, Puppen oder Kokons zu beschaffen. Die Zucht tropischer Saturniiden erfuhr so in den letzten zehn Jahren einen Boom. Eine sehr große Anzahl verschiedenster Arten konnte erfolgreich gezüchtet werden. Leider wurden die meisten Zuchten der letzten Jahre aber nicht dokumentiert oder noch nicht publiziert. Ein wissenschaftlich bedeutender Fundus an Fotos von Präimaginalstadien oder Futterpflanzenlisten mehrerer Hundert Saturniiden-Arten schlummert entweder im Verborgenen oder ist verloren gegangen. Im Internet gibt es viele sporadische Ansätze, Zuchten zu dokumentieren. Wie so häufig auf Internetplattformen werden große Datenmengen in Textform oder an Bildern frei zur Verfügung gestellt, aber oft wenig strukturiert. So ist es fast unmöglich, schnell an die gewünschten Informationen zu kommen, zum Beispiel auf der Suche nach einer Futterpflanze für eine bestimmte Art. So suchen dann viele Züchter lieber in den ihnen zur Verfügung stehenden Standardwerken oder Zeitschriften. Sehr hilfreich sind hier die umfangreichen Zuchtbeschreibungen von Paukstadt & Paukstadt Anfang bis Ende der 1990er-Jahre in »Galathea« und der »Entomologischen Zeitschrift«. Publiziert wurden vorwiegend erstmals erfolgreich durchgeführte Zuchten von asiatischen Saturniiden. Die Zuchten wurden sehr detailliert beschrieben, genauso wie die einzelnen Präimaginalstadien. Auch die von Kirby Wolfe, USA (einem der weltweit bedeutenden Züchter tropischer Saturniiden), aufgelegte CD ist in Qualität und Umfang bemerkenswert. Auf der sehr guten Internetseite von Bill Oehlke in den USA (http://www3.islandtelecom.com/~oehlkew/) findet man eine große Menge von Abbildungen der derzeit bekannten Arten an Saturniiden mit ihren Raupenstadien, aber wenige verwertbare Zuchthinweise. Zumindest werden von der überwiegenden Zahl der abgebildeten Arten Angaben zu den richtigen Futterpflanzen gemacht. Auch auf den bedeutendsten deutschen Seiten www.actias.de oder www.saturniidae-breeder.de kann man sich über Futterpflanzen einzelner Arten informieren. Auf einigen weiteren hier nicht erwähnenswert erscheinenden Internetseiten widmet man sich seit Jahren dem Thema der Zucht. Leider aber begnügt man sich allzu oft nur mit der Abbildung der Präimaginalstadien.

culty finding concrete advice about procedures. I have made many attempts to find such advice on the internet, but it is almost non-existent.

In this book, the attempt will be made to describe a method of propagation that can lead to success for most breeders. For this purpose, I have had discussions with breeders around the world whose experience and knowledge contribute to this work. Here is made available clear advice about the handling of the saturniids in connection with information gathered from around the world about the plants eaten by larvae in nature and in conditions of propagation. Therefore, it should be possible to repeat a great many of the successes already achieved in breeding the Saturniidae. This book provides an intensively researched checklist of publications describing saturniids whose names I consider valid. This list was expanded by searching through currently known synonyms, but I make no claim to absolute completion. Therefore the reader gains a good overview of the presently known species of the Saturniidae, whose number is continually changing. More than one hundred entomologists from around the world have advised me about the plants eaten by their caterpillars based in part on observations made in nature and partly from experiences in the breeding of saturniids.

The question of whether these plants are eaten in their native habitat or under the artificial conditions of breeding is intentionally disregarded here. Had it depended on me to investigate the significant issue of identifying the plants eaten by saturniids, then thousands of reports would not have been processed. Many breeders in various parts of the world obviously have a diverse supply of tropical plants in their gardens or greenhouses. An example is the avocado, *Persea americana*, which proved to be a very good food for the genus *Copaxa* Walker, 1855. Breeders from all over describe the avocado as a reliable source of fodder. Larvae of the genus *Copaxa* are to be found on avocado trees in their homeland of Middle and South America. Because of the enormous number of neophytes in Europe and the species transported to all parts of the world, distinguishing native plants from imports is often difficult for laypersons. Several breeders describe the false acacia, *Robinia pseudoacacia*, as a native European suitable for feeding larvae of the genus *Automeris* Hübner, [1819]. It is well-known how-

Futterpflanzen werden nur sporadisch genannt. Zusammenfassende Zuchthinweise findet man nur ansatzweise. Sehr oft wird in Foren rege zum Thema diskutiert. Leider verlieren sich diese Diskussionen sehr oft in Einzelheiten oder stark subjektiven Erfahrungen, ohne Beweis einer gewissen Gültigkeit für die überwiegende Zahl an Saturniiden-Zuchten. Einsteiger haben es oft sehr schwer, konkrete Handlungsanweisungen zu finden. Ich habe in mehreren Tests versucht, im Internet konkrete Handlungsanweisungen zu einem Thema der Zucht von Saturniiden zu finden. Es ist fast aussichtslos.

In diesem Buch wird nun der Versuch unternommen, eine Methodik der Zucht zu beschreiben, die bei den meisten Zuchten zum Erfolg führen könnte. Dazu wurde mit einer Vielzahl von Züchtern in der ganzen Welt gesprochen und diskutiert. Somit wurde es möglich, deren Erfahrungen und Kenntnisse mit in diese Arbeit einfließen zu lassen. Mit den hier zur Verfügung gestellten konkreten Handlungsanweisungen, in Verbindung mit den weltweit gesammelten Informationen zu den Futterpflanzennachweisen, in der Natur und unter Zuchtbedingungen, sollte es möglich sein, eine Vielzahl der bereits geglückten Zuchten von Saturniiden selbst zu wiederholen. Das Buch liefert eine sehr intensiv recherchierte Checkliste aller bis zum Publikationsdatum beschriebenen Saturniiden, deren Namen meiner Auffassung nach noch Gültigkeit haben. Diese Checkliste wurde ergänzt durch die recherchierten heute bekannten Synonyme, und erhebt nicht den Anspruch der absoluten Vollständigkeit. Somit bekommt der Leser einen guten Überblick über die bis heute bekannten Arten an Saturniiden, deren Zahl sich ständig ändert. Mehr als 100 Entomologen in der ganzen Welt haben mir ihre Futterpflanzen gemeldet, teilweise durch Beobachtungen in der Natur, teilweise durch Erfahrungen beim Züchten von Saturniiden.

Es wurde bewusst darauf verzichtet zu unterscheiden, ob es sich um Futterpflanzennachweise aus dem Habitat der Falter selbst handelt oder um Futterpflanzengaben unter Zuchtbedingungen. Hätte ich diese durchaus diskussionswürdigen Anforderungen an die Meldung von Futterpflanzen an mich gestellt, wären Tausende Futterpflanzenmeldungen nicht zu verarbeiten gewesen. Eine Vielzahl von Züchtern in der gan-

Fig. 5. Final instar larva of *Copaxa rufotincta* Rothschild, 1895. – Photograph courtesy of Horst Käch.
Abb. 5. Raupe von *Copaxa rufotincta* Rothschild, 1895. – Foto: Horst Käch, mit freundlicher Erlaubnis.

ever, that this tree was transplanted to Europe from America, a typical neophyte. In books like Stone's, all the plant species are listed in scientific order by families and higher taxa. That will not be intentionally ignored in this book. It is very easy today for any entomologist to explore the internet with his search machine and learn the family to which a genus of plants belongs. Higher classifications can likewise be found. By entering the genus and species names, one quickly reaches a rich trove of information about almost any plant that often includes illustrations. In this way it is possible to find closely related species. This is the simplest and irreplaceable way of obtaining information today. This book is intended as a source of support and assistance, setting breeders on the right path to the correct food plants. Correspondingly, I wish all entomologists enjoyment and success.

zen Welt besitzt offensichtlich einen gut sortierten Bestand an tropischen Gewächsen in ihren Gärten oder Gewächshäusern. Als Beispiel seien hier Avocado-Bäume, *Persea americana*, genannt, die als sehr gutes Futter für Raupen der Gattung *Copaxa* Walker, 1855 dienen. Züchter auf der ganzen Welt meldeten mir Avocado als Futterpflanze für selbst durchgeführte Zuchten. Die Raupen der Gattung *Copaxa* werden auch in ihrer Heimat Mittel- und Südamerika auf Avocado-Bäumen gefunden. Durch die enorme Anzahl von Neophyten in Europa und verschleppten Pflanzenarten in allen anderen Teilen der Welt ist die Entscheidung, was als einheimische Pflanzenart gilt und was nicht, für den Laien oft nicht mehr zu entscheiden. So meldeten viele Züchter Robinie *Robinia pseudoacacia* als einheimische, lokal gern genutzte Futterpflanze für Raupen der Gattung *Automeris* Hübner, [1819] in Europa. Dieser Baum wurde aber bekannterweise aus Amerika nach Europa eingeschleppt, ein typi-

scher Neophyt. In Büchern wie denen von Stone werden alle Pflanzenarten mit der übergeordneten Klasse, der Pflanzenfamilie, ergänzt. Auch hierauf wurde in diesem Werk bewusst nicht verzichtet. Es ist heute für jeden Entomologen sehr leicht über Suchmaschinen im Internet zu recherchieren, welcher Familie eine Pflanzengattung untergeordnet ist, auch höhere Klassifizierungen sind herausfindbar. Oder anders herum, durch Eingabe der Gattung und des Artnamens bekommt man sehr schnell umfangreiche Informationen über fast jede Pflanze, oftmals sogar mit Bildern im Internet. Es ist so auch ohne Weiteres möglich, nah verwandte Pflanzen zu finden. Diese Informationsbeschaffung ist heute die einfachste und nicht mehr ersetzbar. Dieses Buch kann unterstützen und es kann Hilfestellung sein, Züchter auf die richtige Fährte bei der Suche nach einer bestimmten Futterpflanze zu bringen. Dabei wünsche ich allen Entomologen Spaß und Erfolg.

Fig. 6. Larva (L3) of *Eacles tyrannus* Draudt, 1930. – Photograph courtesy of Horst Käch.
Abb. 6. Raupe (L3) von *Eacles tyrannus* Draudt, 1930. – Foto: Horst Käch, mit freundlicher Erlaubnis.

At this point I ask all the breeders in the world who read this book to share with me their own experiences. Further, I ask you to report on the plants used as food, whether based on observations in nature or gained through your own experience as a breeder. In this way it will be possible to expand this book in years to come. (E-mail: fmeister66@gmx.de).

Ich bitte an dieser Stelle alle Züchter in der ganzen Welt, die dieses Buch lesen, mir die selbst gemachten Zuchterfahrungen mitzuteilen. Ich rufe Sie dazu auf, alle Futterpflanzen zu melden, sei es durch Beobachtungen in der Natur oder durch selbst gemachte Erfahrungen beim Züchten. So wird es möglich sein, dieses Buch über die Jahre zu ergänzen. (E-Mail: fmeister66@gmx.de).

Acknowlegements

I am grateful to all the entomologists who contributed to the success of this work by their advice about edible plants or through reports of their experiences:

Danksagungen

Ich danke allen Entomologen, die durch Futterpflanzenmeldungen oder Erfahrungsberichte zum Gelingen dieses Werkes beigetragen haben:

Peggy Ackermann, Germany; Annette Aiello, Panama; Carsten Anderssohn, Germany; Dirk Bayer, USA; Patsy Bayer, USA; Martin Beeke, Germany; Thierry Bouyer, Belgium; Georg Brenner, Germany; Rainer Brenner, Germany; Ullrich Brosch, Germany; Christopher Conlan, USA; Matt Curtis, Canada; Bostjan Dvorak, Germany; Ronald Brechlin, Germany; Ian J. Kitching, U.K.; Carlos Mielke, Brazil; Ulf Drechsel, Paraguay; Otto Feldner, Kurt Fricke, Germany; Reinhard Foerster, Paraguay; Hector Garcia, Mexico; Caesar "The Tortuga", Mexico; Steve Ife, USA; Martin Jagelka, Germany; Prof. Daniel Janzen, USA; Taylor Jones, USA; Horst Kaech, Ecuador; Andreas Knorke, Germany; Steve Kohll, Luxemburg; Peter Kölzer, Germany; Rudolf E. J. Lampe†, Germany; Renate Lampe, Germany; Jeremy Laratro, USA; Robert Lemaitre, France; Bernd Liese, Germany; Timo Liese, Germany; Laela H. Paukstadt, Germany; Ulrich Paukstadt, Germany; Rainer Plontke, Germany; Alan Marson, U.K.; Hubert Mayer, Austria; Rainer Marx, Germany; André Meister, Germany; Christine Meister, Germany; Werner Meister, Germany; Franka Meister, Germany; Anika Meister, Germany; Bärbel Mönch, Germany; Hans Mönch, Germany; Heinz Niederleitner, Austria; Rolf Oberprieler, Namibia; Bill Oehlke, Canada; Richard Peigler, USA; Helmut Probst, Germany; Andreas Riekert, Germany; Dave Rolfe, U.K.; Pedro Rodrigo, Chile; Rodolphe Rougerie, France; Jan-Peter Rudloff, Germany; Johann Saksida, Germany; Pierre Schmit, France; Robert Schulze, Germany; Leroy Simon, USA; Michael Singer, Argentina; Victor Sinjaev, Russia; Werner Spiess, Germany; Felix Stumpe, Germany; Hubert Thöny, Brazil; Eric Van Schayck, Germany; Michael Weiss, Germany; Bernhard Wenczel, Switzerland; Ulrich Weritz, Germany; Kirby L. Wolfe, USA; Franz Ziereis, Germany;

and all other entomologists worldwide.

Special thanks are due to Frank Haase for the translation of the German text into English, to Ronald Brechlin for many years of valuable collaboration, to Ulrich Weritz for revising the theoretical portions of the text and for many pieces of helpful advice, Rodolphe Rougerie as representative for the collective team of the BOLD project, to the publisher, Friedrich H. Pfeil, and to his coworker, Hubert Hilpert, for their professional services in putting this work into literary form, and of course, to my entire family for their many years of understanding and tolerance that a saturniid breeder so urgently needs from his family, and which I have always received.

und allen anderen Entomologen weltweit.

Besonderer Dank gilt Herrn Frank Haase für die Übersetzung des deutschen Textes ins Englische, Ronald Brechlin für die vielen Jahre der guten Zusammenarbeit, Ulrich Weritz für die Überarbeitung des theoretischen Teils und die vielen guten Ratschläge, dem gesamten Team des BOLD-Projektes, stellvertretend namentlich Rodolphe Rougerie, dem Verleger Friedrich H. Pfeil und seinem Mitarbeiter Hubert Hilpert für die professionelle Betreuung und Umsetzung dieser Arbeit und natürlich meiner ganzen Familie für die vielen Jahre des Verständnisses und der Toleranz, die man als Familienangehöriger eines Saturniiden-Züchters in hohem Maße benötigt und die ich immer erfahren habe.

I The fundamentals of Breeding

1. Sources of material for breeding

There exist a manageable number of possibilities for obtaining breeding material. Eggs, pupae, cocoons and even larvae can be purchased from the internet sites indicated above. In recent years the price of eggs has risen enormously, double what it was twenty years ago. As a rule, eggs are sold by the dozen. In 2010 one paid eight to twelve Euros the dozen depending on the rarity of the species. Prices will also depend on whether the species is a new offering or one seldom available. In 2009 for example, there was no difficulty in selling eggs of *Coscinocera hercules* (Miskin, 1876) for 5 Euros per egg. As in other matters, prices respond to circumstances. A blatant example of this is the decline in prices for *Actias dubernardi* Oberthür, 1897 and *Actias chapae chapae* (Mell, 1950), both from China. The first adults of *Actias dubernardi* appeared on the German market around 1999. At that time, the going price for a male moth was 750 DM. In 2007, the price of a male *Actias chapae chapae* was about 1500 Euros. By 2009, a male *Actias dubernardi* cost about 5 Euros and an *Actias chapae chapae* about 2 Euros. Eggs and cocoons of both species can be had today from the sources mentioned above. Today, the sale of breeding material is carried on almost exclusively through the internet. Advertising in periodicals has almost disappeared. In recent years, the exchange of material among breeders has also declined. Most of the offerings seen on the internet derive from collectors and breeders in Europe and the USA. Caution is needed when ordering directly from tropical countries. I advise beginners to avoid such offerings, even when they seem cheaper. Unfortunately there are very few reliable dealers in these countries. From Africa alone come very few sincere offerings. One always pays but rarely receives the ordered material. Shipping eggs from these countries requires a lot of experience and is also very expensive because of various express charges. Additionally there is the danger that pupae or cocoons are not free of parasites. I myself have received shipments from Africa that contained flies instead of the expected moths. In recent decades, attempts have been made in Europe to make the breeding and sale of Lepidoptera, including the Saturniidae, both practical and economically feasible. A contemporary example is a well-known farm in

I Die Grundlagen der Zucht

1. Die Beschaffung von Zuchtmaterial

Für die Beschaffung von Zuchtmaterial gibt es eine überschaubare Anzahl von Möglichkeiten. Eier, Puppen oder Kokons sowie Raupen kann man heute auf den oben erwähnten Internet-Plattformen kaufen. Der Preis für Eier ist in den letzten Jahren enorm gestiegen und hat sich in 20 Jahren etwa verdoppelt. In der Regel werden Eier im Dutzend angeboten. Im Jahr 2010 zahlte man für 12 Eier, je nachdem wie selten eine Art ist, zwischen 8 Euro und 12 Euro. Der Preis für noch nie oder selten angebotene Arten kann auch deutlich darüber liegen. So wurden im Jahr 2009 für Eier von *Coscinocera hercules* (Miskin, 1876) ohne Probleme 5 Euro pro Ei gezahlt. Wie in anderen Bereichen auch regelt die Nachfrage den Preis. Ein krasses Beispiel für einen Preisverfall sind die Arten *Actias dubernardi* (Oberthür, 1897) und *Actias chapae chapae* (Mell, 1950) aus China. Die ersten Falter von *Actias dubernardi* kamen etwa 1999 auf den deutschen Markt. Für einen männlichen Falter wurde damals auf einer Börse die stolze Summe von 750 DM gezahlt. Der Preis für ein Männchen von *Actias chapae chapae* lag 2007 noch bei 1500 Euro. 2009 kostete ein gezüchtetes Männchen von *Actias dubernardi* etwa 5 Euro und von *Actias chapae chapae* etwa 2 Euro. Von beiden Arten kann man heute Eier und Kokons im oben beschriebenen Rahmen kaufen. Der Verkauf von Zuchtmaterial erfolgt heute fast ausschließlich über das Internet. Anzeigen in Zeitschriften sind diesbezüglich weitgehend ausgestorben. Auch der Tausch von Zuchtmaterial unter Züchtern ging in den letzten Jahren immer mehr zurück. Die meisten Angebote auf den Plattformen im Internet stammen von Sammlern und Züchtern aus Europa und den USA. Vorsicht ist bei direkten Angeboten aus tropischen Ländern angebracht. Laien rate ich vom Bezug solchen Zuchtmaterials ab, auch wenn es vermeintlich billiger ist. Leider gibt es nur sehr wenige seriöse Anbieter in diesen Ländern. Gerade aus Afrika kommen immer wieder viele nicht ernst gemeinte Angebote. Man wird zur Vorkasse gebeten und erhält fast nie das gewünschte Material. Der Versand von Eiern aus solchen Ländern erfordert viel Erfahrung und ist auch wegen diverser Expresszuschläge sehr teuer. Zudem besteht die Gefahr, dass Puppen oder Kokons nicht frei von Parasiten sind. Ich selbst habe aus einigen Sendungen aus Afrika

England that propagates Lepidoptera. However, taking a global view, the few such farms are probably of secondary importance to breeders of saturniids.

Another possibility for collecting saturniid breeding material is traveling to tropical lands oneself. But there are many things to consider when adopting such a course of action. It's not easy to collect larvae in these places and then bring them home. Since the plants on which the larvae are found are not available at home, such collecting is usually futile. If one keeps his larvae in completely different climatic conditions and tries to substitute an alternative food that the caterpillars then reject, then he should expect to lose the entire brood. The evening provides the best opportunity for capturing breeding material. If one is in a position to professionally set up lights for lengthy periods at night, then he can usually expect to obtain saturniid breeding material. In this way, I have sometimes collected breeding material of more than thirty saturniid species in about fourteen days. The females attracted to the light have in most cases already been fertilized, and one has only to allow them to lay their eggs. One should begin using lights as soon as twilight falls. In the last ten years or so, using this method, I have recorded the times at which more than six thousand saturniids were captured in flight. The data reveal a clear pattern. The females of a great many species arrive early at the source of light, sometimes directly with the onset of twilight. My tentative opinion is that females who mated the previous night are captured at the lights early the next evening while en route to laying their eggs. By contrast, it often happened that females of the same species arriving later had not been fertilized. It seems likely that these females were resting in the vicinity of the lights or had emerged nearby before they came to the light. In connection with this, it is possible that in the majority of species the females do not fly at all after emergence but immediately proceed to release the pheromones that attract males. Perhaps future studies will elucidate this question. If one cannot set up his own lighting system, then he can make use of pre-existing sources. Pre-eminent for this purpose are gas stations, since moths flock to them at night on account of their strong spotlights. These gas stations can also be visited during the day. Now and then, thorough searching can discover

statt der erhofften Falter nur Fliegen schlüpfen sehen. Es gab in den letzten Jahrzehnten einige Versuche, in Europa den Verkauf und die Zucht von Schmetterlingen und damit auch Saturniiden kommerziell zu praktizieren. Ein Beispiel ist eine sehr bekannte Schmetterlingsfarm in England, die es auch heute noch gibt. Global gesehen spielen aber diese wenigen Farmen für Saturniiden-Züchter eine eher untergeordnete Rolle.

Eine weitere Möglichkeit an Zuchtmaterial von Saturniiden zu gelangen, ist die Reise in die tropischen Länder selbst. Dazu sind aber sehr umfangreiche Erfahrungen notwendig. Raupen im Ausland zu sammeln und diese dann nach Hause mitzunehmen, ist schwierig. In der Regel ist das zwecklos, weil man zu Hause nicht die Pflanzen zur Verfügung hat, an denen man die Raupen gefunden hat. So muss man umfüttern, die Raupen nehmen dann das neue Futter nicht an und in Verbindung mit völlig anderen klimatischen Bedingungen führt das Ganze im Normalfall zum Totalverlust. Die beste Möglichkeit zur Beschaffung von Zuchtmaterial ist der Nachtfang. Ist man in der Lage nachts über einen längeren Zeitraum professionell zu leuchten, dann bekommt man in der Regel immer Zuchtmaterial von Saturniiden. Ich selbst habe auf diversen Exkursionen in Südamerika in etwa 14 Tagen manchmal Zuchtmaterial von über 30 Saturniiden-Arten auf diese Weise bekommen. Die ans Licht kommenden Weibchen sind in der Regel immer befruchtet und man lässt sie nur noch ihre Eier ablegen. Aber man sollte bereits in der Dämmerung mit dem Leuchten beginnen. In den letzten etwa zehn Jahren habe ich die Flugzeiten von mehr als 6000 am Licht gefangenen Saturniiden dokumentiert. Es zeichnet sich ein klares Muster ab. Die Weibchen sehr vieler Arten kommen sehr früh an die Lichtquellen, manchmal unmittelbar nach Einbruch der Dämmerung. Nicht endgültig belegt ist meine Auffassung, dass Weibchen, die sich in der Nacht zuvor gepaart haben, in der darauf folgenden Nacht sehr früh an die Lichtquellen kommen, weil sie bereits unterwegs sind um Eier abzulegen. Weibchen der gleichen Arten, die später die Lichtquellen besuchen, waren oft noch nicht befruchtet. Es könnte allerdings auch sein, dass diese Weibchen in der Nähe der Lichtquellen saßen, oder in der Nähe geschlüpft sind und deshalb später an die Leuchtquellen kamen. Zudem dürften bei der überwiegenden Mehrzahl der Arten die Weibchen nach dem Schlupf gar female saturniids. In doing this, it is important to search not just the station itself but also the vegetation in the vicinity of the spotlights. This approach always produces results even if one is using his own lights. Many females can be found in shrubs and trees or in the grass. It is beneficial to carry out this search before sundown, because birds are usually more active then. Another, but far more difficult, method is the collecting of cocoons. In this case, it is important to have a good deal of experience. If one knows the plants normally eaten by a species and seeks them out, there is a chance of finding cocoons. It is senseless to just go looking for pupae or cocoons in a tropical biotope, but help may be available. One should never travel without empty cocoons or pupae. Local farmers and children can be shown these specimens and they will frequently know on which trees such cocoons may be found. In the mountains of Sumatra, I showed a cocoon of *Attacus atlas* (Linnaeus, 1758) to a farmer, whereupon he led me to his garden. There stood a tree which in my estimate held several thousand cocoons. Another way to obtain breeding material is to restrain captured females in a natural setting. This method is often used to diversify the breeding stock or to capture a male active by day when the female of the species is nocturnal in habit. Examples include *Rothschildia tucumani* (Dognin, 1901) or species of the genus *Polythysana* Walker, 1855. The ultimate method of increasing the population is, of course, reproduction. For many breeders, the epitome of successful breeding is the achievement of repeated matings. An extensive chapter of this book will be devoted to this subject.

2. The successful treatment of eggs

2.1 Choosing the right substrate for egg-laying

After a successful mating, the next step is arranging for the female to lay her eggs. More will be said on the theme of achieving copulation in a later section. The choice of the appropriate substrate for egg-laying is the decisive issue in the further care of the eggs. First of all, what kind of container is suitable for the female to lay her eggs, and secondly, whether and what beneficial materials should be placed therein?

nicht fliegen, sondern sofort damit beginnen, die Männchen mit Pheromonen anzulocken. Vielleicht werden zukünftige Studien hier für mehr Klarheit sorgen. Besteht keine Möglichkeit professionell zu leuchten, dann kann man Lichtquellen absuchen. Hervorzuheben sind Tankstellen, an denen auf Grund der starken Scheinwerfer nachts viele Saturniiden anfliegen. Diese Tankstellen kann man auch tagsüber absuchen und wird bei gründlicher Suche ab und zu auch Weibchen von Saturniiden finden. Dabei ist es wichtig, nicht nur die Tankstellenanlagen selbst abzusuchen, sondern vor allem die Vegetation in der Nähe der Scheinwerfer. Das gilt immer, auch wenn man selbst leuchtet. Viele Weibchen findet man in den Sträuchern und Bäumen, oder im Gras. Es lohnt sich hier, vor Sonnenaufgang tätig zu werden, weil sonst die Vögel schneller sind. Eine andere weitaus schwerere Methode ist das Sammeln von Kokons. Auch hier ist sehr viel Erfahrung nötig. Kennt man die natürlichen Fraßpflanzen der Arten, die man gezielt sucht, so ist die Chance Kokons zu finden gegeben. Einfach so in einem Biotop in den Tropen nach Puppen oder Kokons zu suchen, ist sinnlos. Aber es gibt vielleicht Hilfe. Man sollte nie eine Reise ohne einige leere Kokons oder Puppen machen. Die leeren Kokons kann man einheimischen Bauern oder Kindern zeigen, die dann häufig wissen, an welchen Pflanzen die Kokons zu finden sind. In den Bergen Sumatras führte mich ein Bauer, nachdem ich ihm einen Kokon von *Attacus atlas* (Linnaeus, 1758) gezeigt hatte, in seinen Garten. Dort stand ein Baum, auf dem sich meinen Schätzungen zu Folge mehrere Tausend Kokons befunden haben dürften. Eine weitere Möglichkeit zur Erlangung von Zuchtmaterial ist, gezüchtete Weibchen in der Wildnis auszubinden. Diese Methode wird häufig verwendet, um Zuchtstämme aufzufrischen, oder um am Tag fliegende Männchen einer Art zu fangen, bei denen die Weibchen nachts aktiv sind, zum Beispiel bei *Rothschildia tucumani* (Dognin, 1901) oder Arten der Gattung *Polythysana* Walker, 1855. Die letzte Methode zur Erlangung von neuem Zuchtmaterial ist natürlich die Vermehrung. Die Krönung einer erfolgreichen Zucht ist für viele Züchter eine erneute Verpaarung. Diesem Thema widmet sich ein eigenes umfangreiches Kapitel in diesem Buch.

Ordinarily, flight cages and other expensive containers are not available when one is on an expedition to capture females in the wild. In such circumstances one's best resort is to simply bag the living females (Fig. 7). The female is placed in an ordinary paper bag in the unnatural attitude of having the wings appressed. The open end of the bag must then of course be closed. The method sounds brutal, but it almost always results in egg-laying. This is likely to happen on the same night in which the female is captured. In this way, a great many females can be taken on expeditions without difficulties. Mated females can be handled the same way in Europe as well. This method has a great advantage. There is no difficulty in removing eggs from the bag. Many females attach their eggs to the substrate with a secretion that works like strong glue. A paper bag is easy to cut. This leads us directly to the matter of the flying cage. The necessity of providing a flying cage in which natural mating can occur will be described later in this section. Following copulation, the male will separate himself from the female. The female will usually begin to lay eggs on the following night. Then the question arises of what to do with the female. There are rather different answers to this question. The female could be left in the cage. In principle, that is possible. But in that case, it is impossible to put another female into the same cage. It can happen that an unmated female will lay sterile eggs that are not readily distinguishable from the fertile eggs. Egg-laying behaviors vary considerably from species to species. Some females lay eggs in symmetrical formations and usually require a secure support for this. For this purpose, such females avoid using the gauze walls of the cage as a place to lay eggs. Then again, others find the gauze an acceptable substrate. Some will simply let the eggs fall. There are breeders who consider it necessary to place plants for the larvae to eat in the container with the eggs. I can recall hardly a single case in the last twenty-five years when that made sense. A possible exception is the genus *Hemileuca* Walker, 1855 from the USA that lays so-called egg-rings on twigs. Other breeders place the female in a shoe box or cardboard carton. My concept is the following: Mated females almost always lay their eggs. When they don't, either they did not mate, or the sex cells had become defective through incest. Therefore it is really more important to pay attention to the medium on which the female lays her eggs.

2. Die erfolgreiche Behandlung der Eier

2.1 Die Wahl des richtigen Ablagemediums

Nach einer geglückten Kopula geht es darum, die Weibchen zur Eiablage zu bringen. Mehr zum Thema der Erzielung einer Kopula wird in einem gesonderten Abschnitt ausgeführt. Ganz entscheidend für die weitere Behandlung der Eier ist die richtige Wahl des Ablagemediums. Gemeint ist zum einen, in welchem Behältnis man die Weibchen ablegen lassen möchte und zweitens, ob und was man in dieses Behältnis als Zulage gibt. Für auf Expeditionen gefangene wilde Weibchen hat man in der Regel keine Flugkästen oder andere aufwändige Behältnisse zur Verfügung. Die besten Ergebnisse erzielt man, wenn man solche Weibchen einfach lebend eintütet (Abb. 7). Man sollte dabei die Weibchen in unnatürlicher Stellung mit nach oben zusammengefalteten Flügeln in eine ganz normale Papiertüte stecken. Die Tüte muss natürlich am offenen Ende verschlossen werden. Die Methode klingt brachial, führt aber fast immer zu einer Eiablage. Diese erfolgt oftmals noch in derselben Nacht, in der die Weibchen gefangen wurden. Selbst eine große Anzahl an Weibchen kann so ohne Probleme auf Expeditionen mitgeführt werden. Auch in Europa könnte man so mit begatteten Weibchen verfahren. Diese Methode hat zudem einen großen Vorteil. Die Eier können ohne jegliche Probleme aus der Tüte entnommen werden. Viele Weibchen platzieren ihre Eier mit einem sehr festen Klebesekret auf dem Ablagemedium. Eine Tüte kann man zerschneiden. Das führt uns nun direkt zum Thema Flugkästen. Wie später im Abschnitt zu lesen sein wird, wie man eine natürliche Kopula erzielt, benötigt man zur Verpaarung einen Flugkasten. Gehen wir nun davon aus, dass die Kopula erfolgt ist. Das Männchen hat sich inzwischen vom Weibchen getrennt. Das Weibchen wird in der Regel in der nächsten Nacht mit der Eiablage beginnen. Stellt sich also die Frage, was mit dem Weibchen zu tun ist? Hierüber gehen die Meinungen weit auseinander. Man kann das Weibchen im Paarungskäfig belassen. Prinzipiell ist das möglich. Es sollten sich dann aber keine anderen Weibchen im Käfig befinden. Es kann vorkommen, dass auch unbefruchtete Weibchen beginnen abzulegen und die befruchteten Eier sind dann von den unbefruchteten nicht sofort zu

The breeder can influence this and he should. As previously mentioned, the eggs must be handled with special care and transferred to a Petri dish. If the eggs are simply lying in the cage, then one picks them up. Eggs laid on gauze are difficult to remove. But when the flight cage has a wooden framework, then one really learns about the problems of removing eggs. Sometimes, the secretions binding eggs to the substrate are so hard that the eggs can't be removed without crushing them. Eggs laid on paper or cardboard can be separated from these substrates with a razor blade. Eggs are in a dangerous situation when laid on plants. If these eggs are put into a Petri dish and sprayed, then the leaf to which they are attached slowly decays, releasing carbon dioxide. This can cause the failure to hatch by otherwise mature larvae. Another result of spraying is the growth of fungus which deprives the eggs of moisture. Consequently, their loss cannot be avoided. For this reason, I never put food plants in the container with eggs. Perhaps this sounds too easy. Using the method described above of bagging fertilized females, I and some other breeders have achieved successful egg-laying in almost one hundred per cent of the cases. Removing the eggs from paper bags is quite easy, with the bonus that most females remain completely unharmed after laying their eggs.

2.2 The handling of eggs until the larvae hatch

Following a successful copulation, the females of most species will begin to lay on the next night. The eggs should be carefully collected and placed in a Petri dish. These dishes are available in both glass and plastic (Fig. 8). An absorbent material such as paper kitchen towels or other paper may line the base of the dish. But I don't see this as necessary. Scented or chemically treated paper should not be used. The problem with using paper usually begins at the time of hatching. The larvae cannot be picked up with a fine brush from this surface, because they strongly adhere to it. Too much pressure with the brush can rupture and kill the larvae. Since not all eggs hatch at once, and must be separated from the lining material, sections of paper holding larvae are cut away from the rest of the lining and transferred to fodder. However, the hatchlings of some species will remain on the paper lining, because they cannot be directly transferred to

unterscheiden. Das Ablageverhalten der Arten ist völlig verschieden. Manche Weibchen legen ihre Eier im Spiegel ab und benötigen dafür meistens eine feste Unterlage. Die Weibchen solcher Arten meiden in der Regel die Gaze des Käfigs zur Ablage. Andere wiederum legen bevorzugt an der Gaze ab, manche lassen die Eier einfach fallen. Es gibt Züchter, die es für notwendig erachten, in den Gazekäfig oder Ablagebehälter Zweige der Raupenfutterpflanze zu legen. Ich kann mich in den letzen 25 Jahren an kaum einen Fall erinnern, in dem das sinnvoll gewesen wäre. Ausnahmen bilden vielleicht Arten der Gattung *Hemileuca* Walker, 1855, aus den USA, die so genannte Eiringe um Zweige legen. Andere Züchter verbringen die Weibchen in Schuhkartons oder Pappschachteln. Meine These ist folgende. Befruchtete Weibchen legen ihre Eier fast immer ab. Weibchen, die nicht richtig legen, sind entweder nicht befruchtet oder das Zuchtmaterial ist durch Inzucht degeneriert. Also ist es doch eigentlich viel wichtiger darauf zu achten, auf welchem Medium die Weibchen ihre Eier ablegen. Das kann ein Züchter beeinflussen und er sollte es. Wie schon erwähnt, müssen die Eier nach der Ablage besonders behandelt werden und in eine Petrischale gegeben werden. Wurden die Eier nun im Käfig abgelegt, sammelt man sie ein. Von Gaze sind sie noch leicht zu entfernen. Aber wenn der Flugkasten einen Holzrahmen hat, dann beginnen die Probleme mit der Entfernung. Manche Klebesekrete, mit denen die Eier an der Unterlage befestigt sind, sind so hart, dass man sie von festen Untergründen nicht entnehmen kann ohne sie zu zerdrücken. Auch auf Karton oder Papier abgelegte Eier sind quasi nur mit Rasierklingen zu entnehmen. Eier, auf Pflanzen abgelegt, sind gefährdet. Bringt man solche Eier in eine Petrischale und besprüht diese, dann verrottet das Laub, auf dem die Eier kleben, allmählich. Kohlendioxid wird freigesetzt. Das kann zum Abtöten der zum Schlupf reifen Raupen führen. Durch das Befeuchten kann sich Schimmel an den Blättern bilden, der den Eiern die Feuchtigkeit entzieht. Der Verlust ist nicht zu vermeiden. Schon aus diesem Grund würde ich niemals Futterpflanzen in die Ablagebehälter legen. Es klingt vielleicht zu einfach. Mit oben beschriebener Methode, der Eintütung der befruchteten Weibchen, haben ich und einige andere Züchter Ablagequoten von nahezu 100 % erzielt. Die Entnahme der Eier ist sehr leicht und zu guter Letzt sind die meisten Weibchen sogar

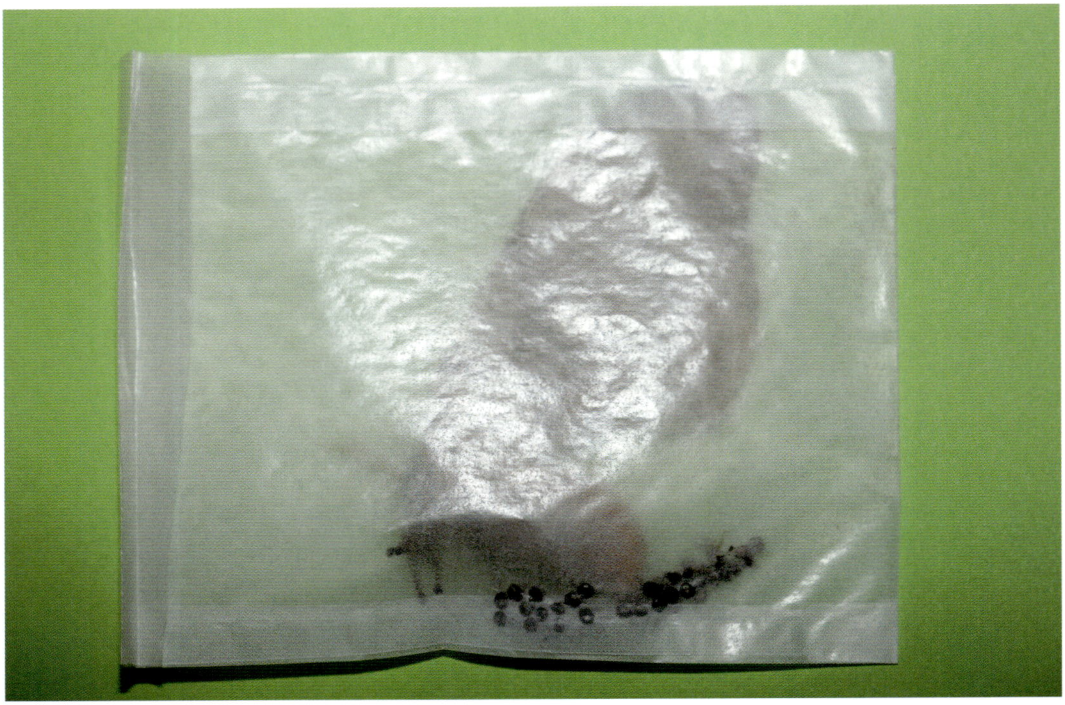

Fig. 7. Female with eggs laid in a bag. – Photograph by Frank Meister.
Abb. 7. Weibchen bei der Eiablage in der Tüte. – Foto: Frank Meister.

fodder. Therefore, I have never put paper linings in my Petri dishes.

Food should never be put into Petri dishes that contain eggs. Eggs and food do not belong together in closed containers. Plants release noticeable amounts of carbon dioxide when they decay. In addition to reports from other breeders, I have myself observed that the out-gassing of carbon dioxide seems to harm eggs, especially just before hatching. When nibbled on, some plants will release defensive or repellant chemicals. So it can happen that larvae will begin to feed and then die, and no more eggs hatch. When this is kept in mind, then only plants completely suitable as food will be placed in the containers. In this way, neither more nor less saturniid eggs are obtained from mated females. Once captured, the females will lay almost all of their eggs in one clutch. It is extremely rare for an undoubtedly mated female to lay no eggs. I can't recall a single instance of this in twenty years of breeding saturniids. Almost certainly, such a female was not mated. Eggs that have been transferred to a Petri dish should be moistened once or twice

noch völlig unbeschädigt, nachdem sie ihre Eier abgelegt haben.

2.2 Eibehandlung bis zum Schlupf der Raupen

Nach einer erfolgreichen Kopula beginnen die Weibchen der meisten Arten in der darauf folgenden Nacht mit der Eiablage. Die Eier sollten umgehend gesammelt und in eine Petrischale gegeben werden. Im Handel gibt es solche Petrischalen aus Glas und aus Plastik (Abb. 8). Man kann diese Petrischalen mit einer saugfähigen Lage aus Küchenrolle oder ähnlichen Papier auslegen. Eine Notwendigkeit dafür sehe ich aber nicht. Es sollte kein parfümiertes oder chemisch behandeltes Papier verwendet werden. Das Problem dieser Papierauslage beginnt meist mit dem Schlupf der Eiraupen. Raupen können nicht mehr mit einem Feinpinsel von dieser Unterlage aufgenommen werden, weil sie sich daran festklammern. Bei zu starkem Druck mit dem Pinsel können die Raupen einreißen und verenden. Da aber nicht alle Raupen gleichzeitig schlüpfen, muss man die Eier von dieser

Fig. 8. Plastic petri dish with eggs. – Photograph by Frank Meister.
Abb. 8. Petrischale aus Plastik mit Eiern. – Foto: Frank Meister.

a day. An atomizer is used to spray a fine mist in the air, and the open dish is passed through the mist. Water is not sprayed directly onto the eggs and neither are they allowed to sit in drops of water. The advantage of the Petri dish is that it is not air-tight, and the mist will gradually evaporate. Standing water is to be avoided. Most of the tropical species need warmth, and room temperature is sufficient. The direct rays of the sun must be avoided and also air-conditioned rooms. Air conditioners in rooms and automo-

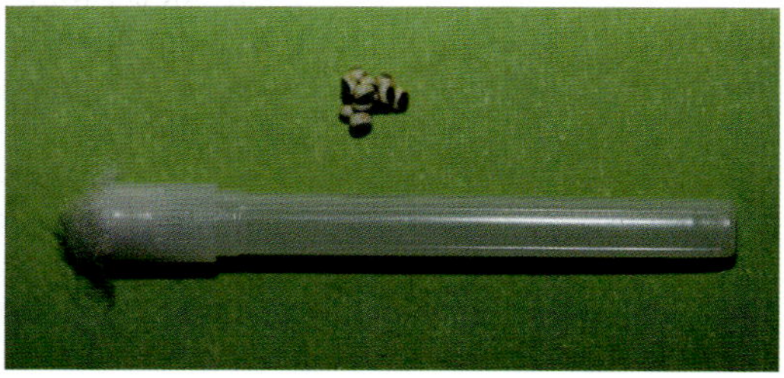

Fig. 9. Small tube for mailing eggs as a letter. – Photograph by Frank Meister.
Abb. 9. Röhrchen für den Eierversand in Briefen. – Foto: Frank Meister.

Unterlage entfernen, die Raupen ausschneiden und mit der Unterlage zum Futter bringen. Die Eiraupen mancher Arten aber gehen auf solch einer Papierunterlage ein, weil sie nicht direkt auf das Futter gesetzt werden können. Ich habe deshalb nie mit Papierauslagen gearbeitet.

Man sollte niemals Futter in diese Petrischalen legen, wenn noch Eier darin liegen. Eier und Futter gehören nicht zusammen in geschlossene Behältnisse. Pflanzen geben bekanntlich CO_2 ab, wenn sie verrotten. Ich habe mehrfach die durch andere Züchter bestätigte Erfahrung gemacht, dass diese CO_2-Vergasung den Eiern bzw. den kurz vor dem Schlupf stehenden Raupen zu schaden scheint. Zudem geben manche Pflanzen, wenn sie angenagt werden, Boten- oder Abwehrstoffe ab. Das führt häufig dazu, dass Eier, die man in die Behälter mit den schon geschlüpften und schon fressenden Eiraupen gibt, eingehen und keine Raupen mehr schlüpfen. Folgt man dieser Betrachtung, dann ist es also auch unzweckmäßig, in die Ablagebehältnisse Zweige oder Blätter der Futterpflanze zu geben. Man wird dadurch bei Saturniiden nicht mehr oder weniger Eier biles remove too much moisture from the air, and eggs obviously can't tolerate that.

2.3 The shipping of eggs by mail

There are a number of suitable methods for shipping eggs. Since eggs can easily be crushed, it is necessary to use crush-proof materials. Items mailed both within Germany and internationally are subject to automated handling and may be compressed by rollers that can squeeze things and tear them apart. A few years ago, it was possible to send eggs in the grooves of corrugated cardboard or in tubes like the fuel lines of mopeds. But today one receives eggs in exotic containers whose construction must have taken a long time. The foremost priorities are the sturdiness of the containers, the cost of their construction, and the expense of their transport. In Germany, it is possible to mail packages at the rate for letters, if they don't exceed 5 mm in thickness. The most economical containers that fulfill all requirements are the barrels of hypodermic syringes, also known as disposable syringes. In 2010, one hundred syringes cost exactly 3.50 Euros in a German pharmacy. That's enough to mail one hundred letters with eggs. These syringes are available in every pharmacy. An alternative is hard, plastic tubes, not more than 5 mm thick. These are available from stores that deal in building supplies and are sold by the meter. The tubes can be cut with a fine-toothed saw or with garden shears into lengths of 2–3 cm. They are very stable and will even resist the blows of a hammer. One end of the tube is plugged with cotton. This is then covered with adhesive tape. For ventilation, the tape is punctured with a needle (Fig. 9). Next, the eggs are inserted into the tube. How to do this without dropping or damaging the eggs is a much-discussed subject. Also, there is considerable disagreement about which method works best. Whether one believes it or not, this is the method that works best: One moistens the tip of his middle finger and picks up the eggs from the Petri dish. With his other hand, he places the open end of the tube over the finger tip. The eggs will spontaneously fall into the tube. Doing this by hand has the advantage of finely balanced, muscle control. Then the open end of the tube is sealed as described above. In the summer, ordinary envelopes are sufficient for mailing eggs. If additional security is desired, then insulated envelopes may be used. In winter,

aus befruchteten Weibchen erhalten. Weibchen, die anfangen zu legen, legen fast immer alle Eier ab. Dass eindeutig befruchtete Weibchen keine Eier legen, ist eher sehr selten. Mir sind aus 20 Jahren Zucht von Saturniiden keine Fälle in Erinnerung. Fast immer sind solche Weibchen nicht befruchtet. Hat man also nun die Eier in eine Petrischale verbracht, sollten diese ein- bis zweimal am Tag befeuchtet werden. Dazu sprüht man feinen Sprühnebel mit einer Sprühflasche in die Luft und führt die offene Schale durch diesen Nebel. Nicht direkt auf Eier sprühen und die Eier dürfen auch nicht in Wassertropfen liegen. Der Vorteil von Petrischalen ist, dass diese nicht luftdicht schließen und der Sprühnebel allmählich wieder verdunstet. Staunässe ist zu vermeiden. Die meisten tropischen Arten benötigen Wärme. Zimmertemperatur ist völlig ausreichend. Direkte Sonneneinstrahlung ist unbedingt zu vermeiden, genauso wie die Aufbewahrung in klimatisierten Räumen. Klimaanlagen in Räumen oder in einem Auto entziehen der Luft sehr viel Feuchtigkeit, was die Eier offensichtlich nicht vertragen.

2.3 Der Versand von Eiern

Für das erfolgreiche Versenden von Eiern gibt es mehrere gute Methoden. Da Eier prinzipiell leicht zerdrückt werden können, benötigt man druckfeste Materialien. Da im internationalen und auch innerdeutschen Postverkehr Briefe weitestgehend durch automatische Verteilzentralen laufen und die Briefe starkem Pressdruck durch Walzen ausgesetzt sind, scheiden alle Materialien aus, die zerdrückt werden können. Noch vor wenigen Jahren war es möglich, Eier in den Rillen von Wellpappe zu versenden oder in Benzinschläuchen von Mopeds. Auch heute noch bekommt man Eier in abenteuerlich gebauten Behältnissen, deren Bau sehr viel Zeit gekostet haben muss. Neben der Stabilität müssen aber aus Kostengründen für die Beschaffung und die Beförderung die zu wählenden Behältnisse auch billig sein. Und sie dürfen, um sie noch als Brief versenden zu können, derzeit in Deutschland 5 mm Dicke nicht überschreiten. Die absolut preisgünstigsten Behältnisse, die alle Anforderungen erfüllen, sind die Schutzhüllen von Spritzenkanülen, auch als Einwegspritzen bekannt. Einhundert Kanülen kosteten in einer deutschen Apotheke im Jahr 2010 exakt 3,50 Euro. Damit könnte man also 100 Briefe mit Eiern versenden.

these will provide some protection against cold. A sheet of paper is folded to fit inside the envelope. The tube containing eggs is placed within the fold of the paper and held in place with adhesive tape. This is inserted into the envelope, always with the tube toward the lower side of the envelope. This is important for a number of reasons. First of all, envelopes are franked near the top. Striking the tube could force the envelope open and expel the tube. Also, if the post office stamp hits the tube, the cancellation will be imperfect and the package may not reach its destination.

2.4 Maintaining eggs over long periods of time

For some species of Saturniidae, it is necessary to maintain the eggs over a long period of time or to overwinter them. This can only succeed when conditions similar to the climate of the natural biotope are provided. For example, the eggs of certain species whose biotope lies at high altitudes may not hatch for several months. Other species will go through their diapause in freezing conditions. Climate diagrams are helpful in working out these requirements. They provide all the information one needs to provide these species with conditions similar to those of their habitats, like temperature, humidity and rainy season. Such diagrams exist for thousands of localities around the world. If the eggs come from a climate with very low temperatures, then they must be kept cool. This sort of thing can be really difficult when the seasons of the hemispheres are opposed. Eggs of *Polythysana cinerascens* (Philippi, 1859) from Chile are available from March to May. In Chile, that is late autumn, but in Europe it is the beginning of spring. So they will certainly die if kept warm. Eggs of *Antheraea yamamai yamamai* (Guérin-Méneville, 1861) arrive in the European fall and must be overwintered. In this species, the larvae develop in the fall and spend the entire winter in the egg, hatching in the spring. Therefore, to overwinter these eggs, it is necessary to maintain continuously cold temperatures with sufficient humidity for about six months. One possibility would be to keep the eggs out-of-doors through the winter in a place protected from extremes of freezing. The container is closed with very fine gauze to provide ventilation. The mesh must be fine enough to exclude potential predators. The container must

Diese Kanülen sind in jeder Apotheke frei erhältlich. Wer es anders machen möchte, der kauft sich runde Hartplastikröhrchen, maximal 5 mm dick, die es in allen Baumärkten als Meterware zu kaufen gibt. Diese Hartplastikröhren können mit einer Feinsäge oder einer Gartenschere in zwei bis drei Zentimeter lange Stücke geschnitten werden. Sie sind sehr stabil und können selbst mit einem Hammer nicht zerstört werden. Zunächst pfropft man in ein Ende dieses Röhrchens ein wenig Watte. Das Ende wird dann über Kreuz mit Klebeband verklebt. Zur Belüftung sticht man die das Röhrchen überklebende Fläche mit einer Nadel ein. Das dient zur Belüftung (Abb. 9). Danach gibt man die Eier in das Röhrchen. Es wurde schon viel darüber diskutiert, wie man am besten Eier in ein Röhrchen verbringt, ohne dass sie herunterfallen oder zerdrückt werden. Auch wenn man es nicht glauben will, am besten funktioniert folgende Methode. Man befeuchtet die Kuppe seines Mittelfingers, wie auch immer, und nimmt dann die Eier aus einer Petrischale auf. Dann streift man das Röhrchen in der einen Hand über die Fingerkuppe. Die Eier fallen von selbst in das Röhrchen. Die Arbeit mit den Fingern hat den Vorteil einer sehr ausgewogenen Feinmotorik. Danach verschließt man das zweite Ende des Röhrchens, wie oben beschrieben. Im Sommer reichen ganz normale Briefe für den Versand. Wer sichergehen will, der nimmt Polsterbriefe, die gerade im Winter einen gewissen Schutz vor Kälte bieten. Man nimmt ein Blatt Papier und faltet es so, dass es in den Brief passt. In die Falz des Briefpapiers legt man das Röhrchen mit den Eiern und fixiert es dort sehr fest mit Klebeband. Dann nimmt man das Papier und positioniert es so im Briefumschlag, dass das Röhrchen immer an der unteren Seite des Briefes liegt. Das ist aus mehreren Gründen wichtig. Erstens werden Briefe ja oben frankiert und sind zudem nach oben offen. Durch den Stempel könnten die Röhrchen aus dem Brief gedrückt werden, oder nicht richtig gestempelt werden und den Adressaten nicht erreichen.

2.4 Die Aufbewahrung von Eiern über einen längeren Zeitraum

Es gibt bei einer Reihe von Saturniiden-Arten die Notwendigkeit, Eier über einen längeren Zeitraum aufzubewahren oder die Eier zu überwintern. Das gelingt nur, wenn man es schafft, ähnliche klimatische Bedingungen anzubieten

Fig. 10. Container for overwintering saturniid eggs in a refrigerator. – Photograph by Frank Meister.
Abb. 10. Eierbehälter für die Überwinterung von Saturniiden-Eiern im Kühlschrank. – Foto: Frank Meister.

also be shielded from direct sunshine. Convenient locations may be found in the garden or on a balcony. If neither is available, then the refrigerator can be used. Refrigerators have the advantage of maintaining a stable temperature, but the disadvantage of very dry air. This can be dealt with by spraying the eggs every day, although this becomes very time-consuming over a period of six months. The construction of a small container can remedy this. Water is placed into a plastic box. A block of Styrofoam is cut to fit the box. A hole, 3 × 3 cm, is made through the block. The block is then divided horizontally into two halves. A layer of gauze is laid between the halves, which are then fastened together with strapping tape. Eggs are placed upon the gauze, and the upper opening of the hole is also covered with gauze. The whole construction is then placed into the box containing water, where it is allowed to float freely. The eggs benefit from the humidity provided by evaporation without coming into contact with the water itself (Fig. 10).

Eggs that remain dormant for several months may need special consideration. In such cases it becomes tempting to abandon hope that the larvae will ever hatch. One should not give up too soon. When one is uncertain as to whether the eggs are alive or dead, he can proceed as follows: One selects an egg and a razor blade. With the

wie in den jeweiligen Biotopen. Die Eier von bestimmten Arten, deren Biotope in großen Höhen liegen, benötigen zum Beispiel bis zum Schlupf der Raupen mitunter mehrere Monate. Andere Arten verbringen ihre Diapausen im Eistadium. Hier kann man sich mit Klimadiagrammen behelfen. In Klimadiagrammen findet man alle Informationen, die man benötigt, um diesen Arten ähnliche Bedingungen bieten zu können wie in ihrem Verbreitungsgebiet, wie die Temperaturen oder Trocken- und Regenzeiten. Diese Klimadiagramme gibt es für Tausende Orte auf der ganzen Welt. Kommen Arten aus Klimazonen mit sehr niedrigen Temperaturen, so muss man die Eier auch kühl aufbewahren. Das kann schwierig sein, wenn die Jahreszeiten vertauscht sind. Eier von *Polythysana cinerascens* (Philippi, 1859) aus Chile erhält man von März bis Mai. Dann ist in Chile Spätherbst und in Europa beginnt der Frühling. Würde man solche Eier nun warm halten, gingen sie mit Sicherheit ein. Eier von *Antheraea yamamai yamamai* (Guérin-Méneville, 1861) bekommt man im europäischen Herbst und die Eier müssen überwintert werden. Bei dieser Art entwickeln sich die Räupchen bereits im Herbst und verbringen dann den ganzen Winter im Ei, um im Frühling zu schlüpfen. Für die Überwinterung dieser Eier ist es also notwendig, über etwa 6 Monate gleichbleibende kalte Temperaturen und genügend Feuchtigkeit zu bieten. Eine Möglichkeit wäre, die Eier an einem sicheren, vor zu starkem Frost geschützten Platz im Freien zu überwintern. Dazu gibt man die Eier in einen luftigen, mit sehr feiner Gaze umspannten Behälter. Es darf sich keine Staunässe bilden. Die Gaze muss so fein sein, dass sie alle in Frage kommenden Fressfeinde abwehrt. Der Behälter darf keinem direkten Sonnenschein ausgesetzt sein. Am besten eignet sich ein Platz im Garten oder auf einem Balkon. Wer beides nicht zur Verfügung hat, der kann einen Kühlschrank nutzen. Kühlschränke haben den Vorteil, dass die Temperatur konstant gehalten werden kann, aber den Nachteil der sehr trockenen Luft. Hier kann man sich behelfen, indem man die Eier jeden Tag besprüht, allerdings ist das sehr zeitaufwändig über sechs Monate lang. Der Bau eines kleinen Behälters kann Abhilfe schaffen. Man nimmt eine Plastikdose und gibt Wasser hinein. Aus einem Styroporklotz schneidet man sich ein Stück heraus. In die Mitte dieses rechteckigen Quaders schneidet man ein etwa 3 × 3 cm großes durchgängiges Loch. Nun teilt man den razor blade, he carefully opens the eggshell and examines the contents with a microscope or a magnifying glass. If the contents are completely dehydrated, then the egg/embryo is dead. If the egg is full of uniformly colored, transparent, yellow-tinted fluid, this egg is unfertilized. Such eggs usually die after only a few days. If the egg is filled with a firm, thread-like mass, or structures of a small larva are visible, then the egg is still developing. The approach of hatching is signaled by darkening of the egg's color. Daily observation will demonstrate changes in form and color as the embryo develops. For example, eggs of the genus *Automeris* develop a black spot on the head end a few days after fertilization – a clearly diagnostic sign. On the whole, a great deal of experience is needed to draw the right conclusions as to whether the eggs are fertilized or not.

3. The successful care of larvae

3.1 Choosing the proper food

The selection of the right food is a decisive factor for successful breeding. Before ordering or obtaining living material for breeding, everyone should be informed about the availability of edible plants of sufficient quality and quantity. Stone's book, "Foodplants of the world Saturniidae", is a standard reference for feeding Saturniid larvae. Further possibilities may be found through the internet. As previously mentioned, Bill Oehlke has a very good website in Canada. It can be found at http://www3.islandtelecom.com/~oehlkew/. The biggest and best German site with a huge market for the purchase and exchange of breeding material is www.actias.de. Presently there is no work dealing with breeding or the feeding of saturniid larvae. Many breeders are disinclined to try out alternative plants that might be appropriate for breeding a species. After years of breeding and exchanges with a series of notable breeders from around the world, I came to the following conclusion: For a huge number of previously bred saturniid species there is a relatively manageable number of plants that are always useful as fodder. When one lacks definite information about which plant might be correct, he generally tries one of the so-called standard food plants. The majority of saturniid larvae is polyphagous, and finds a variety of plants acceptable as food. Those on which

Quader in zwei horizontale Hälften. Zwischen die Hälften legt man Gaze. Nun werden die Quader wieder zusammengeklebt, am besten mit Paketklebeband. Die Eier legt man auf die Gaze und deckt auch noch die obere Öffnung des Quaders mit Gaze ab. Nun legt man den so gebauten Quader in die Plastikdose mit Wasser. Der Quader schwimmt nun frei auf dem Wasser. Das Wasser verdunstet und die Eier liegen in ihrem Gazebett über der verdunstenden Wasserfläche, aber ohne sie zu berühren (Abb. 10).

Eier von Arten, die mehrere Monate überliegen, fallen mitunter sehr stark ein. Man gibt dann die Hoffnung meist schon auf, dass noch Raupen schlüpfen. Man sollte solche Eier nicht zu früh aufgeben. Ist man dennoch unsicher, ob die Eier bereits tot oder noch lebendig sind, kann man folgendermaßen vorgehen. Man nimmt ein Ei und eine Rasierklinge. Mit der Rasierklinge öffnet man vorsichtig die Eischale und betrachtet den Inhalt des Eies unter dem Mikroskop oder mit einer Lupe. Ist der Inhalt völlig trocken, ist das Ei oder die kleine Raupe tot. Ist das Ei mit einer gleichmäßig gefärbten, durchscheinenden meist gelben Flüssigkeit gefüllt, handelt es sich um ein unbefruchtetes Ei. Solche Eier fallen meist bereits nach wenigen Tagen sehr stark ein. Besteht der Inhalt des Eies aus einer sehr zähen, vielleicht auch Fäden ziehenden Masse, oder erkennt man Strukturen einer kleinen Raupe, befinden sich die Eier noch in ihrer Entwicklung. Der Schlupf der Raupen kündigt sich meist dadurch an, dass die Eier sich dunkel verfärben. Bei genauer täglicher Betrachtung der Eier wird man feststellen, ob sie sich in Form und Färbung verändern. Eier der Gattung *Automeris* zum Beispiel bekommen, wenn sie befruchtet sind, nach wenigen Tagen einen schwarzen Punkt am Kopfende, ein klares diagnostisches Zeichen. Insgesamt ist viel Erfahrung nötig, um die richtigen Schlussfolgerungen zu ziehen, ob Eier befruchtet sind oder nicht.

3. Die erfolgreiche Behandlung der Raupen

3.1 Die Wahl des richtigen Futters

Ein ganz entscheidender Faktor für eine erfolgreiche Zucht ist die Auswahl des richtigen Futters. Jeder sollte sich vor der Bestellung oder Beschaffung von Zuchtmaterial informieren, welche Futterpflanzen auch in ausreichender tropical saturniids are most commonly fed in Europe include: oaks *Quercus*, thick-leaf privet *Ligustrum ovalifolium*, willow *Salix*, false acacia *Robinia pseudoacacia*, liquidambar *Liquidambar styraciflua*, fire-thorn *Pyracantha* und blackberry *Rubus fructicosus*. The last two are especially suitable as winter food in Europe. If one is uncertain about the best plant to feed the species he is breeding, then he should offer the young larvae a selection from this list.

Successful breeding depends not only on choosing the proper plants, but also on the quality of the food. One of the commonest errors in breeding is the use of unsuitable fodder. Therefore it is appropriate when searching for a certain tree or bush that one proceeds carefully and does not automatically use the very first leaves he finds. It often happens that trees and bushes may be sick or the leaves show clear indications of deficiencies. Such plants are totally unsuitable for purposes of breeding. Soiled leaves are not edible and must first be washed and dried. This is often the case with old oak leaves in late summer. In early summer, many leaves are infested with plant lice. Using such leaves will most likely cause the total loss of the brood. It is just as important to examine leaves for ants, predatory bugs, or the larvae of carnivorous beetles. Just a few bugs can destroy an entire brood in only a few days. Once a tree or bush has been selected as a source of food, one should take a close look at its surroundings. The foliage along streets or the edges of fields may be contaminated with chemicals. Wilted or limp fodder is unsuitable. One should not use fresh, young leaves, if he as already fed the larvae with older fodder, because doing so carries the serious danger of causing dysentery. This is also true, if the food is wet or sprinkled with dew drops. It makes sense always to take leaves from the tree or bush that has demonstrated its quality as nourishment. The best times for this are early morning and late evening.

3.2 The treatment of hatchlings

If the larvae hatch out of the eggs, one should absolutely give them the opportunity to eat the eggshells. This first ingestion of protein can have greater significance later, if the larvae have difficulty in reaching their normal food. For this reason one observes the hatching in a Petri dish

Menge und Qualität zur Verfügung stehen. Ein sehr nützliches Standardwerk für Raupen und ihre Futterpflanzen für die Familie der Saturniiden ist das Buch von Stone: »Foodplants of the world Saturniidae«. Eine weitere gute Möglichkeit sich über Raupen und deren Futterpflanzen zu informieren, bietet das Internet. Eine sehr gute Website in Kanada bietet wie schon erwähnt Bill Oehlke. Zu finden ist diese unter http://www3.islandtelecom.com/~oehlkew/. Die in Deutschland größte und bekannteste Seite im Internet mit großem Börsenteil zum Kauf oder auch Tausch von Zuchtmaterial findet man unter www.actias.de. Ein aktuelles Werk über die Zucht oder über Raupen und ihre Futterpflanzen für Saturniiden gibt es derzeit nicht. Vielen Züchtern bleibt also oftmals nichts anderes übrig als auszuprobieren, welche Futterpflanzen bei einer neu zu züchtenden Art die richtigen sein könnten. Nach vielen Jahren der Zucht und dem Austausch mit einer Reihe namhafter Züchter weltweit kommt man zu folgendem Schluss. Für eine sehr große Anzahl bisher gezüchteter Arten von Saturniiden gibt es eine relativ überschaubare Anzahl immer wieder verwendeter Standardfutterpflanzen. Findet man also überhaupt keine Hinweise, welche Futterpflanze die richtige sein könnte, probiert man zunächst die so genannten Standardfutterpflanzen aus. Die Mehrheit der Saturniiden-Raupen lebt polyphag, akzeptiert also verschiedene Fresspflanzen. Die mit Abstand am häufigsten akzeptierten Futterpflanzen, mit denen tropische Saturniiden in Europa gezüchtet wurden, sind: Eiche *(Quercus)*, dickblättriger Liguster *(Ligustrum ovalifolium)*, Weide *(Salix)*, Robinie *(Robinia pseudoacacia)*, Amberbaum *(Liquidambar styraciflua)* Feuerdorn *(Pyracantha)*, und Brombeere *(Rubus fructicosus)*. Die beiden zuletzt genannten sind insbesondere als Winterfutter in Europa gut geeignet. Ist man sich also nicht sicher, welche Futterpflanzen von einer zu züchtenden Art gefressen werden, sollte man den Jungraupen zunächst diese Pflanzen zur Auswahl geben.

Entscheidend für eine erfolgreiche Zucht ist aber nicht nur die Wahl der richtigen Futterpflanze, sondern auch die Futterqualität. Einer der häufigsten Fehler bei der Zucht ist die Verwendung ungeeigneten Futters. Es ist deshalb zweckmäßig, bei der Suche nach einem bestimmten Baum oder einem Strauch sorgfältig vorzugehen und nicht das erstbeste Blattwerk zu verwenden.

and leaves the newly hatched larvae alone for at least half a day. The first watering is important for nearly all species. For this, I use luke-warm tap water and with one hand make a fine mist with an atomizer. With the other hand, I pass the open Petri dish through the mist just enough to give the interior a barely visible film of moisture. The larvae can absorb this water, and together with the nourishment of the egg shell, that will suffice them for up to three days. This spraying of the larvae may be repeated during the day. However, water must not be allowed to build up into puddles or even drops. The larvae can drown. When the larvae have drunk some water and eaten most of their eggshells, then they may be moved to their normal food. This has repeatedly shown itself to be a valid principle. There is no point in putting food into Petri dishes with unhatched eggs. When leaves and plant material wilt or are partly eaten, they can release gases. These gases seemingly cause the death of larvae within the eggs before they have a chance to hatch. This interpretation has been supported by several breeders. Normally, one picks up the small caterpillars with a fine brush and sets them on their fodder. In connection with this, there are a number of details to attend to and which need a bit of experience. A great many hatchlings will immediately spin security threads when picked up. In nature, these silk threads most likely serve the purpose of securing the young larva to the leaf on which it hatched in spite of wind or other external forces. The spinning of such threads tests the breeder's patience when he tries to shift hatchlings to fresh food. One should not permit several larvae to entangle each other in their threads, because they will be unable to free themselves and die. In the case of species that produce large amounts of silk, for example *Aglia tau* (Linnaeus, 1758), or several of the Ceratocampinae, it is perhaps advisable to limit the number of larvae allowed to hatch in the same Petri dish. One approach to dealing with these species is to remove unhatched eggs from the dish. Then one gives fresh food to the remaining hatchlings. The leaves onto which they climb can then be transferred to a breeding chamber without trying to pick them up. To begin with, the container for raising the larvae should not be too big. A cassette for photographic film or other really small container will suffice. One should exercise care in choosing this container. Soft plastics should be avoided. Breeders

Viele Bäume oder Sträucher sind oftmals krank oder zeigen schon im Blattwerk eindeutige Mangelerscheinungen. Solche Pflanzen sind für Zuchtzwecke völlig ungeeignet. Auch stark verschmutztes Blattwerk kann nicht verfüttert werden und müsste erst gewaschen und getrocknet werden. Häufig ist älteres Laub von Eichen im Spätsommer sehr stark verschmutzt. Gerade im Frühsommer sind viele Blätter von Läusen befallen. Die Verwendung solcher Blätter führt meist zum Totalverlust der Zucht. Genauso wichtig ist, die Untersuchung des Blattwerkes auf Ameisen, Raubwanzen oder zum Beispiel Larven von Raubkäfern. Wenige Raubwanzen können binnen weniger Tage eine ganze Zucht abtöten. Man sollte sich zudem die Umgebung anschauen, in welcher der Baum oder Strauch steht, den man für die Futterbeschaffung ausgewählt hat. Laub von Straßen- oder Feldrändern kann chemisch belastet sein. Welkes oder angewelktes Futter ist ungeeignet. Man sollte kein junges, frisches Laub verfüttern, wenn man zuvor Altlaub verwendet hat. Die Gefahr von Durchfallerkrankungen ist dann sehr hoch. Auch die Gabe von nassem Futter oder mit Tau bedeckten Blättern kann zu Durchfallerkrankungen führen. Sinnvoll ist es, das Futter immer vom selben Baum oder Strauch zu entnehmen, wenn es sich als qualitativ gut erwiesen hat. Die beste Zeit dafür ist der Morgen oder Spätabend.

3.2 Die Behandlung der Eiraupen

Wenn die Raupen aus den Eiern schlüpfen, sollte man ihnen unbedingt die Gelegenheit geben, ihre Eierschale fressen zu können. Dieser erste Proteinschub könnte später von großer Bedeutung sein, wenn die Raupen schwer an Futter gehen. Beobachtet man also den Schlupf der ersten Raupen in der Petrischale, so lässt man den Eiräupchen mindestens eine halben Tag lang Zeit. Wichtig ist zudem bei fast allen Arten die Erstversorgung mit Wasser. Ich nehme dazu handwarmes Wasser aus der Wasserleitung und sprühe mit einem Zerstäuber feinen Nebel in die Luft. Mit der anderen Hand führe ich nun die Petrischale mit den Eiraupen durch den Nebel, so dass sich ein kaum sichtbarer Sprühwasserfilm in der Petrischale bildet. Die Raupen können nun dieses Wasser aufnehmen, was ihnen zusammen mit der Eierschale einen guten Start für die nächsten bis zu drei Tage gibt. Das Besprühen der Raupen kann mehrmals am tell me that such containers apparently release gases fatal to larvae that die seemingly without cause. The most important task at this point is providing the larvae with food. If one knows the appropriate plants and has them available, then favorable environmental conditions become the next priority. For the first larval stage, I do not place edible plants in water, which could be possible. The breeding chamber should be kept neither too cold nor too warm. The direct rays of the sun are definitely to be avoided. In fact, attempts to raise several species have demonstrated that it is better to keep the container in complete darkness for the first two days. There is the assumption that the larvae will begin to feed more easily. Some breeders share in this interpretation. Many species, like those in the genus *Automeris*, are very gregarious. So it is important that larvae of these species be kept together. Beginning with their purchase, one should acquire a certain number of eggs. Large containers that allow larvae to wander and even escape can cause the death of many larvae. It has often been observed that isolated larvae do not eat, and then they die. The larvae that congregate move to their food sooner, grow better and also eat together. It is of course more difficult to feed larvae of species that have not previously been bred. This book should help with that problem. The best course is to follow the examples that already exist. One begins by finding out what other species in his genus of interest have already been bred. There is a high probability that species within the same genus will favor similar plants as food. Definite examples of this tendency are already known. Species of the genus *Rothschildia* Grote, 1897, have been successfully bred in Europe feeding mainly on privet *Ligustrum* species of the genus *Antheraea* Hübner, 1819 [»1816«], have been raised using oaks *(Quercus)*. Nevertheless, a nutritional specialist like *Rothschildia tucumani* (Dognin, 1901), whose larvae feed on species of *Eupatorium*, does not readily accept privet. In the final analysis, the breeder must experiment. At most, he has two or maybe three days for this before the larvae die of starvation despite repeated misting. It is possible to offer young caterpillars several kinds of plants using a process of elimination. As a rule, larvae usually avoid leaves they are not interested in eating. However, if a great many caterpillars take up positions on the same leaf but have not begun to eat, the probability is high that this will yet

Tag wiederholt werden. Es ist darauf zu achten, dass sich keine Tropfen oder Wasserlachen in der Schale bilden. Die Raupen könnten ertrinken. Haben die Raupen ihre Eischale zum größten Teil gefressen und Wasser getrunken, kann man sie nun auf das Futter setzen. Auch dabei gilt wieder der Grundsatz: Das Futter hat in der Petrischale zusammen mit den noch nicht geschlüpften Raupen nichts zu suchen. Wenn Blätter und Pflanzen verwelken oder angefressen werden, können Gase freigesetzt werden. Diese Gase führen offensichtlich sehr häufig zum Tod der noch in den Eiern befindlichen Raupen, die noch nicht geschlüpft sind. Dieser Umstand wurde mir von mehreren Züchtern bestätigt. Vielmehr nimmt man die kleinen Raupen einzeln mit einem feinen Pinsel auf und setzt sie auf das Futter. Hierbei sind mehrere Aspekte zu beachten und man benötigt einige Erfahrungen. Sehr viele Eiräupchen spinnen sofort Sicherheitsfäden, wenn man sie aufnimmt. Diese Seidenfäden dienen in der Natur mit großer Wahrscheinlichkeit dazu, dass die Raupen nach dem Schlupf bei Wind oder anderen äußeren Einflüssen nicht von den Blättern fallen, auf denen sie geschlüpft sind. Das Spinnen solcher Fäden macht es oft zur Geduldsprobe, die kleinen Raupen auf das Futter zu überführen. Es ist darauf zu achten, dass sich mehrere Raupen nicht gegenseitig mit Fäden überspinnen und dann später nicht freikommen und eingehen. Bei Arten mit extrem spinnfreudigen Raupen, wie zum Beispiel *Aglia tau* (Linnaeus, 1758) oder vielen Ceratocampinae, ist es vielleicht ratsam, nicht zu viele Raupen in einer Petrischale schlüpfen zu lassen. Oder man nimmt bei diesen Arten die Raupen überhaupt nicht auf und entnimmt lieber die noch nicht geschlüpften Eier aus der Petrischale. So kann man dann das Futter zu den Raupen bringen, die darauf klettern und so in den Zuchtbehälter gebracht werden können. Der Zuchtbehälter sollte am Anfang nicht zu groß sein. Es genügt eine Filmdose oder ein anderer sehr kleiner Behälter. Man sollte den Zuchtbehälter sehr sorgsam auswählen. Weichplastikdosen sind zu vermeiden. Züchter berichteten mir, dass diese Dosen offensichtlich gasen und die Raupen ohne Grund eingingen. Die wichtigste Aufgabe am Anfang ist es, die Raupen zum Fressen zu bekommen. Kennt man die Futterpflanze und hat diese auch zur Verfügung, dann sorgt man anfangs vor allem für günstige Umweltbedingungen. Ich stelle im ersten Raupenstadium das Futter nicht ins Wasser, was aber möglich wäre.

prove to be acceptable food. Ultimately, experience counts. I know of no successful use of artificial food for the Saturniidae, and I have made many attempts myself. Caterpillars of *Rothschildia tucumani* have reached the third larval stage on artificial food, and then they all died.

3.3 The shipping of larvae

If one wants to ship larvae, there are a few important things to attend to. The shipping of eggs and even cocoons is relatively problem-free, because the consignment is in transit no more than ten days. The situation of shipping larvae is quite different. Here the shipment must reach the recipient in three or at most four days. Within Europe this may be possible. Shipment overseas takes longer and sending larvae makes no sense. Furthermore, it is important to keep climatic conditions in mind. Eggs and cocoons are relatively robust. Larvae react more sensitively to heat and cold. The ideal phase for shipping larvae is the second day or at best the third day before molting to the third pre-imaginal stage. At this point the larvae have reached a steady level in their feeding and need very little. So it is important to include enough food in the package for up to four days. Older larvae are not suitable for shipping. They need a great deal of space. They experience a higher degree of stress. And it is difficult to provide the amount of food they need for the duration of the trip. During this time, a large amount of fecal material will accumulate. This can dissolve in water that condenses in the package and also give rise to infestation by molds. Either way, the larvae are in danger of infection. All of this is avoided with the first pre-imaginal stage. Such larvae cease feeding and defecating shortly before their next molt. The following method has proved itself the best. One takes the usual sort of bag used for carrying frozen foods with a volume of about two liters. The bag is folded several times and punctured repeatedly with a needle. When the bag is unfolded, it has hundreds of tiny holes. Small twigs of the food plant are placed in the bag. The fodder should be absolutely fresh but never wet. Finally the larvae are added, no more than twenty-five to a two-liter bag. The bag is then securely sealed and placed loosely in a carton. The size of the carton is selected so that the freezer bag just about fills it, leaving little space for play (Fig. 11). The carton is then closed and taken to the post office. I have

Der Zuchtbehälter sollte weder zu kalt noch zu warm stehen. Direkte Sonneneinstrahlung ist unbedingt zu vermeiden. Im Gegenteil hat es sich bei der Zucht von vielen Arten als günstig erwiesen, den Zuchtbehälter in den ersten zwei Tagen völlig dunkel zu stellen. Es besteht die Annahme, dass die Raupen so leichter zu fressen beginnen. Einige Züchter teilen diese Meinung. Viele Arten, zum Beispiel Raupen der Gattung *Automeris*, leben sehr gesellig. Es ist also sehr wichtig, dass man die Raupen dieser Arten zusammensetzt. Schon beim Kauf sollte man eine gewisse Anzahl von Eiern erwerben. Große Zuchtbehälter, in denen sich die Raupen verlaufen und somit auch verlieren, können zum Verlust vieler Raupen führen. Es wurde sehr oft beobachtet, dass solche vereinsamten Raupen nicht zu fressen beginnen und eingehen. Die Raupen, welche Raupenspiegel bilden können, gehen eher an das Futter, wachsen besser und fressen auch zusammen. Schwieriger ist es natürlich, Raupen von Arten zum Fressen zu bringen, die noch nie gezüchtet wurden. Dieses Buch soll dabei helfen. Am besten geht man wie folgt vor. Man schaut zunächst nach, ob andere Arten innerhalb der Gattung schon einmal gezüchtet wurden. Die Wahrscheinlichkeit ist sehr hoch, dass Arten innerhalb einer Gattung ähnliche Futterpflanzen bevorzugen. So gibt es schon klare Tendenzen. Arten innerhalb der Gattung *Rothschildia* Grote, 1897, wurden zum überwiegenden Teil in Europa erfolgreich mit Liguster *(Ligustrum)* gezüchtet, oder Arten der Gattung *Antheraea* Hübner, 1819 [»1816«], mit Eiche (*Quercus*). Trotzdem wird man einen Nahrungsspezialisten wie *Rothschildia tucumani*, (Dognin, 1901), dessen Raupen *Eupatorium*, fressen, nicht leicht an *Ligustrum* bekommen. Am Ende bleibt das Ausprobieren. Man hat dafür maximal zwei bis drei Tage Zeit, dann sterben die Raupen ohne Nahrungsaufnahme in der Regel, selbst bei mehrmaligem Besprühen. Es ist möglich, den Jungräupchen die Blätter mehrerer verschiedener Pflanzen anzubieten und im Ausschlussverfahren wieder zu entziehen. In der Regel meiden die Raupen die Blätter am meisten, die sie am wenigsten fressen würden. Wenn also sehr viele Raupen auf einem Blatt sitzen, aber noch nicht gefressen haben, dann ist die Wahrscheinlichkeit hoch, dass dieses Futter doch noch angenommen wird. Am Ende zählt die Erfahrung. Die erfolgreiche Verwendung von Kunstfutter bei Saturniiden ist mir nicht bekannt. Ich selbst habe mehrere Versuche unternommen.

used this method for several years, and I recall no negative reports about problems associated with shipping in this way. The advantages are obvious. Water does not condense within the bag. The packet weighs very little. It is possible to send the packet as a 'Maxibrief' (to 1 kg) or a 'Großbrief' (to 500 g). In Germany, these travel considerably faster than packages. There is nothing in the packet that can squash the larvae. If the carton itself is compressed, the freezer bag is completely flexible. Should the temperature rise, the interior of the packet is less affected than would be the case of plastic or glass containers.

3.4 The choice of a container for breeding and its correct placement

Once the larvae have begun to feed, it's time to set up their final breeding chamber. First of all, the container's size must accommodate the number and size of the larvae. Every two days (three days maximum) the larvae are moved to fresh food. Pruning shears are useful for this. With the shears, one cuts off the old branch and separates the branches or leaves on which larvae are sitting from those without larvae. The old fodder is completely removed. Next, using smaller shears, the part of the plant holding larvae is trimmed. No more of the old leaves should remain than is necessary. The breeder then sets them aside and lightly mists them with water so that the larvae may drink. Now the breeding chamber is thoroughly cleaned until no plant or fecal material remains. The floor of the chamber is then covered with an absorbent material such as paper towels. Now the chamber can be filled with fresh food. If branches are standing in water to prolong their freshness, then the container of water must also be renewed. When all this has been accomplished, the larvae are moved to the fresh supply of food. The bits of old leaf or twig supporting larvae are attached with needles or clothes pins to new ones. After a few hours, the old bits can be removed once the larvae have moved onto fresh leaves. One should always avoid using force when attaching larvae to or removing them from branches or leaves. Larvae are very sensitive and will move on spontaneously.

Larvae can be raised in a wide variety of containers. One popular method is keeping larvae in aquaria or various kinds of glassware. Glass

Es gelang, Raupen von *Rothschildia tucumani*, an Kunstfutter bis zum dritten Raupenstadium zu ziehen, danach verendeten alle Tiere.

3.3 Der Versand von Raupen

Wenn man Raupen versenden möchte, dann sind einige wichtige Dinge zu beachten. Der Versand von Eiern oder auch Kokons ist relativ unproblematisch, weil die Sendung bis zu 10 Tagen unterwegs sein kann. Das ist beim Versand von Raupen anders. Hier muss die Sendung den Empfänger spätestens nach drei bis vier Tagen erreichen. Innerhalb Europas ist das vielleicht noch möglich. Nach Übersee dauern die Transferzeiten in der Regel länger und der Versand von Raupen macht keinen Sinn. Weiterhin ist es wichtig, die klimatischen Verhältnisse zu beachten. Eier und Kokons sind relativ robust. Raupen reagieren auf zuviel Hitze oder Kälte viel empfindlicher. Das ideale Raupenstadium für den Versand ist das zweite oder dritte, am besten vor der Häutung zum dritten Präimaginalstadium. Die Raupen fressen so schon sehr stabil am Futter, verbrauchen aber noch sehr wenig. Das ist wichtig, um genügend Futter für bis zu vier Tage in das Versandbehältnis geben zu können. Erwachsene Raupen eignen sich nicht gut für den Versand. Sie verbrauchen sehr viel Platz, reagieren gestresster und es ist schwierig abzuschätzen, wie viel Futter während des Transfers verbraucht wird. Außerdem entstehen große Mengen Kot. Dieser Kot kann anfangen zu schimmeln, oder sich in Verbindung mit Kondenswasser verflüssigen. Somit wären die Raupen einer Infektionsgefahr ausgesetzt. All das umgeht man in den ersten Präimaginalstadien. Stehen die Raupen zudem noch kurz vor der nächsten Häutung, hören sie auf zu fressen und koten auch nicht mehr. Am besten hat sich folgende Versandmethode bewährt. Man nimmt einen gewöhnlichen Gefrierbeutel mit einem Volumen von bis 2 Litern. Man faltet diesen Gefrierbeutel mehrmals und sticht dann mit einer Nadel mehrmals hinein. Man entfaltet den Beutel wieder und hat nun einen Beutel mit Hunderten von kleinen Löchern. Nun gibt man zuerst kleine Zweige der Futterpflanze in den Beutel. Das Futter sollte absolut frisch und darf nicht nass sein. Zuletzt kommen nun die Raupen in den Beutel, nicht mehr als 25 Raupen pro 2-Liter-Beutel. Nun wird der Beutel fest verschlossen und wird lose in einen Karton gelegt. Die Größe des Versandkartons sollte so containers have the advantage of transparency so that the larvae are visible and they can be illuminated. Unfortunately they are also quite heavy and can be difficult to handle. When using containers of glass, one must be alert to their propensity for promoting humidity and consequently the rapid growth of molds. Raising larvae in an aquarium has much to recommend it and in many respects is quite practical. I attribute the majority of my successes in breeding to the use of aquaria. They are easy to clean and keep sterile. The contents are visible, and the height of the walls makes it easy to water the fodder. Aquaria are also convenient to use for pupation when filled with soil or peat. The temperature can be controlled by inserting heating elements. Light fixtures can be installed and the humidity adjusted. However, aquaria have really high purchase prices. Plastic boxes in many styles offer an alternative to glass. The advantages of plastic boxes are that they are very light, stackable, available in many sizes, and their purchase is cheap. So, they offer a really good alternative to glass. They prove their worth when obtained from a store free of charge as the packaging or containers of goods. Caution must be exercised in the case of soft plastics. Under the influence of heat, such boxes release definitely noxious gases. This results in the loss of all the larvae and the breeder is left wondering why they died.

Another kind of breeding chamber is a cage made of gauze. This material can be used to make boxes or bags. These containers are suitable for species that do not require a reservoir of water or high humidity, for example the genus *Hyalophora* Duncan, 1841. Of course, one must spray the food with water. Otherwise it would dry out within a few hours. Containers of gauze make a lot of sense, if I want to confine saturniids to a potted plant or raise the larvae out-of-doors. In fact there are species, like the genus *Attacus* Linnaeus, 1767, whose larvae are so lethargic that that they really don't need a container. Food, like a bouquet of flowers, is put into a vase and the larvae are placed on the plants. The placement of breeding chambers always deserves consideration. The direct rays of the sun are to be avoided a priori, because water can condense and accumulate. The danger is that the larvae will seek to drink this water on the walls or floor of the chamber. Any container of water used to keep the fodder moist becomes a hazard for the larvae. Regardless of

gewählt werden, dass der Raupenbeutel den Karton mehr oder weniger ausfüllt und wenig Spiel hat (Abb. 11). Der Karton wird verschlossen und zur Post gebracht. Diese Versandmethode wird von mir seit vielen Jahren praktiziert. Mir sind keine negativen Rückmeldungen über Versandprobleme in Erinnerung. Die Vorteile liegen auf der Hand. Innerhalb des Beutels bildet sich kein Kondenswasser. Das Päckchen hat nur ein sehr geringes Eigengewicht. Ein Versand als Maxibrief oder Großbrief ist möglich. In Deutschland werden diese Briefe wesentlich schneller transportiert als Pakete. Es gibt nichts im Behältnis, was die Raupen zerquetschen könnte. Wird der Karton selbst gequetscht, ist der Beutel absolut flexibel. Bei hohen Temperaturen staut sich die Hitze in einem Karton nicht so sehr wie in einer Plastikschachtel oder einem Glas.

3.4 Die Wahl des Zuchtbehälters und der richtige Standort

Haben die Raupen angefangen zu fressen, so kommen sie in ihren endgültigen Zuchtbehälter. Auch hier gilt, die Größe des Behälters an die Menge und Größe der Raupen anzupassen. Alle zwei bis maximal drei Tage sind die Raupen umzusetzen. Man nimmt dazu am besten eine Pflanzenschere. Mit dieser Schere zerlegt man die alten Zweige und trennt die Zweige oder Blätter, auf denen Raupen sitzen, von denen, auf denen sich keine Raupe befindet. Das alte Futter wird komplett entsorgt. Nun schneidet man mit einer kleineren Schere die Pflanzenteile mit den darauf befindlichen Raupen aus. Es sollte nichts mehr vom alten Futter übrig bleiben. Die Raupen legt man beiseite und kann sie gleich mit einem Sprühnebel überziehen, damit sie trinken können. Nun wird der Zuchtbehälter gründlich gereinigt, bis kein Kot und keine alten Pflanzenteile mehr übrig sind. Den Boden des Behälters belegt man mit einer saugfähigen Unterlage, zum Beispiel Küchenrolle. Nun befüllt man den Behälter mit frischem Futter. Reicht man gewässertes Futter, ist natürlich das Wasser im Wässerungsbehälter zu erneuern. Ist das alles erledigt, dann setzt man die Raupen wieder auf das frische und nun neue Futter. Mit Nadeln oder Wäscheklammern fixiert man die Blätter oder Zweige, auf denen die Raupen noch immer sitzen. Diese Altteile entfernt man nach ein paar Stunden, wenn die Raupen auf das neue Futter gewechselt sind. Man sollte es stets vermeiden,

Fig. 11. Packing larvae for shipping. – Photograph by Frank Meister.
Abb. 11. Versand von Raupen. – Foto: Frank Meister.

whether bottles or vases are used, larvae must be allowed absolutely no opportunity to reach the water. Entire broods have been destroyed that way in a single day. Apparently larvae can sense the presence of water and develop a strong urge to drink. Otherwise, it is difficult to explain why larvae drown themselves. Larvae will follow the stems of their food plants into the container of water where one finds them lying dead on the bottom. The larvae will reach their goal, passing through the smallest spaces between the stems. One can try to prevent this by using folded paper towels to fill up the spaces between stems and the walls of the container. Hopefully, the larvae will no longer be able to reach the interior of the reservoir. Alternatively, the breeder may choose containers with covers, through which he makes holes just large enough to pass the stems of the food plants. Seemingly drowned larvae can sometimes be saved. One massages the larva's body to expel water. Next the larva is laid on an absorbent paper towel and dried with a hair-dryer. Then one waits for a few hours. In this way, some larvae recover and resume their development without difficulty.

Raupen anzufassen oder zu versuchen, sie mit Gewalt von Ästen oder Blättern zu entfernen. Raupen sind sehr empfindlich und reißen dann ein.

Es gibt eine große Vielzahl von Behältnissen, in denen man Raupen züchten kann. Eine weit verbreitete Methode ist das Züchten in Gläsern aller Art und Aquarien. Gläser haben den Vorteil, dass sie durchsichtig sind, den Raupen also viel Helligkeit bieten. Leider sind sie sehr schwer und teilweise unhandlich. Bei Glaszuchten ist auf die Feuchtigkeit zu achten und den damit sehr schnell entstehenden Schimmel. Die Zucht in Aquarien nimmt zwar sehr viel Platz in Anspruch, ist aber in jeder Hinsicht sehr praktisch. Der Großteil meiner Zuchten erfolgt nach wie vor in Aquarien. Sie sind leicht zu säubern und steril zu halten. Man hat alles im Blick und durch die Hochwandigkeit kann man das Futter auch gut wässern. Man kann Aquarien zudem gut für Erdverpuppungen nutzen, wenn man sie mit Erde oder Torf befüllt. Aquarien kann man mit Heizdrähten oder Heizmatten bestücken, man kann so die Temperaturen regeln, man kann sie beleuchten und man kann für Luftfeuchtigkeit sorgen. Die Anschaffung ist aber recht teuer. Eine Alternative bieten Plastikschachteln aller

3.5 Breeding tropical saturniids in an outdoor enclosure

Breeding in the open environment is a somewhat neglected subject. Only in the older literature is there much mention of propagation out-of-doors (c.f. Voelschow 1902). However, tropical saturniids can be raised very well in the open, although such efforts are restricted to the months from May to October in Germany. The conditions are natural enough if one has his own garden or even a balcony and the appropriate plants for food are available. If a greenhouse is considered, then attention must be given to regulating its temperature and humidity. Anyone who wants to propagate the Saturniidae in the long term should have his own garden or a suitable chamber that he can stock with the plants need-

Fig. 12. Garden oak tree wrapped in mosquito netting. – Photograph by Frank Meister.
Abb. 12. Mit Moskitonetzen eingebundene Eiche im Garten. – Foto: Frank Meister.

Art. Der Vorteil von Plastikschachteln ist, dass sie sehr leicht sind, sogar stapelbar, es gibt sie auch in allen Größen und die Anschaffung ist billig. Sie bieten eine sehr gute Alternative zu Glas. Bewährt haben sich zudem Plastikeimer aller Art, die man auch kostenlos im Handel erhält. Vorsicht ist aber vor Weichplastikdosen geboten. Gerade unter Einfluss von Hitze gibt es Dosen, die offensichtlich schädliche Gase freigeben. Das Resultat ist ein Verlust aller Raupen und man fragt sich dann, woran die Raupen eigentlich gestorben sind? Ein weitere Art von Zuchtbehältern sind Gazekäfige, Raupenzuchtkästen oder auch Zuchtsäcke aus Gaze. Diese Gazekäfige eignen sich zur Zucht von Arten, die Staunässe oder hohe Luftfeuchtigkeit nicht vertragen, zum Beispiel aus der Gattung *Hyalophora* Duncan, 1841. Hier muss man das Futter natürlich wässern, weil es sonst in wenigen Stunden vertrocknen würde. Gazekäfige machen mehr Sinn, wenn ich Saturniiden auf eingetopften Pflanzen züchten möchte oder die Raupen ausgebunden werden sollen. Es gibt natürlich auch Arten, wie die der Gattung *Attacus* Linnaeus, 1767, deren Raupen so träge sind, dass sie eigentlich kein Gefäß zur Zucht benötigen. Man stellt das Futter, wie einen Blumenstrauß in eine Vase und setzt die Raupen darauf. Der Standort des Zuchtgefäßes sollte immer genau bedacht werden. Direkte Sonneneinstrahlung ist prinzipiell zu vermeiden, weil sich sonst Kondenswasser bilden kann. Gerade kleine Raupen können so an den Gefäßwänden oder auf dem Boden ertrinken. Eine große Gefahr für die Raupen sind alle Wassergefäße, die man nutzt, um das Futter zu wässern. Egal ob Flaschen oder Vasen, es darf absolut keine Möglichkeit geben, dass die Raupen in die Behälter gelangen können. Ganze Zuchten können so innerhalb eines Tages ihr Ende finden. Die Raupen verspüren offenbar einen starken Drang zu trinken. Anders ist es kaum erklärbar, warum Raupen sich ertränken. Die Raupen klettern an den Stängeln der Futterpflanzen in die Wasserbehälter und man findet sie dann dort tot auf dem Boden liegen. Es reichen kleinste Hohlräume zwischen den Stängeln. Man behilft sich, indem man Küchenrollen zerkleinert und alle Zwischenräume zwischen den Stängeln und den Gefäßwänden damit zustopft, so dass keine Raupe mehr in das Innere der Wasserbehälter gelangt. Oder man wählt Gläser bzw. Dosen mit Deckel, in die man Löcher stanzt, gerade so groß, dass die Zweige der Futterpflanzen hineinpassen. Vermeintlich ed as food. These include *Ligustrum ovalifolium* hedges, winter-hardy oak trees, false acacia trees and lilac bushes, to name a few. They should be pruned to keep them small – that makes it easier to wrap them with materials like gauze. It is also important here to be aware of the species and their climatic requirements. The following method has proved to be the best. One begins by observing the young larva for a few days under normal room conditions. The best time for taking them outside comes when they have begun to eat leaves to which the breeder wants to attach them. He has already sewn a bag of gauze. The gauze must have a very fine mesh. In case of doubt, a double layer may be used. Next one places the gauze sack over a branch of an edible plant situated in partial shade. The branch to provide fodder has already been searched and any harmful or predatory insects removed. One should be especially alert to remove all ants. Now the young larvae may be placed in the sack. To do this one either uses a fine brush or takes a leaf bearing a larva from the breeding container. The leaf is carefully trimmed around the larva and stapled or pinned to the branch inside the gauze sack. During the process, larva must be prevented from falling out of the bag. As a rule, larvae will desert a wilting leaf and move onto a fresh one. What is now the feeding sack is briefly misted with water and the open side is tied up. Great care must be given this. No aperture must remain where the bag joins the branch. Should a larva wish to leave the sack, it will creep along the branch. Furthermore, injurious insects will attempt to enter the bag from the other direction. Therefore, everything must be tightly closed. This should be kept in mind if it becomes necessary to open the sack and move the larvae. Now the larvae may be left to their own devices. It is advisable to periodically spray the bag from the outside. Depending on climatic influences, the larvae are likely to develop more slowly than they would indoors. Meanwhile, the breeder has almost nothing to do. When the branch supporting the larvae becomes denuded, one simply relocates the sack of larvae. It's practical to have bags of various sizes available. With very large species or dealing with large broods, one can use mosquito netting (Fig. 12). With mosquito netting one can enclose entire trees. The breeder is advised to take precautions against birds; they will take advantage of any opening to reach the larvae. I have myself had very good results with

ertrunkene Raupen können manchmal gerettet werden. Man massiert den Raupenkörper und drückt damit Wasser aus der Raupe. Man legt dann die Raupe auf eine saugfähige Küchenrolle und trocknet sie mit einem Föhn und wartet einige Stunden ab. Schon so manche tot geglaubte Raupe erholte sich wieder und setzte ihre Entwicklung ohne Probleme fort.

3.5 Die Zucht von tropischen Saturniiden durch Ausbinden der Raupen im Freien

Etwas in Vergessenheit geraten ist die Freilandzucht. Gerade in älteren Berichten wird viel über die Zucht im Freiland gesprochen (s. Voelschow 1902). Tropische Saturniiden kann man sehr gut im Freien züchten. Die Zucht beschränkt sich dann hier in Deutschland auf die Monate Mai bis Oktober. Die Vorraussetzung ist natürlich, dass man einen eigenen Garten besitzt oder auch einen Balkon und die geeigneten Futterpflanzen zur Verfügung hat. Auch ein Gewächshaus kommt in Frage, muss aber im Bezug auf Temperatur und Feuchtigkeit regelbar sein. Wer sich langfristig mit der Zucht von Saturniiden beschäftigen will, sollte seinen Garten oder entsprechende Kübel mit den Standardfutterpflanzen bestücken. Dazu würden *Ligustrum-ovalifolium*-Hecken, winterharte Eichenbäume oder Robinien sowie Fliederbüsche zählen, um nur einige zu nennen. Man sollte diese Fraßpflanzen durch Verschnitt klein halten. Dadurch wird es einfacher, sie bei Bedarf einzubinden. Auch hier ist es wieder wichtig, die Arten und ihre Anforderungen an klimatische Bedingungen zu kennen. Als beste Methode hat sich die folgende erwiesen. Man zieht die kleinen Raupen unter normalen Zimmerbedingungen ein paar Tage an. Die Raupen haben zu fressen begonnen, am besten genau an den Blättern der Pflanze, an die man sie nun ausbinden möchte. Zuvor hat man einen Gazesack genäht. Die Gaze muss sehr feingliedrig sein. Im Zweifelsfall kann man auch eine doppelte Lage Gaze verwenden. Nun nimmt man einen im Halbschatten befindlichen Zweig einer Futterpflanze und stülpt den Gazesack darüber. Vorher sollte man den Futterzweig auf Schad- und Raubinsekten prüfen und diese entfernen. Besonderes Augenmerk gilt allen Ameisen. Nun verbringt man die kleinen Raupen in den Futtersack. Man kann dazu entweder einen feinen Pinsel verwenden, oder man nimmt die Blätter aus dem Zuchtbehälter, schneithis technique, and in one summer I had more than three hundred larvae of *Antheraea paphia* (Linnaeus, 1758) enveloped on birch trees. The final step is the collecting of the cocoons from the trees (Fig. 13). *Telea montezuma* (Sallé, 1856), *Copaxa lavendera* (Westwood, [1854]) and *Automeris* species have been successfully raised in this way. Still, there are great differences in the factors affecting particular species. Species that spin a free-hanging cocoon are problem-free. Those that pupate in the soil must of necessity be collected. These larvae have such a powerful urge to reach the ground that they will eat their way through the gauze. Very gregarious species like those of the genus *Automeris* wander very far. They too will eat through the gauze and disappear. One should take precautions that foreign species, species that don't belong there, do not expand into the local environment. It seems unlikely that tropical species of saturniids could survive the European winter, but how certain is that? A montane species, one that is frost-resistant, can also survive in Europe. Species from faunas similar to ours would also have a good chance of surviving. Breeders must recognize their responsibility and protect our natural environment from foreign invaders. "Die Ameise als Tramp", written by Bernhard Kegel in 1999, is a very interesting book about invasive plants and animals.

3.6 The issue of sanitation

Before starting out, one should consider how much time he is willing to devote to the activities of a breeder. The breeding of the Saturniidae is very time-consuming and can easily demand several hours a day. He should also consider which species he intends to raise and how many he can handle and what each species eats. Breeding different species simultaneously can take up a great deal of time, since each one must feed on its own preferred plant. If such plants are not immediately available, then the effort of searching for and obtaining them can take up half of one's time. The quantity of fodder should not be underestimated. In their final stage, one hundred growing larvae of the genus *Antheraea* will consume the leaves of a three-meter birch tree every day. This leads to not only the problem of obtaining fodder, but also the problem of providing sufficient space. Unfortunately, there are far too many breeders who take on more than their time allows, which brings us to the problems of sanita-

det die Spiegel mit den Raupen so gut wie möglich ab und fixiert diese Blattraupenspiegel mit einer Klammer oder einer Nadel am Futterzweig. Achtung, dass keine Raupen herunterfallen. In der Regel werden die Raupen das nun welkende Blatt verlassen und auf die frischen Blätter hinüberwechseln. Noch kurz mit Wassernebel besprühen und den Futtersack an der offenen Seite zubinden. Hier ist große Sorgfalt gefragt. Es darf keine einzige offene Stelle am Ast geben. Wenn eine Raupe den Sack verlassen möchte, dann wird sie es am Ast entlanglaufend versuchen. Umgekehrt werden auch Schadinsekten so in den Futtersack gelangen. Also muss alles sehr dicht verschlossen werden. Man sollte allerdings beachten, dass man den Sack auch wieder öffnen muss, um die Raupen umzusetzen. Nun überlässt man die Raupen sich selbst und kontrolliert sie. Regelmäßiges Besprühen von außen ist ratsam. Die Raupen entwickeln sich je nach klimatischen Verhältnissen in der Regel langsamer als bei Zimmerzucht. Dafür hat man so gut wie keine Arbeit. Ist der Futterzweig abgefressen, setzt man die Raupen um. Unterschiedliche Sackgrößen sind sinnvoll. Bei sehr großen Arten oder Massenzuchten verwendet man Moskitonetze (Abb. 12). Mit diesen Moskitonetzen kann man ganze Bäume einbinden. Vorsicht ist vor Vögeln geboten, die jedes Loch nutzen werden, um an die Raupen zu gelangen. Ich selbst habe sehr gute Erfahrungen mit dem Ausbinden gemacht und in einem Sommer über 300 Raupen von *Antheraea paphia* (Linnaeus, 1758), an Birken ausgebunden. Am Ende sammelt man dann nur noch die Kokons von den Bäumen ab (Abb. 13). Auch *Telea montezuma* (Sallé, 1856), *Copaxa lavendera* (Westwood, [1854]) oder *Automeris*-Arten wurden erfolgreich ausgebunden. Dennoch gibt es schon große Unterschiede, was die Eignung der einzelnen Arten betrifft. Arten, deren Raupen einen frei hängenden Kokon spinnen, sind unproblematisch. Arten, deren Raupen sich zur Verpuppung in die Erde begeben, muss man unbedingt abfangen. Die Raupen haben einen derart großen Drang, den Erdboden zu erreichen, dass sie sich durch die Gaze fressen. Sehr gesellig lebende Arten wie aus der Gattung *Automeris* wandern extrem viel. Auch hier fressen sich die Raupen gern durch die Gaze und sind dann verschwunden. Achten Sie darauf, dass keine nicht heimischen Arten in die freie Natur gelangen, die dort nicht hingehören. Zwar überleben tropische Saturniiden-Arten den europäischen Winter

Fig. 13. Tree denuded of leaves with cocoons of *Antheraea paphia* (Linnaeus, 1758). – Photograph by Frank Meister.

Abb. 13. Abgefressener Baum mit Kokons von *Antheraea paphia* (Linnaeus, 1758). – Foto: Frank Meister.

tion and hygiene. The most successful breeders known to me completely change the fodder every two days (three days at most). The containers used for breeding are thoroughly cleaned and the water used for sprinkling fodder is changed. Every larva is examined. Wilted leaves lead to intestinal problems. These are recognized when the larvae stop feeding and become completely inactive. Within a few hours, most of them will empty their intestines. The feces may appear thread-like, or even worse, the larva may seem to have completely emptied itself. Along with this they expel a clear or sometimes black fluid. As a rule, such larvae are beyond saving. A word of caution however, such an event may not spell the end of the larva. Larvae about to pupate also empty their intestines. The sick animals must immediately be separated from the healthy ones in the hope that the infection will not destroy the

nicht, aber wie sicher ist das? Eine hochalpine Art, die auch in ihrer Heimat Frost ausgesetzt ist, kann auch in Europa überleben. Arten aus Faunen, die unserer ähnlich sind, hätten auch gute Überlebenschancen. Werden Sie also Ihrer Verantwortung bewusst und schützen Sie unsere Natur vor fremden Invasoren. Ein sehr interessantes Buch über Neophyten und Neozoen stammt von Bernhard Kegel, 1999 »Die Ameise als Tramp«.

3.6 Das Thema Sauberkeit

Man sollte sich vorher darüber im Klaren sein, ob man die Zeit hat und auch aufbringen möchte, um sich dem Thema der Zucht zuzuwenden. Die Zucht von Saturniiden ist sehr zeitaufwändig und kann schnell mehrere Stunden pro Tag in Anspruch nehmen. Man sollte sich vorher auch überlegen, welche Arten man vorhat zu züchten, welche Mengen an Raupen man bewältigen kann und was die einzelnen Arten fressen. Es kann sehr viel Zeit kosten, verschiedene Arten gleichzeitig zu züchten, die auch noch verschiedene Futterpflanzen benötigen. Hat man diese Futterpflanzen nicht im unmittelbaren Zugriff, dann kann die Futtersuche und Futterbeschaffung die Hälfte der Zeit in Anspruch nehmen. Die Futtermenge ist nicht zu unterschätzen. 100 erwachsene Raupen der Gattung *Antheraea* benötigen im letzten Stadium etwa die Blätter einer drei Meter hohen Birke pro Tag. Das führt nicht nur zu Problemen bei der Futterbeschaffung, sondern wird auch schnell zum Platzproblem. Leider gibt es viel zu viele Züchter mit Zeitproblemen und das bringt uns zum Thema Sauberkeit und Hygiene. Die erfolgreichsten Züchter, die mir bekannt sind, wechseln das Futter alle zwei bis maximal drei Tage komplett aus. Die Zuchtbehälter werden gründlich gereinigt und das Wasser zum Wässern des Futters erneuert. Jede einzelne Raupe wird untersucht. Verwelktes Futter führt zu Darmproblemen. Darmprobleme erkennt man daran, dass die Raupen aufhören zu fressen. Sie werden vollkommen inaktiv. Innerhalb weniger Stunden kommt es dann meist zu einer Darmentleerung. Es bildet sich Schnurkot oder noch schlimmer, die Raupe scheint völlig auszulaufen. Dabei gibt sie eine klare aber manchmal auch schwarze Flüssigkeit von sich. In der Regel sind solche Raupen nicht mehr zu retten. Aber Vorsicht, diese Darmentleerung darf nicht fehlgedeutet werden. Zur Verpuppung rei-

Fig. 14. Final instar larva of *Automeris postalbida* Schaus, 1900, photograph courtesy of Horst Käch.
Abb. 14. Letztes Raupenstadium von *Automeris postalbida* Schaus, 1900. – Foto: Horst Käch, mit freundlicher Erlaubnis.

Fig. 15. Final instar larva of *Automeris oberthurii* (Boisduval, 1875). – Photograph courtesy of Ulf Drechsel.
Abb. 15. Erwachsene Raupe von *Automeris oberthurii* (Boisduval, 1875). – Foto: Ulf Drechsel, mit freundlicher Erlaubnis.

entire brood. If the containers are not cleaned every two days, the fecal material will support the growth of molds. These fungi spread throughout the container and can wipe out the entire brood. Additionally, feces contaminate the fodder eaten by other larvae. Sometimes, all of the larvae will die within a few days, killed by fungal infections. Larvae so infected develop black spots. The powder used against athlete's foot (*Tinea* sp.) can be used to treat sick larvae. The powder is applied to the appropriate places with a cotton swab. Entomological suppliers offer a chemical named Thymol, which is effective against mildew. Thymol has a powerful odor, but it doesn't seem to harm larvae. The best course is to avoid the conditions that allow fungi to grow. If the entire brood is lost through infection, then the container must be thoroughly disinfected.

3.7 Allergic reactions

Caution is advisable in handling species whose larvae are extremely irritating. This advice should not be underestimated. Several larvae in the genera *Automeris* Hübner, [1819], *Lonomia* Walker, 1855, *Periga* Walker, 1855, *Leucanella* Lemaire, 1969, or *Dirphia* Hübner [1819] have poisonous hairs that resemble the spines of the stinging nettle. Contact with these hairs causes varying degrees of inflammation and pain similar to the effects of stinging nettles. The severity of allergic reactions differs from person to person. If there are children in the house, one should exercise care that they do not come into contact with these larvae. Unfortunately, these larvae are very active. Therefore it is important to prevent their escape from the container used for breeding. Very little is known about the mechanisms of toxicity and the effects these larval poisons have on humans. Sadly, these toxins can also cause death. A few years ago, a woman died from this cause. Walking barefoot in the jungle, she stepped on larvae of a *Lonomia* species. There was no help for her, and she died in a few days from internal bleeding.

fe Raupen entleeren auch ihren Darm. Man muss die erkrankten Tiere sofort von den gesunden trennen, in der Hoffnung, dass die Infektion nicht den ganzen Zuchtstamm vernichtet. Reinigt man nicht alle zwei Tage die Zuchtbehälter, so fängt der Kot an zu schimmeln. Diese Schimmelsporen verteilen sich im Zuchtbehälter und können ganze Zuchten vernichten. Außerdem verdreckt der Kot die Futterpflanzen und andere Raupen nehmen dieses Futter auf. Manchmal gehen dann alle Raupen innerhalb weniger Tage ein. Zudem kann es zu Pilzerkrankungen der Raupen kommen. An den Raupen bilden sich dann schwarze Flecken. Man kann diese erkrankten Raupen an den entsprechenden Stellen mit Fußpilzpuder behandeln, indem man mit einem Ohrtupfer den Puder auf die erkrankten Stellen aufträgt. Gegen Schimmel gibt es im entomologischen Handel ein Mittel namens Thymol. Thymol riecht zwar stark, scheint Raupen aber nicht zu schaden. Besser man schafft Bedingungen, die erst gar keinen Schimmel verursachen. Kommt es zu einem Totalverlust der Zucht wegen Infektion, dann muss der Zuchtbehälter gründlich desinfiziert werden.

3.7 Allergien

Vorsicht ist im Umgang mit Arten geboten, deren Raupen sehr stark nesseln. Das Thema ist nicht zu unterschätzen. Viele Raupen der Gattungen *Automeris* Hübner, [1819], *Lonomia* Walker, 1855, *Periga* Walker, 1855, *Leucanella* Lemaire, 1969, oder *Dirphia* Hübner [1819] haben stark nesselnde Brennhaare. Kommt man mit diesen Raupen in Kontakt, so führt das zu leichten bis extremen Hautrötungen, verbunden mit leichten bis starken Schmerzen, ähnlich denen beim Kontakt mit einer Brennnessel. Die allergischen Reaktionen sind von Mensch zu Mensch unterschiedlich. Hat man Kinder im Haushalt, so sollte man darauf achten, dass diese nicht in Berührung mit diesen Raupen kommen. Leider sind die Raupen der oben aufgeführten Gattungen sehr aktiv. Es ist also wichtig, dass die Raupen die Zuchtbehälter nicht verlassen können. Es ist sehr wenig über die toxischen Wirkungen dieser Raupen auf den Menschen bekannt. Es sind leider aber auch schon Todesfälle bekannt geworden. So starb vor einigen Jahren eine Frau, als sie barfuß durch den Dschungel gehend auf Raupen einer *Lonomia* Art trat. Es gab keine Rettung und nach wenigen Tagen erlag sie ihren inneren Blutungen.

4. The successful treatment of cocoons and pupae

4.1 Pupation

The larva will always exhibit clear indications that it is about to enter pupation. Many species, usually those that pupate in the earth, develop distinctive coloration. Shortly before pupation, larvae show signs of a more or less strong wanderlust. This mobile phase can last up to three days. Larvae that make their cocoons among foliage travel on average less than those that enter the soil. Little is known and much is speculated about the reasons for the impressive changes in color and the wandering behavior. The following is a highly probable hypothesis proposed by a few breeders. The color scheme of the larva is associated with its wandering behavior. In their feeding phase, many larvae of non-poisonous species take on the color of the foliage they eat and become virtually invisible to their predators. Larvae of a few species of the genus *Salassa* Moore, 1859, exemplify this. To the human eye, they are hardly distinguishable from the plants they eat, even within a breeding container (Fig. 16). Other species deceive their enemies by using more or less effective mimicry. In contrast, poisonous species exhibit highly conspicuous colors. If the larvae must leave the foliage to pupate in the earth, they creep along the branches and stems of their food plants until they reach the ground, or simply let themselves fall to the ground. A green larva on brown bark would be easily visible to a predator but not a brown one. I am not aware of any saturniid species that becomes brightly colored when ready to pupate. Except for the larvae of species having a distinctive color scheme, they always become darker, with a definite tendency toward brown and red tones. Larvae of the genus *Attacus* neither demonstrate such wandering behavior nor do they change color. As a rule, the larvae of this genus spin a web enclosed by an envelope of edible leaves. The wandering behavior may serve as an additional defense mechanism. This may be the case in some species about to spin a cocoon, when they abandon the tree on which they were feeding and proceed to attach themselves to a different tree. This may have something to do with the attraction trees infested with webs have for certain predators, so the larvae seek out a different, non-edible tree. Evidence is provided

4. Die erfolgreiche Behandlung der Kokons und Puppen

4.1 Die Verpuppung

Steht die Verpuppung einer Raupe bevor, gibt es dafür immer klare Anzeichen. Bei manchen Arten verfärben sich die Raupen, meist bei den Arten, die sich in der Erde verpuppen. Die Raupen entwickeln kurz vor der Verpuppung einen mehr oder weniger starken Wandertrieb. Dieser Wandertrieb kann bis zu drei Tage dauern. Raupen, die im Blattwerk Kokons bilden, wandern im Durchschnitt weniger als jene, die sich in der Erde verpuppen. Über die Gründe dieses Wanderverhaltens und der starken Verfärbungen der zur Verpuppung reifen Raupen ist wenig bekannt und wird viel spekuliert. Sehr wahrscheinlich ist folgende Theorie, die von einigen Züchtern vertreten wird. Die Verfärbung der Raupen steht im Zusammenhang mit ihrem Wanderverhalten. In ihrer Fressphase passen sich viele Raupen der ungiftigen Arten in ihrer Verfärbung dem Blattwerk der Fraßpflanze an, um so vor ihren Fressfeinden so gut wie möglich unsichtbar zu bleiben. So sind die Raupen einiger Arten von *Salassa* Moore, 1859, selbst ausgewachsen, mit dem menschlichen Auge in den Zuchtgefäßen auf den Futterpflanzen kaum zu erkennen (Abb. 16). Andere Arten täuschen ihre Feinde durch eine mehr oder weniger ausgeprägte Mimikri. Giftige Arten dagegen sind extrem auffällig gezeichnet. Wenn die Raupe nun das Blattwerk verlassen muss, um sich in der Erde zu verpuppen, dann geht das nur, indem sie an den Ästen und Stämmen des Fraßbaumes zum Boden kriecht, oder sich auf den Boden fallen lässt. Eine grüne Raupe wäre auf brauner Rinde oder Erde für Fressfeinde leichter zu erkennen, eine braune Raupe nicht. Ich kenne keine Saturniiden-Art, bei der die zur Verpuppung reife Raupe heller wird, sondern die Raupen der Arten mit deutlicher Verfärbung werden immer dunkler, mit klarer Tendenz zu braunen und roten Tönen. Raupen ohne ausgeprägtes Wanderverhalten vor der Verpuppung verfärben sich auch nicht, wie bei der Gattung *Attacus*. Raupen dieser Gattung verspinnen sich in der Regel in der Hülle eines Fraßblattes. Das Wanderverhalten kann zudem auch ein weiterer Schutzmechanismus sein. So wurde bei einigen Arten, die einen Kokon spinnen, beobachtet, dass sie zunächst den Fraßbaum verließen und sich einen neuen Baum suchten, auf dem sie sich dann

Fig. 16. Five larvae of *Salassa lola* (Westwood, 1847) feeding on a plant. – Photograph courtesy of Ronald Brechlin.
Abb. 16. Fünf Raupen von *Salassa lola* (Westwood, 1847) an der Futterpflanze. – Foto: Ronald Brechlin, mit freundlicher Erlaubnis.

by acacias in the Kalahari that release chemicals that regulate the strength of the larva's poison. Perhaps these chemicals are also perceived by the predators. A very definite indication that the larva is nearing pupation is the emptying of the intestine. This is recognized by the fact that the fecal material is no longer firm and stable, but fluid and even transparent. This shows that the larva has expelled the total contents of its intestine and is ready for pupation. Based on their manner of pupation, it is possible to describe three varieties of saturniids which will be treated separately in the following sections.

verspannen. Das könnte damit zu tun haben, dass bestimmte Fressfeinde gezielt von Raupen befallene Bäume nach Gespinsten absuchen und die Raupe sich gezielt einen Baum sucht, der nicht Fraßbaum war. Nachgewiesen ist, dass Akazien in der Kalahari Botenstoffe aussenden können und so den Grad ihrer Giftigkeit regulieren. Vielleicht können diese Botenstoffe auch von Fressfeinden wahrgenommen werden. Ein sehr eindeutiges Symptom für eine nahende Verpuppung ist die Darmentleerung der Raupen. Man erkennt das daran, dass der Kotballen nicht mehr fest und stabil ist, sondern flüssig, manchmal sogar glasklar. Die Raupe hat sich nun ihres gesamten Darminhaltes entleert und ist zur Verpuppung bereit. Prinzipiell kann man bei Saturniiden von drei Verpuppungsformen sprechen, die im Folgenden separat behandelt werden.

4.2 Strongly-woven cocoons and silken webs spun on leaves or bark

This includes those saturniids weaving cocoons that are strong enough in many species to withstand external mechanical forces. Examples include species of *Antheraea* and *Attacus*. The strongest are probably those of *Austrocaligula eucalypti* (Scott, 1864), which seem almost indestructible. These cocoons are so hard that a fine-toothed saw is needed to open them. Those of the largest web-spinning species, like *Hyalophora cecropia* (Linnaeus, 1758) or *Coscinocera hercules*, can be as large as a fist. Larvae about to spin must be given the opportunity to lay out their silken threads. It is appropriate to separate such a larva from those that are not ready to start spinning. There are a few species that need one or more leaves that they can roll up to serve as a substrate for their cocoons. This behavior is exhibited by species of *Attacus* and *Samia* Hübner, 1819 [»1816«]. Other species, like *Antheraea paphia*, construct pendant cocoons and need a strong branch to anchor the suspending cable of silk. It is noteworthy that species constructing pendant cocoons or webs suspend them above the ground and not in contact with the soil. It is best to offer larvae about to spin a variety of possible substrates for a cocoon. There are a number of materials that may be placed in the container with the larvae. Such things should correspond to natural conditions. Species that enroll themselves in leaves will need a sufficient quantity of foliage, while others may be given a layer of sticks or straw. Species that spin their webs on tree bark, like *Austrocaligula eucalypti* or *Saturnia pyri* ([Denis and Schiffermüller]), 1776, should naturally be given similar material. Paper and cardboard are commonly recommended, but should only be used if they can later be separated from the walls of webs or cocoons. This is not always possible when using newspaper. On the one hand, it fuses with the silk, changing the microclimate within the cocoon, and on the other, it may interfere with the emergence of the imago from the cocoon, even to the point of making this impossible. Should the larvae fail to find suitable conditions for spinning their silk, they may wander about for days, starting several webs only to abandon them. Alternatively, they may spin their silk on the walls of the breeding chamber. At worst, no cocoon is formed at all. In searching for an appropriate location, the lar-

4.2 Feste Kokons oder Seidengespinste im Blattwerk oder an der Rinde

Gemeint sind hier Saturniiden-Arten, die einen mehr oder weniger stabilen Kokon spinnen, der bei vielen Arten fest genug ist, um gegen äußere mechanische Einflüsse zu bestehen, wie bei *Antheraea*- oder *Attacus*-Arten. Die wohl härtesten Kokons findet man bei *Austrocaligula eucalypti* (Scott, 1864), die unzerstörbar erscheinen. Diese Kokons sind so hart, dass man eine Feinsäge benötigt, um sie zu öffnen. Die größten Gespinste bauen Arten wie *Hyalophora cecropia* (Linnaeus, 1758) oder *Coscinocera hercules*, die faustgroß werden können. Der spinnreifen Raupe muss Gelegenheit gegeben werden, ein Gespinst anzulegen. Es ist zweckmäßig, eine solche Raupe vom Rest der noch nicht spinnreifen Raupen zu trennen. Einige Arten benötigen zur Anlage des Kokons nur ein oder mehrere Blätter, die eingerollt werden, wie bei Arten von *Attacus* oder *Samia* Hübner, 1819 [»1816«]. Andere Arten bilden hängende Kokons und benötigen einen dicken Ast, an dem der Halte- und Hauptseidenstrang befestigt wird, wie bei *Antheraea paphia*. Es ist darauf zu achten, dass Arten, die hängende Kokons oder Gespinste anlegen, auch hängend gelagert werden und nicht auf den Boden gelegt werden. Am besten bietet man den spinnreifen Raupen vielfältige Möglichkeiten zur Anlage eines Kokons. Es gibt verschiedene Materialien, mit denen man einen Verspinnungsbehälter füllen könnte. Das Material sollte den natürlichen Bedingungen entsprechen. Arten, die Kokons in eingerollten Blättern anlegen, brauchen genügend Blattwerk, anderen bietet man ein Gestrüpp aus Reisig oder Stroh an. Arten, die sich an Baumrinde verspinnen, wie *Austrocaligula eucalypti* oder *Saturnia pyri* ([Denis & Schiffermüller], 1776), sollte auch Rinde geboten werden. Die Verwendung von Papier und Pappe wird häufig auch empfohlen, sollte aber nur dann zum Einsatz kommen, wenn man sie später auch wieder von den Gespinst- oder Kokonwänden entfernen kann. Gerade bei Zeitungspapier ist das aber nicht immer der Fall. Verbindet sich das Papier mit den Gespinstwänden, so ist einerseits wieder das Mikroklima im Kokon verändert und es kann zu Schlupfproblemen der Imagines führen, die dann nicht aus den Kokons herauskommen. Finden die Raupen keine geeigneten Spinnmöglichkeiten, so laufen sie oft tagelang umher, fertigen mehrere Gespinste, die dann wieder verlassen vae become exhausted. If they waste too much silk in abortive beginnings, the final cocoon will be weak or incomplete. Compared to a normal cocoon, the delicate microclimate within a defective cocoon will in time be destroyed, and the pupa will die during a long diapause or emerge as a crippled adult. There is great danger that the larva will not even reach the pupa stage. Once they begin to spin, the larvae require absolutely calm conditions. Fluctuations of climate, temperature and humidity are to be avoided. The time required for transformation from larva to pupa varies from species to species and may last as much as three weeks. This is most easily determined for species that construct cocoons. If a cocoon is lightly shaken and one hears a faint rattling, then it appears that the larva has dehydrated. Such cocoons are also much lighter than those containing active larvae. If the sound is duller and deeper, that means the transformation is still proceeding. If no sound is heard, then the larva has not yet changed into the pupa. The cocoons and webs may be cleaned after two or three weeks when the pupae are fully developed. Then one carefully removes all foreign materials. The cocoons and webs must not be damaged while doing this. The materials are removed to prevent the growth of fungus and to preserve the optimal microclimate within the cocoons or webs. Branches with attached webbing or suspension cables may be left with the cocoons. One should check whether the escape hatch on the upper end of the cocoon or web is free from obstruction to avoid problems when the imago tries to emerge. The completed cocoons are then placed in the hatching chamber in accordance with their natural positions, preferably hanging freely. In this way, the adults can emerge faster and with less stress. This is essential for females of genera with massive abdomens, like species of *Antheraea* and *Attacus*. Were these cocoons to lie on the ground, the female would have to struggle against their walls in order to rise. Many would not succeed and would become crippled. Imagos of most of the tropical species without diapause exit their cocoons by the end of four weeks, whereas, depending on the species, some require up to six months. Still other species enter diapause, such as the North American genera *Hyalophora* und *Callosamia* Packard, 1864. Species expected to hatch out soon need no special care. Many breeders spray the cocoons with water several times a day. There is no definite evidence

werden, oder legen Gespinste an den Wänden der Zuchtbehälter an. Im schlimmsten Falle wird dann kein Gespinst mehr angelegt. Die Raupen werden durch die Suche nach einem geeigneten Platz für ihr Gespinst geschwächt. Wenn sehr viel Seide für mehrere angefangene Kokons verbraucht wird, dann ist der letztendlich gefertigte Kokon instabil oder halbentwickelt. Das Mikroklima in diesen Kokons, im Vergleich zu einem normalen Kokon, wird auf Dauer empfindlich gestört und bei längeren Diapausen sterben die Puppen in den Kokons ab oder es schlüpfen nur verkrüppelte Falter. Die Gefahr ist zudem groß, dass es gar nicht erst zur Weiterentwicklung bis zu einer Puppe kommt. Die spinnenden Raupen brauchen also absolute Ruhe. Klima- und Temperaturschwankungen sind zu vermeiden, genauso wie Nässe. Die Zeitspanne der Verwandlung von der Raupe zur Puppe ist je nach Art unterschiedlich lang und kann bis zu drei Wochen dauern. Bei Arten, die Kokons bilden, kann man das sehr gut testen. Man schüttelt die Kokons leicht. Hört man ein helles Klappern, so ist davon auszugehen, dass die Raupe im Kokon vertrocknet ist. Diese Kokons sind auch viel leichter als Kokons mit vitalen Puppen. Ist ein dumpfes, viel tieferes Klappern zu hören, so ist die Verpuppung erfolgt. Wenn nichts zu hören ist, hat sich die Raupe noch nicht zur Puppe weiterentwickelt. Sind die Kokons oder Gespinste nach zwei bis drei Wochen mit Puppen fertig entwickelt, werden sie gesäubert. Vorsichtig entfernt man alle fremden Bestandteile. Die Gespinste dürfen dabei nicht beschädigt werden. Somit ist sichergestellt, dass es zu keiner späteren Schimmelbildung kommen kann und dass sich das Mikroklima im Kokon oder Gespinst optimal einstellt. Äste mit Sicherungs- oder Haltesträngen können an den Kokons belassen werden. Man sollte prüfen, ob die am oberen Ende der Kokons oder Gespinste liegende Schlupfreuse freiliegt, um Probleme beim Schlupf zu vermeiden. Die nun fertigen Kokons werden entsprechend ihrer natürlichen Lage am besten freihängend in einen Schlupfkasten gebracht. Freihängend deshalb, weil sich so die später schlüpfenden Falter ohne lange suchen zu müssen, an ihren Kokons entfalten können. Gerade für die Weibchen einiger Gattungen mit ihren schweren Abdomen, wie *Antheraea*- oder *Attacus*-Arten, ist das unerlässlich. Würden diese Kokons am Boden liegen, müssten sich die Weibchen erst an den Wänden des Schlupfkastens nach oben aras to whether that is at all necessary. If such cocoons are stored at room temperature and a humidity of 60 %, the adults will still emerge. Experiments with *Antheraea paphia* have shown that there is no difference with regard to hatching behavior, whether the cocoons are sprayed or not. Of course, the reality is that the longer the cocoons are in storage, the more important the conditions of their environment become. There is no all-encompassing rule about this. Every species has its own requirements and is more or less sensitive to climatic conditions. Species that spin thick, hard cocoons, like the genus *Antheraea*, are less sensitive, since they occasionally experience severe weather conditions in their natural environment. And this is exactly the point. Based on the point of the breeding material's origin, it is possible to obtain information about natural conditions at their source. Using diagrams provided through the internet, we can study climatic conditions in almost any place in the world. From such climatic data, one can learn the natural conditions appropriate to his material and attempt to simulate it. For example, it would make no sense to store webs of *Hyalophora cecropia* from Canada at room temperature, spray them twice a day with water and expect them to hatch out at Christmas when their native environment is experiencing the depths of winter. One repeatedly hears that many breeders open cocoons, by making a slit or by enlarging the escape hatch. This is a very serious mistake and should be avoided. The presumptive and only basis for this practice is the notion that it would simplify and ease the escape of the adult from the cocoon. How did that come about, and who opens cocoons in nature? Healthy and active adults need no assistance in emerging from their cocoons. Artificially opening cocoons has consequences. When one slices into a cocoon, the incision leaves sharp edges. These edges will bend inward during the next few days and can crush the pupa inside the cocoon. Also, the natural microclimate of the cocoon is of course destroyed by such operations. The pupa is then exposed to the climate of the room in which it is stored. The pupae in opened cocoons frequently desiccate, if they fail to hatch within a few days. In addition to the disadvantages described above, dependent cocoons are unlikely to maintain their suspended positions when opened. Clearly, one of the greatest risks is that the pupae of many species cannot withstand direct spraying with water.

beiten. Manche schaffen das nicht und verkrüppeln. Aus den Kokons der meisten tropischen Arten schlüpfen die Imagines ohne Diapause schon wieder nach vier Wochen, je nach Art bis zu sechs Monaten. Andere Arten legen Diapausen ein, wie zum Beispiel Saturniiden der Gattungen *Hyalophora* und *Callosamia* Packard, 1864, aus Nordamerika. Arten, aus deren Kokons mit einem baldigen Schlupf zu rechnen ist, benötigen keine besondere Pflege. Viele Züchter besprühen die Kokons mehrmals am Tag mit Wasser. Es gibt keine klaren Erkenntnisse, ob das überhaupt notwendig ist. Lagert man diese Kokons bei Zimmertemperatur bei einem Luftfeuchtigkeit von 60 %, so werden die Falter schlüpfen. Experimente bei *Antheraea paphia* haben ergeben, dass es keine Unterschiede hinsichtlich des Schlupfverhaltens bei besprühten oder nicht besprühten Kokons gibt. Fakt ist natürlich, dass das Raumklima umso wichtiger wird, je länger die Kokons lagern. Es gibt keine allgemeingültigen Regeln. Jede Art verhält sich unterschiedlich und ist auch mehr oder weniger empfindlich gegen klimatische Verhältnisse. Arten, die dichte und feste Kokons spinnen, wie bei der Gattung *Antheraea*, sind eher unempfindlich, schließlich sind diese auch zum Teil extremen Witterungen in der Natur ausgesetzt. Und das genau ist der Punkt. Anhand der Fundorte lässt sich analysieren, woher das Zuchtmaterial stammt. Im Internet ist es heute leicht möglich, über Klimadiagramme die klimatischen Verhältnisse fast jeden Ortes der Welt zu studieren. Anhand dieser klimatischen Daten bekommt man Informationen über die Verhältnisse vor Ort und kann diese versuchen zu simulieren. Es würde also zum Beispiel keinen Sinn machen, Gespinste von *Hyalophora cecropia* aus Kanada bei Raumtemperatur zu lagern und jeden Tag zweimal zu besprühen und auf einen Schlupf um Weihnachten zu hoffen, wo am Fundort des Zuchtmaterials tiefer Winter herrscht. Immer wieder ist zu hören, dass viele Züchter die Kokons öffnen, aufschneiden, oder an der Schlupfreuse weiten. Das ist ein sehr großer Fehler und sollte unterlassen werden. Begründet wird diese Prozedur einzig und allein damit, dass es so für die schlüpfenden Falter einfacher wäre, die Kokons zu verlassen. Wieso sollte das so sein und wer öffnet Kokons in der Natur? Gesunde und vitale Falter benötigen keine Hilfe beim Schlupf. Das Öffnen der Kokons hat nur Nachteile. Zerschneidet man Kokons, so bilden sich scharfe Kanten an den Schnitten. Die-

Fig. 17. Freshly hatched hybrid male of *Graellsia isabellae* (Graëlls, 1849) ♂ × *Actias dubernardi* (Oberthür, 1897) ♀. – Photograph courtesy of Robert Lemaitre.

Abb. 17. Frisch geschlüpftes hybrides Männchen von *Graellsia isabellae* (Graëlls, 1849) ♂ × *Actias dubernardi* (Oberthür, 1897) ♀. – Foto: Robert Lemaitre, mit freundlicher Erlaubnis.

Particular examples include the genera *Attacus* and *Coscinocera*, also species of the genus *Argema* Wallengren, 1958, whose pupae have relatively thin cuticles. Spraying can lead to absorption of water into the unfolding wings of adults as the wings grow. Other pupae simply die. The extensive spraying of pupae or cocoons that many breeders do must at least to some degree be considered questionable. In my opinion, pupae can neither absorb nor make use of water, because of their hard sclerites of chitin. More likely, water serves at most to alter climatic conditions of the surroundings for a short while. Indeed, it seems very likely that water is the key stimulus for hatching in a few species, for example those saturniid genera in the savannahs of Africa or in regions with clearly defined periods of rain and drought. Accordingly, the majority of new

se Schnittkanten wölben sich dann nach wenigen Tagen nach innen und können die Puppen im Kokon zerdrücken. Durch das Öffnen der Kokons ist das Mikroklima im Kokon natürlich zerstört. Die Puppe ist nun dem Raumklima ausgesetzt. Die Puppen aus geöffneten Kokons vertrocknen nicht selten, wenn sie nicht innerhalb weniger Tage schlüpfen. Zerschnittene Kokons lassen sich kaum noch aufhängen, mit den oben beschriebenen Nachteilen. Eines der größten Risiken allerdings ist, dass die Puppen vieler Arten das direkte Besprühen mit Wasser nicht vertragen. Genannt seien hier die Puppen der Gattungen *Attacus*, *Coscinocera* oder auch Arten von *Argema* Wallengren, 1958, die relativ dünne Puppenhüllen haben. Das Besprühen mit Wasser führt dazu, dass sich Wassereinschlüsse in den sich entfaltenden Flügeln der Falter bilden kön-

Fig. 18. Full-grown larva (L5) of *Eacles ormondei niepelti* Draudt, 1930, before pupation. – Photograph courtesy of Horst Käch.

Abb. 18. Ausgewachsene Raupe (L5) von *Eacles ormondei niepelti* Draudt, 1930, vor der Verpuppung. – Foto: Horst Käch, mit freundlicher Erlaubnis.

nen. Andere Puppen sterben ab. Das weitläufig von vielen Züchtern angewandte Besprühen der Puppen oder Kokons muss zumindest teilweise in Frage gestellt werden. Puppen können meiner Meinung nach überhaupt kein Wasser aufnehmen und verarbeiten, wie auch durch den harten sklerotisierten Chitinpanzer. Vielmehr dient das Wasser maximal dazu, für kurze Zeit das Umgebungsklima zu verändern. Sehr wahrscheinlich dürfte sein, dass das Wasser Schlüsselreize bei einigen Arten hervorruft, zum Beispiel bei Saturniiden-Gattungen in Savannengebieten in Afrika oder in Gebieten mit klar abzugrenzenden Regen- und Trockenperioden. So schlüpft in diesen Klimazonen die Masse der neuen Saturniiden-Generationen mit Beginn der Regenzeit. Setzt der Regen zum Beispiel um Wochen verspätet ein, so verschiebt sich auch der Schlupf der Imagines. Diese Erfahrungen konnte ich auf einigen Reisen in Südamerika machen. Ob diese Klimaverhältnisse bei einer großen Anzahl von Hobbyzüchtern simuliert werden können, bleibt zu bezweifeln. In den vielen Jahren der Zucht von Saturniiden hier in Europa konnte dieses Phänomen auf Nachfrage bei vielen Züchtern

Fig. 19. Larva of *Gynanisa maja maja* (Klug, 1836). – Photograph courtesy of Andreas Riekert.
Abb. 19. Raupe von *Gynanisa maja maja* (Klug, 1836). – Foto: Andreas Riekert, mit freundlicher Erlaubnis.

saturniid generations hatch out at the beginning of the rainy season. Should the rains come a few weeks later, the emergence of the imagos is similarly delayed. I experienced this phenomenon a number of times when traveling in South America. Whether a great many hobbyists are able to simulate these climatic conditions seems doubtful. During the many years of breeding Saturniidae here in Europe, this phenomenon has not been unambiguously observed despite the efforts of several breeders. Therefore it is much more important to provide cocoons or pupae with a stable and appropriate climate. As previously indicated for species without diapause, storage at a room temperature of 20–25 °C is adequate. They should not be exposed to direct sunlight and kept away from both air conditioners and sources of heat. Actually, the best place may be the bathroom, because on average it has the highest relative humidity. Those pupae whose hatching does not follow directly may be kept there. If that is not possible, pupae may be stored in a chamber following the model of Friedrich (c.f. Friedrich 1975, page 46). Cocoons and webs expected to hatch out in the following year may be stored in available space in a cellar or in a cool room with temperatures in the range 8–10 °C.

4.3 Loosely spun silk on the ground

Many species, such as those of the genus *Automeris*, spin neither cocoons nor webs in the trees or bushes that provide nourishment, but weave their silk among plants on the ground or even just below the surface of the earth. Shortly before pupation, these larvae develop an intense wanderlust. It is absolutely necessary to remove the wanderers from the breeding containers and separate them from the larvae not yet ready to spin their webs. Otherwise, they will lay their cocoons among feces and plant remains that encourage the growth of fungi and infections. When a suitable substrate for underground pupation is not available, these mobile larvae will often creep under layers of paper towels lining their box or spin their silk in a corner of the breeding chamber. In such cases, duplicate cocoons with several pupae are common within the same web. To later remove such webs without damaging them, it is better to use a pail or container of plastic. Peat or sand 5 cm deep is placed in such containers and overlaid with leaves, twigs, or

nicht eindeutig beobachtet werden. Also ist es viel wichtiger, den Kokons oder Puppen ein durchschnittlich verträgliches Klima zu bieten. Wie gesagt, bei Arten ohne Diapause reicht die Lagerung bei Zimmertemperatur zwischen 20 und 25 °C an einem nicht direkt von der Sonne bestrahlten Platz und nicht in der unmittelbaren Nähe von Heizquellen oder Klimaanlagen. Der beste Platz dürfte wegen der durchschnittlich relativ höchsten Luftfeuchtigkeit das Badezimmer sein. Dort können auch Puppen gelagert werden, deren Schlupf nicht unmittelbar erfolgt. Ist das nicht möglich, lagert man die Puppen in dem bei Friedrich abgebildeten Modell eines Puppenkastens (s. Friedrich, 1975, S. 46). Kokons oder Gespinste, bei denen mit einem Schlupf erst wieder im nächsten Jahr zu rechnen ist, lagert man, wenn vorhanden im Keller oder einem kühlen Raum bei 8 bis 10 °C.

4.3 Lockere Seidengespinste am Boden

Viele Arten, zum Beispiel der Gattung *Automeris*, verspinnen sich nicht in Kokons oder Gespinsten in den Bäumen oder Sträuchern ihrer Futterpflanzen, sondern verspinnen sich am Boden, zwischen Bodenlaub oder knapp unter der Erdoberfläche. Diese Raupen entwickeln kurz vor der Verpuppung einen starken Wandertrieb. Man sollte sie unbedingt aus den Raupenzuchtbehältern entnehmen und von den noch nicht spinnreifen Raupen trennen. Ansonsten legen sie ihre Kokons zwischen dem Kot und den Pflanzenresten an, was Schimmel und Infektionen begünstigt. Bietet man diesen Raupen keinen geeigneten Verpuppungsuntergrund, so verkriechen sich diese oftmals unter den Einlagen aus Papiertüchern im Raupenkasten oder legen ihre Gespinste in den Ecken der Zuchtbehälter an. Doppelkokons mit mehreren Puppen in einem Gespinst sind dann häufig. Ohne Zerstörung sind diese Gespinste später nicht zu entnehmen. Man nimmt besser einen Eimer oder Plastikbehälter, den man mit Torf oder Sand etwa 5 cm hoch befüllt, darauf gibt man Laub, Zweige oder Rindenstücke. Für gute Belüftung ist zu sorgen, damit das Substrat nicht schimmelt. Dazu können alte Gaze- oder Gardinenstückchen helfen, die man an den Rändern mit Gummisehnen oder auch Paketklebeband fixiert, um ein Entweichen der Raupen zu verhindern. Der Behälter sollte dunkel gelagert werden. Solche gesponnenen pieces of bark. Care must be taken to provide ventilation and avoid mildew. Gauze or lace curtains can promote ventilation and prevent the larvae from becoming wet by fastening the edges with elastic or strapping tape. The containers should then be kept in a dark place. Cocoons spun in this manner are usually not very sturdy and may even be transparent. They should be cleaned as much as possible before being stored in a cellar. The pupae of the species described here are especially vulnerable to climatic conditions, particularly dehydration.

4.4 Pupation in the soil

Species of these saturniid species do not construct cocoons nor do they even weave silk; instead, they proceed to pupate in the soil. Typical genera are *Eacles* and *Citheronia* Hübner, [1819], from the Neotropical faunal region, or *Aurivillius* Packard, 1902, *Gynanisa* Walker, 1855, and *Gonimbrasia* Butler, 1878, from Africa. Their larvae usually develop intense coloration and wander all around before pupating. The best results are seen when one places larvae in a suitable container of glass, clay, or plastic holding a moist mixture of sand and peat at least 20 cm deep. The minimum volume for such a container is one liter. Sand is most appropriate, because it is readily obtainable, reusable, and free of the ants and parasites that are often encountered when peat or soil from the garden is used. Unlike peat and soil, sand does not support the growth of fungi, because it contains no organic matter. The sand should be neither completely dry nor wet. Each container so prepared receives one larva. The container should remain covered in darkness until the larva has completely buried itself. Once the larva has disappeared into the soil, the cover may be removed. The containers should not be emptied sooner than two weeks afterward, and three weeks is even better. Since not all larvae enter the pupa stage simultaneously, it makes sense to date the containers. If space is available, the containers can be placed in a cellar. This is usually done before removing the pupae. Dehydration is a threat to pupae of these genera. Pupae of genera from the savannahs of Africa resist dehydration and are extremely robust, whereas others that pupate in the soil are especially vulnerable, if kept warm after the diapause. This is where one can easily suffer serious losses. The pupae buried in the earth must be maintained in

Kokons sind meist nicht sehr stabil und eher transparent. Auch diese Kokons sollten so gut wie möglich gereinigt werden, bevor sie in den Keller zur Lagerung gebracht werden. Die Puppen der hier beschriebenen Arten sind auf Grund des dünnwandigen Gespinstes wesentlich anfälliger in Bezug auf klimatische Verhältnisse, insbesondere gegen Vertrocknung.

4.4 Erdpuppen

Die Raupen dieser Saturniiden-Arten bauen keinen Kokon oder ein Gespinst, sondern begeben sich zur Verpuppung in die Erde. Typische Gattungen sind *Eacles* und *Citheronia* Hübner, [1819], aus dem neotropischen Faunengebiet oder *Aurivillius* Packard, 1902, *Gynanisa* Walker, 1855, und *Gonimbrasia* Butler, 1878, aus Afrika. Diese Raupen verfärben sich meist stark vor der Verpuppung und wandern umher. Die besten Ergebnisse erzielt man, wenn man den Raupen einen geeigneten Glas-, Ton- oder Plastikbehälter anbietet, den man mindestens 20 cm hoch mit einem feuchten Sand-Torf-Gemisch befüllt. Das Mindestvolumen für solch einen Behälter sollte 1 Liter betragen. Sand ist deshalb gut geeignet, weil er leicht zu beschaffen ist, immer wieder verwendbar ist und frei ist von Parasiten und Ameisen, die man sich oftmals mit Erde oder Torf aus dem Garten mit in die Verpuppungsbehälter holen kann. Sand schimmelt im Vergleich zu Erde oder Torf nicht, da er keine organischen Bestandteile enthält. Der Sand sollte weder staubtrocken noch nass sein. In einen solch gefüllten Behälter gibt man jeweils eine Raupe. Der Behälter sollte bis zum vollständigen Eingraben der Raupe abgedeckt werden und dunkel stehen. Ist die Raupe in der Erde verschwunden, so kann man die Abdeckung entfernen. Man sollte die Behälter frühestens nach zwei Wochen entleeren, besser erst nach drei Wochen. Da sich nicht alle Raupen gleichzeitig vergraben, ist es sinnvoll, die Behälter zu datieren. Wer sehr viel Platz hat, kann die Behälter auch so wie sie sind in den Keller bringen. Üblich ist aber eher eine Entnahme der Puppen. Auch Puppen dieser Gattungen sind gefährdet zu vertrocknen. Puppen von Gattungen aus den Savannengebieten Afrikas vertrocknen kaum und sind extrem robust. Andere Erdpuppen sind besonders gefährdet, wenn man sie nach der Diapause wieder warm hält. Hier verzeichnet man schnell sehr hohe Verluste. Diese Erdpuppen müssen unbedingt ein feuchtes

Fig. 20. Male and female pupae of *Telea polyphemus mexicana* C. C. Hoffmann, 1942. The gender of each pupa is clearly distinguished by the antenna processes of the imago. – Photograph by Frank Meister.
Abb. 20. Männliche und weibliche Puppe von *Telea polyphemus mexicana* C. C. Hoffmann, 1942. Deutlich sind die Geschlechtsunterschiede der beiden Puppen anhand der imaginalen Fortsätze der Antennen zu erkennen. – Foto: Frank Meister.

Abb. 21. Male and female pupae of *Telea polyphemus mexicana* C. C. Hoffmann, 1942. The gender of each pupa is clearly distinguished by the genital processes of the imago. – Photograph by Frank Meister.
Abb. 21. Männliche und weibliche Puppe von *Telea polyphemus mexicana* C. C. Hoffmann, 1942. Deutlich sind die Geschlechtsunterschiede der beiden Puppen anhand der imaginalen Fortsätze der Genitalapparate zu erkennen. – Foto: Frank Meister.

humid surroundings but without direct contact with water.

4.5 Distinguishing the sexes of pupae and cocoons

Since it is impractical to open cocoons or the silk enclosing a pupa, it is not a simple matter to determine its gender in the case of those species that spin a strongly woven cocoon. However, there are methods that offer the possibility of distinguishing the sexes with a high degree of probability. On average, the pupae of female moths are considerably larger and heavier. As explained above, the species that spin complete cocoons can be distinguished by visual inspection as in the case of *Argema mittrei* (Guérin-Méneville, 1847). Species with transparent cocoons, such as those in the genus *Automeris*, are held against a source of light to discern the size of the enclosed pupa. When the cocoon is opaque, only the size can be measured and the weight found using a postage scale. Thus the largest and heaviest are divided from the smallest and lightest with a high probability of effectively separating the females from the males. Saturniid pupae appear mummified,

Umgebungsklima erhalten, dürfen zugleich aber auch nicht direkt besprüht werden.

4.5 Unterscheidung der Geschlechter bei Puppen und Kokons

Da es nicht sinnvoll ist, die Kokons oder Gespinste zu öffnen, ist die Unterscheidung der Geschlechter bei einigen Arten, die geschlossene feste Kokons spinnen, nicht einfach. Dennoch gibt es Methoden, die eine hohe Wahrscheinlichkeit zur sicheren Trennung möglich machen. Die Puppen der weiblichen Falter sind im Durchschnitt wesentlich größer und damit auch schwerer. Wie schon oben erläutert, ist auch bei eingesponnenen Arten eine Unterscheidung durch Sichtprüfung der Puppen möglich, so bei *Argema mittrei* (Guérin-Méneville, 1847). Arten mit transparenten Kokons, wie bei der Gattung *Automeris*, hält man gegen eine Lichtquelle und prüft so die Größe der darin befindlichen Puppe. Bei geschlossenen Kokons ohne mögliche Sichtprüfung bleibt nur, die Größe der Kokons zu messen und die Kokons mit einer Briefwaage zu wiegen. Man trennt dann die größten und schwersten Kokons von den kleinsten und leichthe so-called *Pupa obtecta*. In this kind of pupa, all the imaginal processes involved in the development of the adult moth are enclosed in a cuticle that encloses and strongly adheres to the body of the pupa. The processes within saturniid pupae that lead to those of the imago are readily visible and therefore distinguishable. Thus it is possible to draw conclusions about the gender of saturniid pupae. There are two very typical, differentiating characters. For a great many species, one can distinguish male from female by the form of the antennae (Fig. 20). Many saturniid males possess fan-shaped antennae, for example all the species of the genera *Antheraea* und *Attacus*. These broad antennae are readily visible in their early development within the pupa. As a rule, the corresponding imaginal anlagen of females of these species are very delicate and thin in the pupae. There is another character of external morphology that serves to distinguish the sexes among many pupae (Fig. 21). For this, one needs only a little experience. In a variety of saturniid pupae, one can use the abdominal segments as external morphological characters of the pupae. To do this, one counts the free abdominal segments – the segments without imaginal markings or processes below the visible wing casings to the genital and anal opening. The males usually have one more abdominal segment than the female. Frequently, there are very large differences in the appearance of the male imago's genital and anal processes when compared with those of the female. The termination of the female's abdomen is often rounded, while that of the male is noticeably more pointed.

4.6 Shipping pupae and cocoons

Problem-free, world-wide shipping of pupae and cocoons is common practice. Simplest is the shipping of cocoons, which in many species are strong enough to withstand pressure without damage. Pupae can be packed in cardboard tubes of the sort used with paper towels. These are then placed in a sturdy cardboard box. Pupae should not be placed in direct contact with cotton wool, because as the pupae move, cotton fibers become entangled between the segments of cuticle. Removal of these fibers is usually impossible. Most often, the same procedures apply as with the shipping of eggs and larvae. One must give consideration to climatic conditions. Both high temperatures and exposure to cold must be

testen, und mit sehr hoher Trefferquote hat man nun die weiblichen Puppen von den männlichen getrennt. Saturniiden-Puppen sind Mumienpuppen, die so genannten »Pupa obtecta«. Bei dieser Art von Puppen sind alle imaginalen Fortsätze, die später den Falter ausmachen, von einer einheitlichen Cuticula umgeben und kleben am Puppenkörper fest. Diese imaginalen Fortsätze sind bei Saturniiden-Puppen gut zu sehen und damit unterscheidbar. Somit ist es möglich, anhand der äußeren Morphologie einer Saturniiden-Puppe Rückschlüsse auf das Geschlecht zu ziehen. Zwei sehr typische Unterscheidungsmerkmale sind leicht differenzierbar. Bei einer großen Artenzahl kann man an den Puppen erkennen, ob es sich um eine männliche oder weibliche Puppe handelt, indem man sich die Antennen anschaut (Abb. 20). Viele Männchen bei den Saturniiden besitzen Fächerantennen, zum Beispiel bei allen Arten der Gattung *Antheraea* und *Attacus*. Diese Fächerantennen sind als imaginale Fortsätze an den Puppen gut zu sehen. Die Weibchen dieser Arten haben dann in der Regel sehr feine und dünne imaginale Fortsätze an den Puppen. Eine weitere Sichtprüfung kann an den äußeren Geschlechtsmerkmalen erfolgen, die auch bei vielen Puppen gut zu unterscheiden sind (Abb. 21). Dafür ist einige Erfahrung notwendig. Man kann bei einer Vielzahl von Saturniiden-Puppen die Abdominalsegmente als äußeres morphologisches Merkmal am Puppenkörper untersuchen. Dabei zählt man die freien Abdominalsegmente, also die Segmente ohne erkennbare imaginale Markierungen oder Fortsätze unterhalb der sichtbaren Flügelscheiden bis zur Geschlechts- und Afteröffnung. Die Männchen haben in der Regel ein freies Abdominalsegment mehr als die Weibchen. Häufig gibt es sehr große Unterschiede in der Darstellung der imaginalen Geschlechts- und Afterfortsätze an männlichen Puppen im Vergleich zu den weiblichen. Oftmals ist der Puppenapex am Ende des Abdomens bei weiblichen Puppen rundlicher, bei männlichen Puppen wesentlich spitzer.

4.6 Der Versand von Puppen und Kokons

Der Versand von Puppen und Kokons ist unproblematisch und wird weltweit praktiziert. Am einfachsten ist der Versand von Kokons, die bei vielen Arten so stabil sind, dass sie selbst bei Druck nicht beschädigt werden können. Puppen avoided. If one does not know when the pupa phase of a particular species will end, then the pupae should be forwarded as soon as transformation from the larva is complete. I strongly advise against shipping larvae during the process of transformation. It is simply not possible until the pupa is fully developed. Incompletely formed pupae would be exposed to shaking and changes of climatic conditions in transit. The result is usually dead larvae or pupae whose growth is stunted. For successful pupation, the larvae require absolutely calm conditions, which they don't have when transported.

5. The successful propagation of breeding stock

5.1 Achieving copulation

Within the circle of breeders, hardly any other subject is discussed as thoroughly as the propagation of saturniids and the necessity of achieving successful matings. For many breeders, being able to repeatedly achieve such matings represents the height of success. Success in this enterprise depends almost entirely on the species involved. Several species are easy to mate; others almost never. Thus, there is no single, simple solution to this problem. In this connection, useful experiences and helpful methods will be described in the following sections. Also, the distinction will be made between natural copulation and artificially assisted intercourse.

5.2 Determining the gender of adult moths

In hardly any other family of Lepidoptera is it so easy to recognize the sex of imagos as it is with the saturniids. However, lay persons find this difficult when they begin. There is a whole series of clear, diagnostic characters which make it possible for the layman to distinguish the sexes for himself. The first decisive character is the size of the moth. In the majority of genera, the females are bigger than the males. Next, one examines the abdomen, because the hinder portion of the female's body is markedly larger, considering that in some species it may contain several hundred eggs. Another test based on the external genital organs of the abdomen makes it easy to distinguish the sexes. Males always have

kann man in Haushaltsrollenpapier wickeln und sie dann sicher in stabilen Kartons versenden. Vorsicht mit Watte, man sollte die Puppen nicht direkt in Watte legen, weil sich die Puppen bewegen, durch diese Bewegung die Watte zwischen die einzelnen Segmente der Puppenhülle gerät und diese verstopft. Eine Entfernung ist dann meist nicht mehr möglich. Ansonsten gelten auch hier die gleichen Regeln wie beim Versand von Eiern oder Raupen. Klimatische Verhältnisse beachten. Große Hitze oder Kälteperioden meiden. Ist man sich nicht sicher, wie lange die einzelne Art von der Verpuppung bis zum Schlupf benötigt, dann sollte man die Puppen oder Kokons sofort nach vollzogener Verpuppung versenden. Vom Versand der Raupen im Verpuppungsstadium ist dringend abzuraten. Hat sich die Raupe noch nicht voll zur Puppe entwickelt, so ist kein Versand möglich. Die Verpuppung würde dann während des Transportes erfolgen, begleitet von Erschütterungen und wechselnden klimatischen Verhältnissen. Das Resultat wären dann in der Regel verkrüppelte Puppen oder tote Raupen. Raupen benötigen im Verpuppungsprozess absolute Ruhe, die sie während eines Versandes nicht hätten.

5. Die erfolgreiche Vermehrung von Zuchtmaterial

5.1 Die Erzielung einer Kopula

Kaum ein anderes Thema wird in Züchterkreisen so stark diskutiert wie die Vermehrung von Saturniiden und die damit notwendige erfolgreiche Kopula. Für viele Züchter ist das erneute Verpaaren erst die Krönung einer erfolgreichen Zucht. Die Erzielung einer Kopula ist sehr stark von der jeweiligen Art abhängig. Manche Arten sind leicht zur Kopula zu bringen, andere fast nie. Es gibt also kein Patentrezept. Auch hier können nur die gemachten Erfahrungen geschildert werden und Methoden, die eine Kopula begünstigen. Man unterscheidet zwischen der natürlich herbeigeführten Kopula und der künstlichen Kopula, der so genannten Handpaarung.

5.2 Die Unterscheidung der Geschlechter bei den Faltern

Bei kaum einer anderen Familie von Lepidopteren ist die Unterscheidung der Geschlechter bei

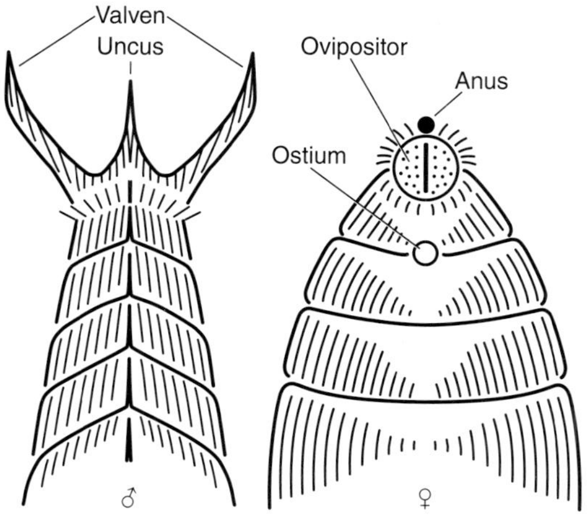

Fig. 22. Schematic drawing of a male and a female abdomen. – After Friedrich, 1975: 25, Fig. 9.
Abb. 22. Schematische Darstellung eines männlichen und weiblichen Abdomens. – Nach Friedrich, 1975: 25, Abb. 9.

small to very large processes which they use to hold onto the female during intercourse (Fig. 22) Light pressure at the end of the abdomen usually stimulates the male to spread these processes, leaving no doubt as to his gender. This method must be used with caution lest the male have a premature ejaculation and be unable to mate later that night. In contrast, females do not have these claspers. Some species have a ring-shaped ovipositor at the end of the abdomen (Fig. 23) which will extend in response to light stroking. In many saturniid species, differentiating the sexes is really easy, because they have completely different color schemes. The shape and size of the antennae form a further typical and diagnostic character. In several species, the masculine antennae are large and bushy with very long branches. Feminine antennae are small, lightly or not at all bushy and with short, hardly noticeable branches (Figs. 24–25).

5.3 Natural copulation under conditions of artificial breeding

By this is meant the conditions in which two moths of different sex are able to accomplish mating by their own means. This has previously been alluded to. The livelier and healthier the available breeding material is, the easier it is to achieve copulation. Willingness to engage in mating declines in many species as inbreeding

den Imagines so einfach wie bei den Saturniiden. Dennoch haben gerade Laien am Anfang Probleme. Es gibt eine ganze Reihe von deutlichen diagnostischen Merkmalen, die es selbst dem Laien ermöglichen, die Geschlechter zu unterscheiden. Erstes Unterscheidungsmerkmal ist die Größe der Falter. Bei der überwiegenden Zahl der Gattungen sind die Weibchen größer als die Männchen. Schaut man sich dazu noch die Abdomina an, dann sind auch hier die Hinterkörper der Weibchen wesentlich größer, enthalten sie doch bei manchen Arten mehrere Hundert Eier. Auch eine Prüfung der äußeren Geschlechtsorgane am Abdomen macht eine Unterscheidung leicht. Männchen besitzen immer kleine bis sehr große Valven (Abb. 22), mit denen sie sich bei der Kopula an die Weibchen klammern können. Bei einem leichten Druck auf das Ende des Abdomens spreizen die Männchen häufig diese Valven und man erkennt sie so eindeutig als Männchen. Dabei ist aber Vorsicht geboten, weil die Männchen bei dieser Methode häufig ihren Samenvorrat abgeben und sich dann in der Nacht nicht mehr paaren. Weibchen dagegen besitzen diese Klammern nicht. Sie haben am Hinterleibsende einen ringförmigen Ovipositor (Abb. 23), den manche Arten bei leichter Druckausübung nach außen stülpen. Bei vielen Saturniiden-Arten ist die Unterscheidung der Geschlechter schon deshalb leicht, weil sie völlig unterschiedlich gefärbt sind. Ein weiteres typisches diagnostisches Merkmal findet man in der Form und Größe der Antennen. Männliche Antennen sind bei vielen Arten größer, sehr stark gefiedert, mit sehr langen Rami. Weibliche Antennen sind schmal, wenig bis gar nicht gefiedert und mit kurzen, kaum erkennbaren Rami (Abb. 24–25).

5.3 Die natürliche Kopula unter Zuchtbedingungen

Gemeint ist hier, Bedingungen zu schaffen, dass zwei Falter unterschiedlichen Geschlechtes einer Art selbstständig in Kopula gehen. Je vitaler und gesünder das vorhandene Zuchtmaterial ist, desto leichter ist es, eine Kopula zu erzielen. Die Bereitschaft freiwillig in Kopula zu gehen, nimmt bei vielen Arten mit zunehmender Inzucht ab oder ist gar nicht mehr möglich. Dagegen ist eine Kopula bei inzuchtfreien Stämmen oftmals sehr leicht zu erreichen. Vertreter bestimmter Gattungen gehen erfahrungsgemäß sehr leicht in Kopula, bei einigen Arten ist eine freiwillige

Fig. 23. Female of *Citheronia mexicana* Grote & Robinson, 1866, with clearly visible ovipositor. – Photograph by Frank Meister.

Abb. 23. Weibchen von *Citheronia mexicana* Grote & Robinson, 1866, mit klar sichtbaren Ovipositor. – Foto: Frank Meister.

increases, even to the point where breeding becomes impossible. On the other hand, it is often quite easy to accomplish matings using stock that is free of inbreeding. From experience, there are certain genera that very easy to breed, but there are a few species with which it is rare or nearly impossible to accomplish this in captivity. An example of this is *Argema mittrei* from Madagascar. The following will describe the methods and factors that promote natural mating.

Many breeders are aware of the uncertainties. One has only a limited supply of pupae to work with and must hope that at least one male and one female will emerge within a short span of time to make mating possible. If one intends to continue breeding, then he needs an appropriate number of pupae of both sexes. Since not all moths hatch at the same time, and because the period of hatching may last up to eight weeks, a supply of twenty to thirty pupae is necessary. The best proportion is 50:50 to 60:40 in favor of male pupae. As a rule, the males emerge a few days earlier than the females. There are different explanations for this; one of which is as follows. In their natural state, the saturniids are at the bottom of the food chain. As adults, they take no nourishment. Their total metabolic energy is therefore limited to what was present in the pupa

Kopula selten oder auch fast gar nicht zu erzielen, wie beispielsweise bei *Argema mittrei* aus Madagaskar. Im Folgenden soll nun beschrieben werden, welche Methoden und Faktoren eine freiwillige Kopula begünstigen können.

Viele Züchter kennen diese Ungewissheit. Man hat nur eine begrenzte Anzahl an Puppen zur Verfügung und hofft nun darauf, dass innerhalb einer kurzen Zeitspanne ein Männchen und ein Weibchen schlüpfen, um eine Paarung zu erzielen. Beabsichtigt man eine Weiterzucht, dann braucht man dafür eine entsprechende Anzahl an Puppen beider Geschlechter. Da nicht alle Falter gleichzeitig schlüpfen und die Schlupfspanne in der Regel bis zu 8 Wochen dauern kann, sind etwa 20 bis 30 Puppen nötig. Das beste Geschlechterverhältnis liegt wohl bei 50:50 oder 60:40 zugunsten von männlichen Puppen. In der Regel schlüpfen die Männchen ein paar Tage früher als die Weibchen. Dazu gibt es verschiedene Thesen. Eine These sieht wie folgt aus. In der freien Natur stehen Saturniiden am unteren Ende der Nahrungskette. Sie nehmen im Falterstadium keinerlei Nahrung auf. Dadurch ist ihr gesamter Energiehaushalt auf die mit dem Schlupf vorhandenen eigenen Energieressourcen begrenzt. Für die Erhaltung der eigenen Art hat das große Auswirkungen. Die Energieressourcen müssen sehr sparsam eingesetzt werden und dienen nur einem Ziel, der Paarung, Eiablage und damit der Fortpflanzung. Alles, was diesem Ziel nicht dienlich ist, wäre damit zu vermeiden. Vielleicht erklärt sich so, warum Weibchen in der Regel sehr träge sind und sofort nach dem Schlupf damit beginnen, Pheromone freizusetzen, um Männchen anzulocken. Weite Flüge würden Energie vergeuden und die Gefahr, von Fressfeinden erbeutet zu werden, wäre sehr groß. Folgt man dieser These, dann wären viele Verbreitungsmuster neu zu überdenken, die noch von riesigen Verbreitungsarealen einzelner Arten über ganze Kontinente ausgehen. Neuere DNA-basierende Untersuchungen von Brechlin & Meister zeigen, dass es ganz klar definierte zoografische Verbreitungsräume gibt. Die Männchen hingegen verhalten sich anders und ihr Lebenszweck dürfte dazu dienen, ein paarungsbereites Weibchen zu finden. Dazu legen sie teilweise sehr weite Strecken zurück. So würden natürlich wesentlich mehr Männchen Fressfeinden zum Opfer fallen. Ein Beleg dafür, dass das Geschlechterverhältnis in der Natur zu Gunsten

Fig. 24. Male of *Telea polyphemus mexicana* C. C. Hoffmann, 1942, with typical, strongly pennate antennae. – Photograph by Frank Meister.
Abb. 24. Männchen von *Telea polyphemus mexicana* C. C. Hoffmann, 1942, mit typischen, stark gefiederten Antennen. – Foto: Frank Meister.

Fig. 25. Female of *Telea polyphemus mexicana* C. C. Hoffmann, 1942, with typical, only weakly pennate antennae. – Photograph by Frank Meister.
Abb. 25. Weibchen von *Telea polyphemus mexicana* C. C. Hoffmann, 1942, mit typischen nur schwach gefächerten Antennen. – Foto: Frank Meister.

stage. This has serious consequences for the preservation of their species. Their energy resources must be used sparingly and directed to one goal: mating and egg-laying, in short, reproduction. Anything that does not serve this goal is to be avoided. Perhaps that explains why females usually seem so lethargic and begin releasing pheromones that attract males as soon as they emerge from the pupa. Extensive flight would waste energy and invite the attention of predators. If one accepts this hypothesis, it may be necessary to reconsider the patterns by which the ranges of moths expand. In a few species this may become a gigantic area, spreading across entire continents. Recent DNA studies by Brechlin and Meister show that there are clearly defined zoogeographic ranges. However, this does not limit the male as much as the female, since his purpose in life is to find and mate with a female. This partly does away with the need for ranges of broad extent. Wide-ranging activity would naturally cause considerably more males to fall victim to predators. I know of no better proof that the proportions in nature tend to favor the production of greater numbers of male moths. A reason for the earlier emergence of the male moths from a clutch of pupae could be that there will be a large number of males available for mating when the first female makes her appearance. I received another interesting hypothesis from Ulrich Weritz, of Adenbüttel, Germany, one of the foremost breeders on the saturniid scene. He expressed the opinion that this schedule of hatching could have something to do with avoiding inbreeding in a state of nature. If all the males in a clutch, males and females alike, were to emerge simultaneously, it would be to the advantage of the males to be the first to reach a female. So, the timing of emergence serves to reduce the possibility of inbreeding. This sounds logical and is not to be dismissed out of hand. This proposition has some repercussions for the breeder. For certain species, it can be practical after the diapause to provide warmth for the female pupae as much as two weeks earlier than for the males. As a rule however, the males are the first to emerge. If no female is available, then one should keep the males cool or continuously exposed to light. Since the imagos of all saturniid species take no nourishment, such treatment will extend the life span of the males until females become available. This method can be effective for as much as five days. Some breeders place the living males in bags and

von mehr männlichen Faltern tendiert, ist mir nicht bekannt. Ein Grund für den frühzeitigeren Schlupf männlicher Falter eines Geleges könnte sein, dass mit dem Schlupf der ersten Weibchen bereits eine große Anzahl von Männchen für die Paarung zur Verfügung steht. Eine andere interessante These bekam ich von Ulrich Weritz, Adenbüttel, Deutschland, einem der besten Züchter innerhalb der Saturniiden-Szene. Er vertritt die Auffassung, dass dieses zeitlich verzögerte Schlupfverhalten etwas mit dem Thema Inzucht in freier Natur zu tun haben könnte. Würden alle Falter eines Geleges, also Männchen und Weibchen gleichzeitig schlüpfen, wären die Männchen innerhalb eines Geleges im Vorteil und wohl immer die ersten bei den schlüpfenden Weibchen. Der zeitlich versetzte Schlupf trägt also zur Verhinderung von Inzucht bei. Das klingt logisch und ist nicht von der Hand zu weisen. Für den Züchter ergeben sich aus dieser Betrachtung wenige Schlussfolgerungen. Es kann bei bestimmten Arten sinnvoll sein, nach der Diapause die weiblichen Puppen bis zu zwei Wochen früher ins Warme zu holen als die Männlichen. In der Regel schlüpfen aber trotzdem meist zuerst männliche Falter. Ist kein Weibchen verfügbar, sollte man die Männchen entweder kühl lagern oder permanent hell. Da die Imagines aller Saturniiden-Arten prinzipiell keine Nahrung aufnehmen können, bleibt auch nur eine lebensverlängernde Lagerung der Männchen, bis Weibchen verfügbar sind von bis zu maximal 5 Tagen. Einige Züchter tüten zu diesem Zweck die lebenden Männchen ein und geben diese in den Kühlschrank. Eine sehr drastische Methode, die aber schon zum Erfolg geführt hat, wenn die Männchen nur ein bis zwei Tage im Kühlschrank verblieben. Die Lagerung in einem Flugkäfig in einem kühlen Raum, der auch nachts beleuchtet wird, ist eine weitere Methode. Man kann die Männchen auch in einem Flugkäfig ins Freie stellen, da die europäischen Nächte ja recht kühl sind. Die beste Methode ist jedoch eine Mixtur aus mehreren Komponenten, die dem Flugverhalten der Arten Rechnung trägt. Untersuchungen haben ergeben, dass die Regulierung des Schlupfzeitpunktes vor allem von hormonellen Mechanismen reguliert und von genetischen Faktoren kontrolliert wird. Diese hormonellen Faktoren spielen die entscheidende Rolle, wann ein Falter schlüpft. Ansonsten müsste es möglich sein, bestimmte Arten mit einer klar definierten Diapause allein durch äußere klimatische Verrefrigerate them. This is really a drastic method, but one that has produced good results when the males are kept only one or two days in the refrigerator. Holding the males in a flight cage in a cool room which is kept illuminated through the night is an alternative method. One can also place a flight cage out of doors, since the nights in Europe are cool. However, the best method is a mixture of several components that take into consideration the flight behavior of the species. Studies have shown that emergence of the adult from the pupa is regulated by hormonal mechanisms, which are in turn controlled by genetic factors. The hormonal factors play the decisive role as to when the moth emerges. Otherwise it must be possible that certain species with a clearly defined diapause are stimulated to shed the pupal casing by external climatic changes alone. That simply does not work. If one takes North American saturniids of the genus *Hyalophora* to Europe in December and gives them warm and moist conditions, they will still not shed their pupal casing. These species emerge in the USA between April and June, and their behavior is essentially similar in Europe. In recent years, the pupae of a series of saturniids and sphingids have undergone treatment with ecdyson (molting hormone). Ecdyson from the prothoracic gland is clearly the essential hormone regulating the completion of pupation. For example, this expensive process was used to inject ecdyson into pupae of species such as *Sphinx ligustri* in a state of diapause in December. The result was astonishing. In a few weeks, the moths shed their pupal casings almost simultaneously at a time of deepest winter (c.f. Plontke and Predel 1999). Since the time of emergence is hormonally controlled, the question is then to what extent this event can be controlled by the breeder. Two factors play very important roles here, the conditions of light and humidity. For most of the Saturniidae, very little is known about their generation time. A few tropical species reproduce constantly; others consistently produce only one generation a year. Some species may remain in the diapause phase as long as five years, while another regularly passes through further generations. It is possible with many species to make use of external stimuli to force diapause. If pupae of neotropical species are placed in a cool, dark cellar, they will probably enter diapause. If some are kept warm and moist, there is a very good chance that the adults will emerge after a few

änderungen zum Schlupf anzuregen. Das funktioniert definitiv nicht. Würde man nordamerikanischen Saturniiden der Gattung *Hyalophora* in Europa im Dezember feuchte und warme Bedingungen bieten, würden sie deshalb nicht schlüpfen. Diese Art schlüpft in den USA etwa von April bis Juni und auch in Europa ist das nicht wesentlich anders. In den letzten Jahren wurden eine Reihe von Saturniiden- und Sphingiden-Puppen einer Ecdyson-Behandlung unterzogen. Das Ecdyson der Prothorakaldrüsen ist demnach offensichtlich das entscheidende Hormon zur Schlupfregulierung. Man spritzte in einem aufwändigen Verfahren Ecdyson in die Puppen von Arten, die sich in Diapause befanden, zum Beispiel in Puppen von *Sphinx ligustri* im Dezember. Das Resultat war verblüffend. Die Falter schlüpften fast gleichzeitig nach wenigen Wochen, also in einer Zeit, in der in der Natur tiefster Winter herrschte (s. Plontke & Predel, 1999). Wenn der Schlupfzeitpunkt also hormonell bedingt ist, stellt sich die Frage, inwieweit ein Züchter den Schlupfzeitpunkt überhaupt beeinflussen kann. Zwei Faktoren spielen hier eine sehr bedeutende Rolle, die Licht und Feuchtigkeitsverhältnisse. Über die Generationsfolgen ist bei den meisten Saturniiden sehr wenig bekannt. Einige Arten in den Tropen reproduzieren sich ständig neu, andere durchlaufen nur eine Generationsfolge pro Jahr, bei manchen Arten macht ein Teil Diapausen von bis zu fünf Jahren und ein anderer Teil durchläuft weiter regelmäßige Generationsfolgen. Bei vielen Arten ist es möglich, äußere Reize zu setzen und damit Diapausen zu erzwingen. Bringt man die Puppen von neotropischen Arten nach der Verpuppung in einen kühlen und dunklen Keller, werden sie wahrscheinlich in Diapause gehen. Würde man einen Teil nach der Verpuppung warm und feucht halten, wäre die Chance sehr groß, dass die Falter nach wenigen Wochen erneut schlüpfen. Weitere Faktoren, die das Paarungsverhalten beeinflussen dürften, sind die Lichtverhältnisse und natürlich die Flugzeit. Die Flugzeit ist bei jeder Saturniiden-Art völlig verschieden. Selbst innerhalb einer Art fliegen in der Natur die Männchen nicht gleichzeitig mit den Weibchen. So konnte bei einer Reihe von Expeditionen nach Südamerika folgendes dokumentiert werden. Bei der Auswertung von über 3000 einzeln notierten Flugzeiten von Saturniiden in Argentinien ergaben sich erstaunlich eng gefasste Flugzeitspannen bei einzelnen Arten, die geschlechterspeziweeks. Further factors that can influence mating behavior are the conditions of light and the time of flight. Every species of saturniid flies at a different time. In a state of nature, the males of a particular species do not fly at the same time as the females. It was possible to document this observation in a series of expeditions to South America. Analysis of more than 3000 recorded flight times of saturniids in Argentina revealed a remarkably narrow span for the flight of individual species whose opposite sex was nevertheless to be found some distance away. Thus the females of some species would take flight very early in twilight, whereas males of the same species would follow suit around two o'clock the next day. Such broad observations were repeatedly supported by more than fifty captures. One could almost set his clock by the appearance of a certain species at night. For example, individuals of the genera *Arsenura* Duncan, 1841, or *Copiopteryx* Duncan, 1841, never appeared before four o'clock. In Europe, the ecdysis from the pupal state and the flying behavior become completely confused in moving across the time zones. As far as I know, this problem has not been studied at all. However, this temporal effect seems unimportant. Additionally, I find no marked differences between moths from imported cocoons and those bred in Europe. The majority of moths here in Europe seem to emerge from the pupa in the afternoon and proceed directly to mating as soon as darkness falls. Their flying behavior can be suppressed, if one exposes them to continuous light. This makes sense for certain species. The males of many species are so active that they will quickly take flight shortly after emerging from the pupa. In theory, this is not a bad thing. Unfortunately, after two days, such males no longer have functional wings and are unable to fly to a female. In several species, the one- or two-day-old males are considerably more active in mating than those newly hatched. To accomplish mating, one places one male with one female or if possible, the ratio 2:1 into a mating chamber of gauze. The mating chamber should have a definite size, from 50 × 30 × 30 cm to 100 × 80 × 80 cm. I advise that if eggs are not to be laid in the same cage, then neither substrate materials nor plants are necessary. The last thirty years have given me the impression that fragrant blossoms or branches of plants eaten by saturniids are useful in promoting mating. The mating chamber is then placed in a closed room

fisch zudem noch weit auseinander lagen. So flogen die Weibchen einiger Arten bereits sehr früh in der Dämmerung und die Männchen der gleichen Art erst gegen 2 Uhr am nächsten Tag. Solche immer wiederkehrenden Beobachtungen konnten in mehr als 50 Fangnächten dokumentiert werden. Man konnte fast die Uhr danach stellen, wann eine bestimmte Art in der Nacht zum Licht kommen würde. Vertreter der Gattung *Arsenura* Duncan, 1841, zum Beispiel oder *Copiopteryx* Duncan, 1841, kamen niemals vor 4 Uhr ans Licht. Das Schlupf- und natürlich das Flugverhalten der einzelnen Arten kommt durch die in Europa verschobenen Zeitzonen völlig durcheinander. Dieses Problem ist meinem Kenntnisstand nach völlig unerforscht. Es scheint aber keine große Rolle zu spielen. Auch sind mir keine Unterschiede zwischen Faltern aus Importkokons und in Europa gezüchteten Tieren aufgefallen. Die überwiegende Zahl der Falter scheint hier in Europa nachmittags zu schlüpfen und geht wenn es dunkel wird unproblematisch in Paarung. Das Flugverhalten kann unterdrückt werden, wenn man Saturniiden permanent Licht aussetzt. Das kann bei bestimmten Arten Sinn machen. Die Männchen mancher Arten sind so aktiv, dass sie sich innerhalb kürzester Zeit abfliegen. Im Prinzip ist das nicht schlimm. Leider führt es aber auch dazu, dass solche Männchen nach zwei Tagen keine erkennbaren Flügel mehr besitzen und dann die Weibchen nicht mehr anfliegen können. Bei vielen Arten sind ein bis zwei Tage alte Männchen wesentlich paarungsfreudiger als frisch geschlüpfte Männchen. Um nun Paarungen zu erzielen, bestückt man einen Paarungskäfig aus Gaze mit einem Männchen und einem Weibchen, oder wenn möglich auch im Verhältnis 2:1. Der Paarungskäfig sollte eine gewisse Größe haben, je nach Art von 50 × 30 × 30 cm bis 100 × 80 × 80 cm. Da ich dazu rate, die Eiablage nicht im Käfig selbst zuzulassen, ist keinerlei Befüllung nötig, auch nicht mit Pflanzen. Ich hatte in den letzten 30 Jahren nie den Eindruck, dass duftende Blumen oder Zweige der Futterpflanzen bei Saturniiden irgendetwas nützen, um eine Paarung positiv zu beeinflussen. Man stellt diesen Paarungskäfig in geschlossenen Räumen an einen Platz, wo Luft zirkulieren kann, vielleicht auch an ein offenes Fenster. Das trägt vielleicht dazu bei, dass sich die Pheromone der Weibchen besser im Raum verteilen und die Männchen aktiver werden. Bewiesen ist das allerdings nicht. Hat man am glei-

Fig. 26. The correct mating position of a male with a female moth during assisted copulatio. – Photograph by Frank Meister.
Abb. 26. Richtige Paarungsstellung der männlichen und weiblichen Falter bei der künstlichen Kopula. – Foto: Frank Meister.

where the air can circulate, perhaps by an open window. The result may be that the female's pheromone diffuses better within such a space and the male becomes more active. But this has not been proved. If one intends to mate different species on the same day, then only couples of the same species should be placed in the same room. If one places a flying cage in the open, then it is necessary to take precautions against predators. Hedgehogs will force an entrance into cages placed on the ground. Birds will tear a way through the gauze to reach the moths. I have myself seen a tit rip a hole in a large mosquito net to seize a female of *Argema mittrei*, a moth with nearly twice the wingspan of the bird. It's not always easy to know whether intercourse has actually taken place. Mating behavior is very variable. Some species will remain in copulation more than 24 hours while for others the period of actual mating is very short. There are however a few indications. Males that have been engaged in intercourse usually die within the next two days. Females that have mated begin to lay eggs either the same night or the next at

chen Tag Paarungen verschiedener Arten in Aussicht, so könnte es sinnvoll sein, nur Paare der gleichen Art in ein und demselben Raum zu deponieren. Stellt man Flugkäfige ins Freie, dann sollte man diese vor Fressfeinden geschützt aufstellen. Igel dringen in Käfige ein, wenn sie auf dem Boden stehen. Vögel hacken selbst durch die Gaze, um an die Falter zu kommen. Ich selbst beobachtete, wie eine Meise ein Loch in ein riesiges Moskitonetz hackte, um sich dann ein Weibchen von *Argema mittrei* zu schnappen, immerhin ein Falter, der fast die doppelte Flügelspannweite hat. Ob eine Kopula stattgefunden hat, ist nicht immer leicht erkennbar. Das Kopulationsverhalten ist sehr unterschiedlich. Manche Arten verweilen über 24 Stunden in Kopula, andere kopulieren nur sehr kurz. Es gibt dennoch einige Anzeichen. Männchen, die in Kopula waren, gehen häufig innerhalb der nächsten zwei Tage ein. Weibchen, die sich gepaart haben, beginnen entweder in der gleichen Nacht mit der Eiablage oder spätestens in der nächsten. Sollte ein Weibchen nach einer vermeintlichen Kopula in der darauffolgenden Nacht wieder zu locken the latest. If, after an apparent mating, a female begins again in the following night to attract males, then no true mating took place. One also recognizes this when the female's ovipositor has not opened widely.

5.4 Assisted copulation

Assisted copulation is a technical method substituting for natural mating and will be thoroughly discussed here, because it is particularly useful in the breeding of the Saturniidae. The first consideration in dealing with a species is that the males have valves (sclerites protecting the genital opening) that are sufficiently large and robustly developed to be noticeable. However, within a species there is considerable variation from one male to another as to how well the valves are developed. Among these, one should select the strongest males with the largest valves and uncus. The uncus is a small hook, which the male uses to secure his position and hold himself upright on the female's back (Fig. 26). In the following, the procedure used by Friedrich (1975, p. 26) will be described: "One grasps the male and the female at the roots of the wings, or rather by the thorax, being careful that the abdomen extends freely from the hind wings. Then one lightly rubs the terminal portion of the abdomen of one after the other in turn for a few minutes. This is done with a circular motion until the valves of the male spread apart and the uncus of the abdomen is fastened to the female's abdomen between the ostium and the ovipositor. At this point, the bodies should be at an angle of about 135° to one another. Should the uncus contact a different area, such as the ovipositor, the bodies are misaligned and must be separated as quickly as possible. After complete union, the abdomens may remain in contact for a short while. If the mating has been successful, the male is rendered lifeless, leaving the female as the surviving partner".

I can almost always observe the death of the male if the assisted mating has been successful as described above. One then places the still conjoined pair on an adherent surface to which the female can cling. Gauze attached to a vertical board is the best for this purpose. The female should always be placed above the male, since several males will remain in the vicinity of the mating, attracted to the tethered female. These

beginnen, hat keine Paarung stattgefunden. Man erkennt das daran, dass die Weibchen ihren Ovipositor weit von sich spreizen.

5.4 Die Handkopula

Die Handkopula ist eine Form der künstlichen Paarungsmethoden, die hier ausführlich besprochen werden soll, weil sie sich besonders gut für die Verpaarung von Saturniiden eignet. Grundvorrausetzung ist, dass es sich um Arten handelt, bei denen die Männchen ausreichend große und robust ausgebildete Valven besitzen. Es ist aber auch innerhalb einer Art sehr unterschiedlich, wie die Valven von Männchen zu Männchen ausgebildet sind. Man sollte sich die kräftigsten Männchen mit den größten Valven und dem größten Uncus heraussuchen. Der Uncus ist ein kleiner Haken, mit dem sich Männchen zusätzlich vertikal am Hinterleib des Weibchen festklammern (Abb. 26). Im Folgenden wird nun das Verfahren der Handkopula nach (Friedrich, 1975, S. 26) beschrieben: »Man fasst ♂ und ♀ an den Flügelwurzeln bzw. am Thorax, achtet darauf, dass die Abdomina der Tiere frei aus den Hinterflügeln herausragen und reibt die Hinterleibsenden, einander zugewandt, einige Minuten leicht und unter annähernd kreisenden Bewegungen so lange, bis das ♂ die Valven spreizt und der Uncus das Abdomen des ♀ zwischen Ostium und Ovipositor fasst; zwischen den Körpern sollte dabei ein Winkel von etwa 135° entstehen. Wenn der Uncus aber z. B. den Ovipositor erfasst, ist die Prozedur misslungen, und die Partner müssen so schnell wie möglich getrennt werden. Die Abdomina können nach vollzogener Vereinigung kurze Zeit etwas gegeneinandergedrückt werden. Ist die Kopula zustande gekommen, wirkt das ♂ bei den Arten leblos, bei denen das ♀ während der Vereinigung der führende Partner ist . . . .)«. Ich konnte diese Leblosigkeit der Männchen fast immer feststellen, wenn die Handverpaarung wie oben beschrieben erfolgreich war. Man setzt nun das miteinander fest verbundene Paar auf eine griffige Unterlage, an der sich das Weibchen festhalten kann. Am besten eignet sich dafür Gaze, die vertikal auf einen Brett fixiert ist. Das Weibchen sollte immer über den Männchen sitzend platziert werden, denn manche Männchen verharren in Paarungsstellung mit eingezogenen Beinen. Diese hängen dann nach unten nur über den Genitalapparat mit dem Partner verbunden. Das Paar sollte nun will then hang above or below the genitals that are still attached to the surviving partner. The mated pair should remain in a location protected from outside influences such as light or movement. The slightest disturbance can terminate the mating. Before beginning the procedure described above, one should rub his hands with sand or earth. The best time to carry out assisted mating is the evening as twilight falls or during a phase of masculine activity. Finally, it requires both skill and much experience if one wants to succeed with this method. Many breeders have tried this for years without success. For others, this method works as well as any other.

5.5 Confining the female moth in an open breeding environment

There is a little regarded, but very productive method to insure successful copulation. This involves binding a female in breeding condition in such a way that she attracts wild males but cannot herself escape. This makes it possible to protect one's stock from the danger of inbreeding. This naturally assumes that I am in a location where the appropriate species occurs in the wild. In Europe I have achieved very good success with *Aglia tau*, species of *Saturnia* or *Graellsia isabellae* (Graëlls, 1849). To restrain the female, one winds a thin thread between the fore- and hindwings so that it goes around the thorax above the abdomen. A knot is made around a leg, and then one places the female on a tree or bush and securely fastens the other end of the thread to a branch. It is important to note that the female is securely restrained. Knots should not bind the female too tightly and not so loosely that she could free herself. An alternative method is to place the female in a flight cage that serves as a trap. Here one makes use of the circumstance that, as a rule, moths will climb upwards rather than the reverse. When a male approaches and tries to reach the female, and sooner or later succeeds in entering the cage, he will be unable to leave afterward. This method can also be used to capture males for one's collection. I know of instances in Chile, where a newly-hatched female of the genus *Polythysana* was used to capture dozens of males.

vor jeglicher Beeinflussung, ob durch Licht oder Bewegung, an einem geschützten Ort verharren können. Kleinste Störungen könnten die Paarung beenden. Vor Beginn der oben beschriebenen Prozedur sollte man seine Hände mit Sand oder Erde verreiben. Der beste Zeitpunkt ist abends mit Einbruch der Dämmerung oder in einer Phase, in der die Männchen aktiv sind. Letztendlich gehört sehr viel Erfahrung und Geschick dazu, will man mit dieser künstlichen Paarungsmethode Erfolg haben. Manche Züchter versuchen das seit vielen Jahren und schaffen es nicht. Andere verpaaren so fast jede Art.

5.5 Das Ausbinden oder Verbringen von weiblichen Faltern im Freien

Eine selten anzuwendende aber recht erfolgreiche Methode der Erzielung von Kopulas ist das Ausbinden von Faltern oder das Anlocken von wilden Männchen mit gezüchteten Weibchen. So ist es möglich, seine Zuchtstämme vor Inzucht zu schützen. Vorraussetzung ist natürlich, dass ich mich in einem Habitat befinde, in dem die jeweilige Art auch wild vorkommt. In Europa wurden mir sehr gute Erfolge mit *Aglia tau*, *Saturnia*-Arten oder *Graellsia isabellae* (Graëlls, 1849) gemeldet. Zum Ausbinden zieht man einen dünnen Faden zwischen dem Vorderflügel und dem Hinterflügel hindurch und umschlingt so den Thorax am Übergang zum Abdomen. Den Knoten fixiert man an den Beinen. Dann setzt man das Weibchen in einen Baum oder Busch und bindet das andere Ende des Fadens an einem Ast fest. Wichtig ist dabei, dass das Weibchen sich selbst festhalten kann. Der Knoten darf nicht zu fest sein, um das Weibchen nicht abzuschnüren und nicht zu locker, so dass sich das Weibchen selbst befreien könnte. Eine andere Methode ist, die Weibchen in einen Anflugkäfig mit Reuse zu setzen. Hier bedient man sich des Umstandes, dass Falter in der Regel von unten nach oben laufen und nicht umgekehrt. Fliegt ein Männchen an, dann versucht es das Weibchen zu erreichen und gelangt früher oder später in den Kasten, kommt aber später nicht wieder heraus. Mit dieser Methode kann man auch Männchen zum Zwecke des Sammelns fangen. Mir sind Fälle aus Chile bekannt, wo mit einem frischen Weibchen der Gattung *Polythysana* Dutzende Männchen gefangen wurden.

II Systematic list of all currently described wild silk moths (Lepidoptera: Saturniidae) and the plants presently known to be eaten by their larvae

II Systematische Auflistung aller bis dato beschriebenen Wilden Seidenspinner (Lepidoptera: Saturniidae) und den bis heute bekannten Fraßpflanzen ihrer Raupen

Taxonomic revisions

Against the background of a large number of related DNA analyses made by the BOLD-Project of the Canadian Center for DNA Barcoding of the Biodiversity Institute of Ontario of the University of Guelph, the following changes in taxonomy have been adopted. In this connection, special thanks are due to the entire team of the BOLD-Project represented by Rodolphe Rougerie, whose work has made substaintial contributions to taxonomy.

The following species are no longer considered synonyms but have been raised to the status of species: *Paradaemonia castanea* Rothschild, 1907 **rev. stat.** (previously a synonym of *platydesmia* Rothschild, 1907); *Titaea raveni* Johnson and Michener, 1948 **rev. stat.** (previously a synonym of *Titaea tamerlan guayaquila* Schaus, 1932); *Citheronioides lamata* Schaus, 1933 **rev. stat.** (previously a synonym of *Citheronioides collaris* Rothschild, 1907); *Eubergia boetifica* Druce, 1899 **rev. stat.** (previously a synonym of *Eubergia caisa* Weymer, 1908); *Antheraea myanmarensis* Paukstadt, Paukstadt and Brosch, 1998 **rev. stat.** (previously a synonym of *Antheraea gschwandneri* Niepelt, 1918).

The following species was considered a synonym but has been raised to the status of species: *Actias truncatipennis* Sonthonnax, 1899 (previously a subspecies of *Actias luna* Linnaeus, 1758).

The following have been reduced in rank to synonymy: *Giacomellia inversa* Giacomelli, 1911 **syn. nov.** (currently a synonym of *Giacomellia bilineata* Burmeister 1878); *Leucanella piura* Lampe, 2004 **syn. nov.** (currently a synonym of *Leucanella*

Taxonomische Änderungen

Im Rahmen der Auswertung einer Vielzahl von vergleichenden DNA-Analysen des BOLD-Projektes des Canadian Centre for DNA Barcoding des Biodiversity Institute of Ontario der University of Guelph werden folgende taxonomische Änderungen vorgenommen. In diesem Zusammenhang geht ein besonderer Dank an das gesamte Team des BOLD-Projektes und hier stellvertretend namentlich an Rodolphe Rougerie, deren Arbeit heute zu einem unverzichtbaren Bestandteil der taxonomischen Arbeit geworden sind.

Folgende Arten werden aus ihren Synonymien gelöst und nun im Artstatus geführt: *Paradaemonia castanea* (W. Rothschild, 1907) **rev. stat.** (zuvor Synonym von *platydesmia* (W. Rothschild, 1907)); *Titaea raveni* (Johnson & Michener, 1948) **rev. stat.** (zuvor Synonym von *Titaea tamerlan guayaquila* (Schaus, 1932)); *Citheronioides lamata* (Schaus, 1933) **rev. stat.** (zuvor Synonym von *Citheronioides collaris* (W. Rothschild, 1907)); *Eubergia boetifica* (Druce, 1899) **rev. stat.** (zuvor Synonym von *Eubergia caisa* (Weymer, 1908)); *Antheraea myanmarensis* Paukstadt, Paukstadt & Brosch, 1998 **rev. stat.** (zuvor Synonym von *Antheraea gschwandneri* Niepelt, 1918).

Folgende weitere bisher als Subspezies geführte Taxa werden in Artrang gehoben: *Actias truncatipennis* Sonthonnax, 1899 **rev. stat.** (bisher Subspezies von *Actias luna* (Linnaeus, 1758)).

Folgende bisher im Artrang geführte Taxa werden synonymisiert: *Giacomellia inversa* (Giacomelli, 1911) **syn. nov.** (nun Synonym von *Giacomellia bilineata* (Burmeister, 1878)); *Leucanella* *memusoides* Lemaire, 1973); *Paradirphia wenczeli* Racheli, 2008 **syn. nov.** (currently a synonym of *Paradirphia latipunctata* Lemaire, 1976).

The following subspecies has undergone synonymization: *Automeris lumbrosa lampei* Lemaire, 2002 **syn. nov.** (currently a synonym of *Automeris naranja* Schaus, 1898).

piura Lampe, 2004 **syn. nov.** (nun Synonym von *Leucanella memusoides* Lemaire, 1973; *Paradirphia wenczeli* Racheli, 2008 **syn. nov.** (nun Synonym von *Paradirphia latipunctata* (Lemaire, 1976).

Folgende Unterart wird synonymisiert: *Automeris umbrosa lampei* Lemaire, 2002 **syn. nov.** (nun Synonym von *Automeris naranja* Schaus, 1898).

Oxyteninae

Homoeopteryx C. Felder, 1874

divisa Jordan, 1924
Foodplants: Unknown

elegans Jordan, 1924
Foodplants: Unknown

major Jordan, 1924
Foodplants: Unknown

malecena malecena (Druce, 1886)
Foodplants: Unknown

malecena prona Jordan, 1924
Foodplants: Unknown

syssauroides C. Felder, [1874]
Foodplants: Unknown

Oxytenis Hübner, [1819]

affinis Hoffmann, 1942
Foodplants: Unknown

albilunulata Schaus, 1912
Foodplants: Unknown

beprea Druce, 1886
Foodplants: *Alibertia edulis, Amaioua pedicellata, Augusta rivalis, Bertiera bracteosa, Chimarrhis parviflora, Coffea arabica, Genipa americana, Bertiera bracteosa, Ixora floribunda, Pentagonia donnell-smithii, Posoqueria latifolia, Psychotria grandis, Randia aculeata, Randia armata, Randia genipoides, Randia grandifolia, Randia monantha, Randia pittieri.*

bicornis Jordan, 1924
Foodplants: Unknown

epiphaea Jordan, 1924
Foodplants: Unknown

erosa Jordan, 1924
Foodplants: Unknown

ferruginea (Walker, 1855)
Foodplants: *Alibertia edulis, Posoqueria latifolia, Randia armata.*

gigantea (Druce, 1890)
Foodplants: *Alibertia edulis, Posoqueria latifolia, Randia armata.*

leda Druce, 1906
Foodplants: Unknown

mirabilis (Cramer, 1780)
Foodplants: Unknown

modestia (Cramer, 1780)
attacina Walker, 1865
bimaculatus (Doumet, 1859)
honesta (Cramer, 1780)
lonomica Druce, 1886
Foodplants: *Alibertia edulis, Amaioua pedicellata, Augusta rivalis, Genipa americana, Lonicera, Posoqueria latifolia, Randia grandifolia.*

naemia aravaca Jordan, 1924
Foodplants: Unknown

naemia naemia Druce, 1906
Foodplants: Unknown

naemia orecta Jordan, 1924
Foodplants: *Arachnothryx buddleioides, Sabicea panamensis, Sabicea villosa, Uncaria tomentosa.*

nubila acuta Jordan, 1924
Foodplants: Unknown

nubila nubila Jordan, 1924
Foodplants: Unknown

peregrina (Stoll, 1780)
Foodplants: Unknown

plettina Jordan, 1924
Foodplants: Unknown

sobrina Jordan, 1924
Foodplants: Unknown

spadix Jordan, 1924
Foodplants: Unknown

zerbina (Cramer, 1780)
angulata (Cramer, 1775)
harpago (Hübner, [1819])
Foodplants: Unknown

Therinia Westwood, 1879

amphira amphira Druce, 1890
Foodplants: Unknown

amphira melini Bryk, 1953
Foodplants: Unknown

buckleyi buckleyi Jordan, 1924
Foodplants: Unknown

buckleyi paraensis Jordan, 1924
Foodplants: Unknown

celata Jordan, 1924
Foodplants: Unknown

diffissa diffissa Jordan, 1924
Foodplants: Unknown

diffissa fortis Jordan, 1924
Foodplants: Unknown

diffissa inca Jordan, 1924
Foodplants: Unknown

diffissa tridens Jordan, 1924
Foodplants: Unknown

geometraria C. Felder, 1862
Foodplants: Unknown

lactucina (Cramer, 1780)
Foodplants: Unknown

paulina Jordan, 1924
Foodplants: Unknown

podaliriaria (Westwood, 1841)
Foodplants: Unknown

spinicauda Jordan, 1924
Foodplants: Unknown

stricturaria (Hübner, [1831])
machaonaria (Guérin-Méneville, [1844])
Foodplants: Unknown

terminalis Jordan, 1924
Foodplants: Unknown

transversaria columbiana Jordan, 1924
Foodplants: Unknown

transversaria salax Jordan, 1924
Foodplants: *Chimarrhis parviflora, Guettarda foliacea, Pentagonia donnell-smithii, Renistipulia costaricensis, Sabicea panamensis, Uncaria tomentosa*

transversaria transversaria Druce, 1887
Foodplants: *Calycophyllum candidissimum, Warscewiczia coccinea.*

Cercophaninae

Cercophanini

Cercophana C. Felder, 1862

frauenfeldi C. Felder, 1862
Foodplants: *Betula, Cryptocarya alba, Quercus.*

venusta (Walker, 1856)
aristoteliae (Philippi, 1860)
daphnea Maassen & Weymer, 1885
rufescens Philippi, 1864
vulpes Butler, 1882
Foodplants: *Aristotelia chilensis, Maytenus boaria.*

Microdulia Jordan, 1924

mirabilis (Rothschild, 1895)
izquierodoi (Silva, 1920)
Foodplants: Unknown

Neocercophana Izquierdo, 1895

philippii Izquierdo, 1895
canenilla Voelschow, 1902
Foodplants: *Hydrangea scandens*

Janiodini

Janiodes Jordan, 1924

bethulia (Druce, 1904)
Foodplants: Unknown

cuscoensis Brechlin & Meister, 2008
Foodplants: Unknown

dognini Jordan, 1924
Foodplants: Unknown

ecuadorensis (Dognin, 1890)
flexuosa Dognin, 1911
monacharia (Maassen, 1890)
Foodplants: Unknown

laverna laverna (Druce, 1890)
laverna pulverosa (Strand, 1912)
Foodplants: *Fagus, Quercus*

laverna nigropunctata (Druce, 1906)
Foodplants: *Fagus, Quercus × hispanica.*

lavernoboliviana Brechlin & Meister, 2011
Foodplants: Unknown

lavernocuscoensis Brechlin & Meister, 2011
Foodplants: Unknown

lavernodiosiana Brechlin & Meister, 2011
Foodplants: Unknown

lavernohuanucensis Brechlin & Meister, 2011
Foodplants: Unknown

lavernopascoensis Brechlin & Meister, 2011
Foodplants: Unknown

lucida Brechlin & Meister, 2011
Foodplants: Unknown

madrediosiana Brechlin & Meister, 2011
Foodplants: Unknown

manzanoi Pinas-Rubio, 2000
Foodplants: Unknown

meridensis Brechlin & Meister, 2011
Foodplants: Unknown

napoensis Brechlin, Meister & Käch, 2009
Foodplants: Unknown

oxapampensis Brechlin & Meister, 2008
Foodplants: Unknown

peruviana Brechlin & Meister, 2011
Foodplants: Unknown

pichinchensis Brechlin, Meister & Käch, 2009
Foodplants: Unknown

piurensis Brechlin & Meister, 2011
Foodplants: Unknown

russea (Dognin, 1912)
Foodplants: Unknown

virgata Jordan, 1924
Foodplants: Unknown

Arsenurinae

Arsenurini

Arsenura Duncan, 1841

albopicta Jordan, 1922
Foodplants: *Tilia cordata*

altocymonia Brechlin & Meister, 2010
Foodplants: Unknown

angulatus Bouvier, 1924
Foodplants: Unknown

archianassa archianassa (Draudt, 1930)
Foodplants: *Ceiba pentandra, Tilia cordata*

archianassa porioni Lemaire, 1980
Foodplants: Unknown

arianae Brechlin & Meister, 2010
Foodplants: *Tilia cordata*

armida (Cramer, 1779)
cassandra (Cramer, 1779)
erythrinae (Fabricius, 1781)
Foodplants: *Annona, Annona rensoniana, Bombax ceiba, Ceiba insignis, Ceiba pentandra, Ceiba speciosa, Goethalsia meiantha Guazuma ulmifolia, Heliocarpus americanus, Heliocarpus appendiculatus, Pachira quinata, Padus avium* var. *avium, Prunus laurocerasus, Salix caprea, Tilia cordata.*

aspasia (Herrich-Schäffer, [1853])
Foodplants: *Ceiba speciosa, Luehea divaricata, Lychnophora, Myrsine umbellata, Rapanea, Tibouchina, Vernonia erigeroides.*

batesii arcaei Druce, 1886
Foodplants: *Apeiba tibourbou, Byttneria catalpifolia, Ceiba, Guazuma ulmifolia, Heisteria, Heliocarpus americanus, Helicteres baruensis, Helicteres guazumifolia, Herrania purpurea, Luehea candida, Luehea seemannii, Luehea speciosa, Malvaviscus palmanus, Mortoniodendron guatemalense, Pachira aquatica, Pachira sessilis, Pseudobombax septenatum, Quararibea asterolepsis, Quararibea funebris, Sida, Theobroma angustifolium, Theobroma bicolor, Tilia platyphyllos, Triumfetta bogotensis, Triumfetta lappula, Waltheria glomerata.*

batesii aurantiaca Lemaire, 1976
Foodplants: Unknown

batesii batesii (C. Felder & R. Felder, 1874)
crenulata Schaus, 1921
Foodplants: *Apeiba tibourbou, Ceiba speciosa, Guazuma ulmifolia, Luehea speciosa, Mortoniodendron guatemalense.*

batesii gadoui Lemaire, 1980
Foodplants: Unknown

beebei (Fleming, 1945)
Foodplants: Unknown

biundulata Schaus, 1906
Foodplants: Unknown

ciocolatina Draudt, 1930
Foodplants: Unknown

cymonia (W. Rothschild, 1907)
Foodplants: *Myriocarpa, Saurauia, Wercklea.*

delormei Bouvier, 1929
Foodplants: Unknown

drucei Schaus, 1906
Foodplants: *Ceiba speciosa, Malvaviscus palmanus, Pachira aquatica, Theobroma angustifolium.*

fuscata Brechlin & Meister, 2010
Foodplants: Unknown

giuglarisi Bènèluz, 2009
Foodplants: Unknown

jennettae Wolfe, Conlan & Kelly, 2000
Foodplants: Unknown

kaechi Brechlin & Meister, 2010
Foodplants: Unknown

meander (Walker, 1855)
Foodplants: Unknown

mossi Jordan, 1922
Foodplants: Unknown

orbignyana (Guérin-Méneville, [1844])
dorbignyana Bouvier, 1924
Foodplants: *Brachychiton discolor, Ceiba speciosa., Luehea divaricata, Luehea paniculata, Tilia cordata, Tilia platyphyllos.*

pandora (Klug, 1836)
Foodplants: *Luehea divaricata*

paraorbignyana Brechlin & Meister, 2010
Foodplants: Unknown

polyodonta (Jordan, 1911)
Foodplants: *Ceiba insignis, Ceiba speciosa, Salix caprea, Tilia cordata, Tilia platyphyllos.*

ponderosa guianensis (W. Rothschild, 1907)
Foodplants: *Salix caprea*

ponderosa ponderosa W. Rothschild, 1895
Foodplants: *Hibiscus*

ponderosa yungasensis Brechlin & Meister, 2010
Foodplants: Unknown

rebeli Gschwandner, 1920
oweni Schaus, 1921
Foodplants: *Tilia*

sylla sylla (Cramer, 1779)
pelias (Jordan, 1911)
Foodplants: *Hirtella racemosa*

sylla hercules (Walker, 1855)
Foodplants: *Hirtella racemosa*

sylla niepelti (Schüssler, 1936)
Foodplants: *Hirtella racemosa, Luehea candida.*

sylla winbrechlini Brechlin & Meister, 2010
Foodplants: Unknown

thomsoni lemairei Racheli, 1998
Foodplants: Unknown

thomsoni thomsoni Schaus, 1906
Foodplants: Unknown

xanthopus (Walker, 1855)
alcmene Draudt, 1930
paranensis (Schüssler, 1936)
Foodplants: *Brachychiton discolor, Luehea divaricata, Urena sinuata.*

yungascymonia Brechlin & Meister, 2010
Foodplants: Unknown

Caio Travassos & Noronha, 1968

championi (Druce, 1886)
columbiana (Rothschild, 1907)
undilinea (Schaus, 1921)
Foodplants: *Ceiba pentandra, Chorisia, Pachira aquatica, Pachira quinata, Tilia.*

chiapasiana Brechlin & Meister, 2010
Foodplants: *Tilia cordata, Tilia platyphyllos.*

hidalgensis Brechlin & Meister, 2010
Foodplants: Unknown

harrietae (Forbes, 1944)
Foodplants: *Chorisia, Pachira quinata, Tilia cordata, Tilia occidentalis, Tilia platyphyllos.*

richardsoni (Druce, 1886)
extrema (Beutelspacher, 1986)
mestiza (Draudt, 1930)
Foodplants: *Ceiba acuminata, Ceiba pentandra, Ceiba speciosa.*

romulus (Maassen, 1869)
brasiliensis (Zikan, 1941)
Foodplants: Unknown

witti Brechlin & Meister, 2010
Foodplants: *Tilia cordata, Tilia platyphyllos.*

Copiopteryx Duncan, 1841

decerto (Maassen, [1872])
Foodplants: *Manilkara chicle*

jehovah (Strecker, 1874)
adaheli Draudt, 1930
Foodplants: *Manilkara chicle*

semiramis andensis (Lemaire, 1974)
Foodplants: *Acanthosyris spinescens, Manilkara chicle, Pouteria caimito.*

semiramis banghaasi Draudt, 1930
Foodplants: *Manilkara chicle*

semiramis gadourum Lemaire, 1971
Foodplants: Unknown

semiramis montei Gagarin, 1934
travassosi May, 1934
Foodplants: *Pouteria caimito, Pouteria ramiflora.*

semiramis phoenix (Deyrolle, 1869)
Foodplants: Unknown

semiramis semiramis (Cramer, 1775)
igoryi Gagarin, 1934
Foodplants: *Acanthosyris spinescens, Manilkara chicle, Pouteria caimito.*

semiramis steindachneri Fassl, 1917
Foodplants: Unknown

sonthonnaxi É. André, 1905
phippsi Schaus, 1932
Foodplants: *Mimusops*

virgo Zikán, 1929
biedermanni Kotzsch, 1930
Foodplants: Unknown

Dysdaemonia Hübner, 1819 [»1816«]

australoboreas Brechlin & Meister, 2009
Foodplants: *Bombax ceiba, Ceiba pentandra, Chorisia.*

boreas (Cramer, 1775)
auster (C. Felder & R. Felder, 1874)
Foodplants: *Bombax ceiba, Ceiba pentandra, Ceiba speciosa.*

brasiliensis W. Rothschild, 1907
Foodplants: *Bombax ceiba, Ceiba pentandra, Chorisia.*

concisa Becker & Camargo, 2001
Foodplants: Unknown

fosteri W. Rothschild, 1906
cortesi Giacomelli, 1925
Foodplants: *Ceiba speciosa, Samanea.*

undulensis Brechlin & Meister, 2009
Foodplants: *Ceiba speciosa*

Grammopelta W. Rothschild, 1907

lineata (Schaus, 1906)
cervina Rothschild, 1907
convergens (Bouvier, 1928)
niepelti Draudt, 1930
Foodplants: Unknown

Loxolomia Maassen, 1869

johnsoni Schaus, 1932
zwechti May, 1934
Foodplants: *Cariniana legalis, Quercus × hispanica.*

serpentina Maassen, 1869
Foodplants: Unknown

Paradaemonia Bouvier, 1925

andensis (W. Rothschild, 1907)
andina (Fassl, 1920)
Foodplants: *Coffea, Cuphea ignea.*

balsasensis C. Mielke & E. Furtado, 2005
Foodplants: Unknown

berlai Oiticica, 1946
Foodplants: Unknown

castanea (W. Rothschild, 1907) **rev. stat.**
Foodplants: Unknown

gravis (Jordan, 1922)
rufomaculata Bouvier, 1929
Foodplants: Unknown

mayi (Jordan, 1922)
Foodplants: Unknown

meridionalis Camargo, O. Mielke & Casagrande, 2007
Foodplants: Unknown

nycteris (Jordan, 1922)
Foodplants: Unknown

orsilochus (Maassen, 1869)
caligula Girard, 1885
Foodplants: Unknown

platydesmia (W. Rothschild, 1907)
despinayi (Bouvier, 1923)
guianenesis (Bouvier, 1925)
Foodplants: *Cuphea*

pluto (Westwood, [1854])
glaucescens (Walker, 1855)
kadenii (Herrich-Schäffer, 1855)
Foodplants: Unknown

ruschii May & Oiticica Filho, 1943
Foodplants: Unknown

samba (Schaus, 1906)
Foodplants: Unknown

terrena (Jordan, 1922)
susannae L. Racheli, 2005
Foodplants: Unknown

thelia (Jordan, 1922)
wagneri (Bouvier, 1924)
Foodplants: *Chrysophyllum marginatum*

Rhescyntis Hübner, [1819]

descimoni Lemaire, 1975
Foodplants: Unknown

hermes (W. Rothschild, 1907)
Foodplants: Unknown

hippodamia colombiana (Bouvier, 1927)
pomposa (Draudt, 1930)
Foodplants: Unknown

hippodamia gigantea (Bouvier, 1930)
Foodplants: Unknown

hippodamia hippodamia (Cramer, 1777)
guianensis (Bouvier, 1924)
martii (Perty, 1833)
Foodplants: *Myristica fragrans*, *Ligustrum ovalifolium*, *Virola bicuhyba*, *Virola koschnyi*, *Virola sebifera*.

hippodamia norax Druce, 1897
septentrionalis Vásquez, 1966
Foodplants: *Ligustrum*, *Myristica fragrans*, *Virola bicuhyba*, *Virola koschnyi*, *Virola sebifera*.

pseudomartii Lemaire, 1976
Foodplants: Unknown

reducta Camargo & Becker, 2001
Foodplants: Unknown

Titaea Hübner, [1823]

lemoulti (Schaus, 1905)
Foodplants: *Brachychiton populneus*, *Ceiba pentandra*, *Ceiba speciosa*, *Pachira quinata*, *Tilia caroliniana* subsp. *floridana*, *Tilia platyphyllos*.

orsinome Hübner, [1823]
latifascia (Walker, 1855)
Foodplants: *Ceiba speciosa*, *Eriotheca gracilipes*, *Pachira quinata*.

raveni (F. Johnson & Michener, 1948) **rev. stat.**
Foodplants: *Pachira quinata*, *Tilia cordata*.

tamerlan amazonensis Lemaire, 1980
Foodplants: *Ceiba speciosa*, *Pachira quinata*, *Tilia cordata*, *Tilia platyphyllos*.

tamerlan guayaquila (Schaus, 1932)
Foodplants: *Pachira quinata*

tamerlan nobilis (Schaus, 1912)
andicola (Bouvier, 1927)
avangareza (Schaus, 1932)
columbiana (Draudt, 1930)
Foodplants: *Ceiba pentandra*, *Ceiba speciosa*, *Goethalsia meiantha*, *Pachira aquatica*, *Pachira quinata*, *Tilia cordata*.

tamerlan tamerlan (Maassen, 1869)
Foodplants: *Bombax*, *Brachychiton populneus*, *Ceiba speciosa*, *Pachira quinata*, *Tilia cordata*, *Tilia platyphyllos*.

timur (Fassl, 1915)
Foodplants: *Pachira quinata*

Almeidaiini

Almeidaia Travassos, 1937

aidae Mielke & Casagrande, 1981
Foodplants: *Eriotheca gracilipes*, *Pseudobombax longiflorum*.

romualdoi Travassos, 1937
Foodplants: Unknown

Ceratocampinae

Adeloneivaia Travassos, 1940

acuta (Schaus, 1896)
Foodplants: *Acacia baileyana, Robinia pseudoacacia.*

bahiana Brechlin & Meister, 2011
Foodplants: Unknown

bellardi (Schaus, 1927)
Foodplants: Unknown

boisduvalii (Doûmet, 1859)
carisma (Schaus, 1904)
guianensis (Bouvier, 1923)
marginata (Bouvier, 1923)
Foodplants: *Acacia baileyana, Acacia dealbata, Gleditsia triacanthos, Robinia pseudoacacia.*

boliviana (Bouvier, 1927)
Foodplants: *Acacia baileyana, Robinia pseudoacacia.*

catharina (Bouvier, 1927)
apicalis (Bouvier, 1927)
Foodplants: Unknown

catoxantha catoxantha (W. Rothschild, 1907)
marioniana (Bouvier, 1927)
Foodplants: *Acacia baileyana*

catoxantha lacrimata (Dognin, 1921)
Foodplants: Unknown

cayennsis Brechlin & Meister, 2011
Foodplants: Unknown

erici Brechlin & Meister, 2011
Foodplants: Unknown

fallax (Boisduval, 1872)
boulleti (Bouvier, 1923)
curvilinea (Schaus, 1900)
Foodplants: *Acacia velutina, Inga sapindoides.*

irrorata (Schaus, 1900)
Foodplants: *Acacia baileyanaleyana*

isara (Dognin, 1905)
pacifica (Schaus, 1911)
packardi (Bouvier, 1927)
Foodplants: *Acacia baileyana, Acacia collinsii, Cercidium microphyllum, Ebenopsis ebano, Lysiloma aurita, Lysiloma divaricatum, Prosopis juliflora.*

jamazonica Brechlin & Meister, 2011
Foodplants: Unknown

jason (Boisduval, 1872)
montezuma (Packard, 1905)
Foodplants: *Cojoba valerioi, Enterolobium schomburgkii, Inga edulis, Inga longispica, Inga mortoniana, Inga oerstediana, Inga punctata, Inga samanensis, Inga sapindoides, Malus »hillieri«, Paraserianthes lophantha, Quercus agrifolia, Quercus dumosa, Quercus kelloggii, Salix caprea, Zygia longifolia.*

jaustralica Brechlin & Meister, 2011
Foodplants: Unknown

kawiana Brechlin & Meister, 2011
Foodplants: Unknown

leopoldina Brechlin & Meister, 2011
Foodplants: Unknown

minuta (Bouvier, 1927)
Foodplants: Unknown

misionesa Brechlin & Meister, 2011
Foodplants: Unknown

muelleri Brechlin & Meister, 2011
Foodplants: Unknown

nigripuncta Lemaire, 1984
Foodplants: *Inga sapindoides*

orientoandensis Brechlin & Meister, 2011
Foodplants: Unknown

orientoboliviana Brechlin & Meister, 2011
Foodplants: Unknown

pallida Lemaire, 1982
Foodplants: Unknown

paracatharina Brechlin & Meister, 2011
Foodplants: Unknown

pelias (W. Rothschild, 1907)
confusa Travassos & Noronha, 1965
Foodplants: Unknown

sabulosa brasiliensis (Bouvier, 1927)
meridionalis (Bouvier, 1927)
Foodplants: Unknown

sabulosa sabulosa (W. Rothschild, 1907)
diluta (W. Rothschild, 1907)
Foodplants: *Acacia baileyana*

schubarti Rêgo-Barros & Mielke, 1970
Foodplants: *Stryphnodendron adstringens*

subangulata (Herrich-Schäffer, [1855])
obscura (Bouvier, 1924)

Foodplants: *Acacia baileyana, Anadenanthera peregrina* var. *falcata, Malus »hillieri«, Mimosa, Robinia pseudoacacia, Salix caprea.*

wellingi Lemaire, 1982
Foodplants: Unknown

Adelowalkeria Travassos, 1941

caeca Lemaire, 1969
Foodplants: Unknown

eugenia (Druce, 1904)
comstocki (Fleming, 1945)
Foodplants: Unknown

eugenicolombiana Brechlin & Meister, 2011
Foodplants: Unknown

flavoboliviana Brechlin & Meister, 2011
Foodplants: Unknown

flavoparvissima Brechlin & Meister, 2011
Foodplants: Unknown

flavosignata (Walker, 1865)
caripitoensis (Fleming, 1945)
philiponi (Bouvier, 1927)
sanctissima D'Abrera, 1995
wardii (Boisduval, 1872)
Foodplants: *Cassia, Crotalaria, Machaerium brasiliense, Mimosa, Myrocarpus frondosus.*

kitchingi Brechlin & Meister, 2011
Foodplants: Unknown

plateada (Schaus, 1905)
amazonica (W. Rothschild, 1907)
Foodplants: *Acacia baileyana*

siriae Brechlin & Meister, 2011
Foodplants: Unknown

torresi Travassos & May, 1941
Foodplants: Unknown

tristygma (Boisduval, 1872)
columbia (Schaus, 1892)
Foodplants: Unknown

vanschaycki Brechlin & Meister, 2011
Foodplants: Unknown

witti Brechlin & Meister, 2011
Foodplants: Unknown

Almeidella Oiticica, 1946

almeidai Oiticica, 1946
Foodplants: Unknown

approximans (Schaus, 1921)
Foodplants: Unknown

corrupta (Schaus, 1913)
Foodplants: Unknown

Anisota Hübner, [1820]

assimilis (Druce, 1886)
Foodplants: *Quercus albocincta, Quercus grisea, Quercus × hispanica.*

consularis Dyar, 1896
Foodplants: *Quercus falcata, Quercus laevis, Quercus myrtifolia, Quercus nigra, Quercus palustris, Quercus phellus, Quercus robur, Quercus virginiana.*

dissimilis (Boisduval, 1872)
suprema H. Edwards, 1884
Foodplants: *Malus »hillieri«, Quercus × hispanica, Quercus robur.*

finlaysoni Riotte, 1969
Foodplants: *Castanea sativa, Quercus alba, Quercus michauxii, Quercus muhlenbergii, Quercus palustris, Quercus rubra.*

kendallorum Lemaire, 1987
Foodplants: *Quercus albocincta, Quercus grisea.*

leucostygma (Boisduval, 1872)
Foodplants: *Quercus dumosa, Quercus lobata.*

manitobensis McDunnough, 1921
Foodplants: *Corylus, Quercus macrocarpa.*

oslari W. Rothschild, 1907
neomexicana Brehme, 1909
skinneri Biedermann 1908
Foodplants: *Acer negundo, Acer rubrum, Acer saccharinum, Quercus arizonica, Quercus emoryi, Quercus gambelii, Quercus kelloggii, Quercus oblongifolia, Quercus palustris, Quercus robur, Quercus rubra, Quercus turbinella, Quercus velutina, Quercus virginiana, Schinus terebinthifolia.*

peigleri Riotte, 1975
Foodplants: *Castanea sativa, Quercus falcata, Quercus gambelii, Quercus marilandica, Quercus nigra, Quercus palustris, Quercus pubescens, Quercus rubra, Quercus velutina.*

pellucida (J. E. Smith, 1797)
Foodplants: *Castanea sativa, Quercus falcata, Quercus imbricaria, Quercus lyrata, Quercus macrocarpa, Quercus marilandica, Quercus myrtifolia, Quercus nigra, Quercus texana, Quercus palustris, Quercus pubescens, Quercus velutina.*

punctata Riotte & Peigler, 1981
Foodplants: *Castanea sativa, Quercus falcata, Quercus gambelii, Quercus marilandica, Quercus nigra, Quercus palustris, Quercus pubescens, Quercus rubra, Quercus velutina.*

senatoria (J. E. Smith, 1797)
Foodplants: *Betula pendula, Carya, Castanea dentata, Castanea pumila, Castanea sativa, Corylus, Fagus, Quercus alba, Quercus bicolor, Quercus coccinea, Quercus ilicifolia, Quercus imbricaria, Quercus macrocarpa, Quercus michauxii, Quercus palustris, Quercus prinoides, Quercus rubra, Quercus stellata, Quercus velutina, Rubus allegheniensis, Rubus idaeus.*

stigma (Fabricius, 1775)
fuscosa Ferguson, 1971
Foodplants: *Castanea dentata, Castanea mollissima, Castanea pumila, Castanea sativa, Corylus, Quercus alba, Quercus falcata, Quercus marilandica, Quercus michauxii, Quercus nigra, Quercus palustris, Quercus pubescens, Quercus rubra, Quercus stellata, Quercus velutina, Quercus virginiana.*

virginiensis (Drury, 1773)
discolor Ferguson, 1971
pellucida (Smith, 1797)
perspicellina (Sherborn, 1898)
sinulus Riotte, 1970
Foodplants: *Betula papyrifera, Carpinus caroliniana, Castanea dentata, Corylus, Fagus grandifolia, Quercus alba, Quercus falcata, Quercus macrocarpa, Quercus marilandica, Quercus nigra, Quercus palustris, Quercus petraea, Quercus michauxii, Quercus robur, Quercus rubra, Quercus stellata, Quercus velutina.*

Bathyphlebia C. Felder & R. Felder, 1874

aglia C. Felder & R. Felder, 1874
gschwandneri Schawerda, 1925
Foodplants: *Quercus*

aglioides Naumann, Brosch & Wenczel, 2009
Foodplants: Unknown

eminentoides Brechlin & Meister, 2009
Foodplants: Unknown

eminens (Dognin, 1891)
Foodplants: *Morella pubescens, Quercus rubra.*

flavior Oiticica Filho & Michener, 1950
Foodplants: *Quercus*

johnsoni Oiticica Filho & Michener, 1950
Foodplants: *Mimosa, Quercus*

rufescens Oiticica Filho & Michener, 1950
Foodplants: *Quercus*

Ceratesa Michener, 1949

hemirhodia (W. Rothschild, 1907)
Foodplants: Unknown

Ceropoda Michener, 1949

tibialis (W. Rothschild, 1907)
johnsoni Michener, 1949
Foodplants: Unknown

Cicia Oiticica, 1964

citrina (Schaus, 1904)
Foodplants: *Gleditsia triacanthos*

crocata (Boisduval, 1872)
eolus (Bouvier, 1927)
invalida (Schaus, 1892)
jucundoides (Bouvier, 1924)
latus (Bouvier, 1927)
typica (Bouvier, 1927)
Foodplants: *Gleditsia triacanthos*

nettia (Schaus, 1921)
callichroma (Le Cerf, 1934)
Foodplants: Unknown

norape Becker & Camargo, 2001
Foodplants: Unknown

pamala (Schaus, 1900)
lineata (Druce, 1904)
tricolor (Bouvier, 1923)
Foodplants: Unknown

pelota (Schaus, 1905)
Foodplants: Unknown

Citheronia Hübner, [1819]

andina Lemaire, 1971
Foodplants: *Malosma laurina, Schinus terebinthifolia.*

aroa Schaus, 1896
brasiliensis Bouvier, 1927
consobrina Rothschild, 1907
mogya Schaus, 1920
Foodplants: *Juglans regia, Ligustrum ovalifolium, Liquidambar styraciflua, Malosma laurina, Prunus domestica, Prunus laurocerasus Prunus serotina, Prunus spinosa, Malosma laurina.*

azteca albescens Lemaire, 1973
Foodplants: *Malosma laurina, Rhus typhina, Schinus terebinthifolia*

azteca azteca Schaus, 1896
Foodplants: *Liquidambar styraciflua, Prunus domestica, Quercus robur, Malosma laurina, Rhus typhina, Schinus terebinthifolia.*

beledonon Dyar, 1912
colimae Draudt, 1930
nayaritensis Bouvier, 1931
Foodplants: *Gossypium, Ligustrum ovalifolium, Ligustrum vulgare, Liquidambar styraciflua, Malosma laurina, Rhus typhina, Schinus terebinthifolia.*

bellavista bellavista Draudt, 1930
Foodplants: *Calycophyllum candidissimum, Cochlospermum vitifolium, Ligustrum ovalifolium, Liquidambar styraciflua, Malosma laurina, Pachira quinata, Phoradendron quadrangulare, Spondias mombin, Uncaria tomentosa, Vernonanthura patens, Zanthoxylum setulosum.*

bellavista cinerea Lemaire, 1982
Foodplants: *Rhus*

brissotii brissotii (Boisduval, 1868)
catharina Draudt, 1930
Foodplants: *Amygdalus, Carya illinoinensis, Euphorbia, Juglans nigra, Ligustrum japonicum, Ligustrum ovalifolium, Liquidambar styraciflua, Prunus laurocerasus, Psidium, Ricinus, Sapium glandulosum.*

brissotii meridionalis Bouvier, 1927
argentina Draudt, 1930
benjamini Schaus, 1933
Foodplants: *Ligustrum japonicum, Ligustrum ovalifolium, Prunus persica.*

equatorialis Bouvier, 1927
Foodplants: *Juglans neotropica, Prunus domestica.*

guayaquila Schaus, 1927
Foodplants: *Ligustrum ovalifolium, Malosma laurina, Prunus domestica, Rosa, Schinus molle.*

hamifera bodoquena Travassos & Rêgo-Barros, 1966
Foodplants: Unknown

hamifera hamifera W. Rothschild, 1907
fuscalis W. Rothschild, 1907
guayanensis Bouvier, 1927
Foodplants: *Ligustrum ovalifolium, Liquidambar styraciflua, Malosma laurina, Pyracantha coccinea, Salix caprea, Terminalia catappa.*

johnsoni Schaus, 1928
Foodplants: Unknown

laocoon (Cramer, 1777)
anassa Hübner, [1819]
cacicus (Walker, 1855)
ixion (Boisduval, 1868)
princeps (Walker, 1855)
Foodplants: *Acalypha, Anacardium occidentale, Campomanesia mediterranea, Caryocar brasiliense, Cupressus, Diospyros kaki, Gossypium herbaceum, Ilex paraguariensis, Joannesia princeps, Juglans nigra, Ligustrum ovalifolium, Liquidambar styraciflua, Malosma laurina, Mangifera indica, Padus avium* var. *avium, Psidium guineense, Psidium guajava, Psidium grandifolium, Psidium guineense, Psidium rufum, Rosa, Ricinus communis, Schinus terebinthifolia, Sebastiania commersoniana, Terminalia catappa, Vernicia fordii, Vernonanthura tweedieana, Vernonia cuneata, Zinnia elegans.*

lichyi Lemaire, 1971
Foodplants: *Prunus spinosa, Ligustrum ovalifolium, Salix caprea.*

lobesis centraliamericana Bouvier, 1927
Foodplants: Unknown

lobesis jordani Draudt, 1930
Foodplants: *Liquidambar styraciflua, Malosma laurina, Prunus spinosa, Rhus typhina.*

lobesis lobesis W. Rothschild, 1907
claveryi Bouvier, 1924
marion Dyar, 1914
Foodplants: *Bursera simaruba, Calycophyllum candidissimum, Cochlospermum vitifolium, Eucalyptus gunnii, Ligustrum ovalifolium, Liquidambar styraciflua, Maclura tinctoria, Malosma laurina, Pachira quinata, Phoradendron quadrangulare, Prunus domestica, Prunus spinosa, Psidium guajava, Pyracantha coccinea, Sapium glandulosum, Spondias mombin, Spondias purpurea, Syzygium salomonense, Zanthoxylum setulosum.*

maureillei Wolfe & Herbin, 2002
Foodplants: *Liquidambar styraciflua, Malosma laurina, Salix caprea, Schinus molle, Schinus terebinthifolia.*

mexicana Grote & Robinson, 1866
Foodplants: *Carya, Juglans regia, Ligustrum ovalifolium, Liquidambar styraciflua, Prunus domestica.*

phoronea (Cramer, 1779)
minutus Fleming, 1945
Foodplants: *Bixa orellana, Ligustrum ovalifolium, Malosma laurina, Prunus domestica, Prunus laurocerasus, Prunus serotina, Prunus spinosa, Pyracantha coccinea.*

pseudomexicana Lemaire, 1974
Foodplants: *Liquidambar styraciflua, Malosma laurina, Rhus typhina, Schinus terebinthifolia.*

regalis (Fabricius, 1793)
infernalis Strecker, 1884
maxima cornuta (Edwards, 1771)
regia (Abbot & Smith, 1797)
Foodplants: *Carya glabra, Carya illinoinensis, Carya ovata, Cephalanthus occidentalis, Corylus, Diervilla, Diospyros virginiana, Ficus sycomorus, Fraxinus excelsior, Gossypium herbaceum, Juglans cinerea, Juglans nigra, Juglans regia, Ligustrum ovalifolium, Liquidambar styraciflua, Malosma laurina, Nyssa sylvatica, Oxydendrum arboreum, Platanus occidentalis, Platanus orientalis, Prunus domestica, Prunus serotina, Quercus robur, Quercus × hispanica, Rhus glabra, Schmaltzia choriophylla, Schmaltzia copallinum, Rhus glabra, Rhus typhina, Salix caprea, Sassafras albidum, Schinus terebinthifolia, Syringa vulgaris.*

sepulcralis Grote & Robinson, 1865
Foodplants: *Pinus caribaea, Pinus echinata, Pinus halepensis, Pinus mugo, Pinus nigra, Pinus ponderosa, Pinus rigida, Pinus strobus, Pinus sylvestris, Pinus taeda.*

splendens queretana Vásquez, 1944
Foodplants: *Carya illinoinensis, Juglans nigra, Mangifera indica, Malosma laurina, Rhus typhina, Schinus molle, Thespesia lampas* var. *thespesioides.*

splendens sinaloensis C. C. Hoffmann, 1942
Foodplants: *Arbutus menziesii, Arctostaphylos pungens, Carya glabra, Juglans nigra, Juglans regia, Ligustrum ovalifolium, Liquidambar styraciflua, Malosma laurina, Pistacia lentiscus, Schmaltzia choriophylla, Rhus trilobata, Rhus typhina, Schinus molle, Schinus terebinthifolia, Thespesia lampas* var. *thespesioides.*

splendens splendens (Druce, 1886)
Foodplants: *Arbutus, Arctostaphylos, Carya illinoinensis, Gossypium hirsutum, Juglans nigra, Juglans regia, Liquidambar styraciflua, Lithraea brasiliensis, Mangifera indica, Malosma laurina, Rhus typhina, Schinus polygama, Schinus molle, Schinus terebinthifolia, Struthanthus flexicaulis, Thespesia lampas* var. *thespesioides.*

vogleri (Weyenburgh, 1881)
leona (Druce, 1890)
Foodplants: *Eucalyptus gunnii, Jodina rhombifolia, Ligaria cuneifolia, Liquidambar styraciflua, Lithraea brasiliensis, Lithraea molleoides, Loranthaceae, Malosma laurina, Prunus spinosa, Pyracantha coccinea, Schinus molle, Schinus polygama, Schinus terebinthifolia, Tripodanthus acutifolius.*

volcan Lemaire, 1982
Foodplants: *Conostegia, Quetzalia occidentalis.*

Citheronioides Lemaire, 1987

collaris (W. Rothschild, 1907)
Foodplants: *Acalypha diversifolia, Ardisia standleyana, Cestrum racemosum, Chimarrhis parviflora, Chione venosa* var. *Venosa, Conceveiba pleiostemona, Conostegia micrantha, Conostegia xalapensis, Cordia stellifera, Hampea appendiculata, Heliocarpus appendiculatus, Hyeronima oblonga, Liquidambar, Melastoma, Miconia affinis, Miconia trinervia, Myrcia splendens, Prunus annularis, Prunus domestica, Quercus, Vitex cooperi.*

lamata (Schaus, 1933) **rev. stat.**
Foodplants: Unknown

Citheronula Michener, 1949

armata (W. Rothschild, 1907)
Foodplants: *Anacardium occidentale, Jatropha cinerea.*

sonyae Breyer, 1957
Foodplants: *Jatropha cinerea*

Citioica Travassos & Noronha, 1965

anthonilis (Herrich-Schäffer, [1854])
analis (Rothschild, 1907)
grisescens (C. C. Hoffmann, 1942)
ocarona (Schaus, 1933)

Foodplants: *Acacia baileyana, Robinia pseudoacacia, Salix caprea.*

guayensis Brechlin & Meister, 2011
Foodplants: Unknown

homoea (W. Rothschild, 1907)
Foodplants: Unknown

rubrocanescens Brechlin & Meister, 2011
Foodplants: Unknown

Dacunju Travassos & Noronha, 1965

jucunda (Walker, 1855)
dimidiata (Herrich-Schäffer, [1855])
smithii (Ehrmann, 1909)
Foodplants: Unknown

Dryocampa Harris, 1833

rubicunda (Fabricius, 1793)
alba Grote, 1874
pallida Bowles, 1875
roseilinea Schaus, 1920
sperryae Bower, 1942
venusta Walker, 1865
Foodplants: *Acer campestre, Acer monspessulanum, Acer negundo, Acer pseudoplatanus, Acer rubrum, Acer saccharinum, Acer spicatum, Betula, Fagus grandifolia, Juglans cinerea, Malus »hillieri«, Quercus coccinea, Quercus ilicifolia, Quercus laevis, Quercus nigra, Quercus velutina.*

Eacles Hübner, 1819 [»1816«]

acuta Schaus, 1905
Foodplants: Unknown

adoxa Jordan, 1910
Foodplants: Unknown

barnesi Schaus, 1905
johnsoniella Oiticica Filho & Michener, 1950
Foodplants: *Eucalyptus gunnii, Fagus grandifolia* subsp. *grandifolia, Schinus terebinthifolia.*

bertrandi Lemaire, 1981
Foodplants: Unknown

callopteris W. Rothschild, 1907
Foodplants: Unknown

camposportoi Mendes, 1937
Foodplants: Unknown

canaima Feige, 1971
ambicolor Lemaire, 1972
Foodplants: *Coriariaceae*

cuscoensis Brechlin & Meister, 2009
Foodplants: Unknown

ducalis (Walker, 1955)
columbiana Draudt, 1929
Foodplants: *Fagus, Juglans, Ligustrum ovalifolium, Quercus robur.*

fairchildi May & Oiticica Filho, 1941
Foodplants: *Ligustrum, Quercus.*

fulvaster fulvaster W. Rothschild, 1907
Foodplants: *Eucalyptus gunnii, Fagus grandifolia* subsp. *grandifolia.*

fulvaster oriecuadoriana Brechlin & Meister, 2011
Foodplants: Unknown

guianensis Schaus, 1905
Foodplants: *Fagus, Juglans, Ligustrum, Quercus.*

guinlei Oiticica, 1941
Foodplants: Unknown

imperialis anchicayensis Lemaire, 1971
Foodplants: *Betula pendula, Castanea sativa, Cupressus, Ficus, Gossypium anomalum, Ligustrum ovalifolium, Prunus laurocerasus, Pyracantha coccinea, Quercus × hispanica, Quercus robur.*

imperialis cacicus (Boisduval, 1868)
approximans Bouvier, 1923
Foodplants: *Betula pendula, Crataegus monogyna, Cupressus, Erythrina crista-galli, Ficus, Gossypium anomalum, Ligustrum ovalifolium, Ligustrum vulgare, Malus »hillieri«, Mangifera indica, Melastoma, Myrtaceae, Quercus ilex, Quercus robur, Pistacia lentiscus, Prunus laurocerasus, Prunus spinosa, Psidium guajava, Pyracantha coccinea, Rhus, Salix babylonica, Salix caprea, Sambucus nigra, Schinus polygama.*

imperialis decoris W. Rothschild, 1907
Foodplants: *Astronium graveolens, Betula pendula, Billia hippocastanum, Bixa orellana, Bursera simaruba, Bursera tomentosa, Calycophyllum candidissimum, Calyptrogyne trichostachys, Cedrela odorata, Cochlospermum vitifolium, Cupressus lusitanica, Dalbergia retusa, Erythrina costaricensis, Eucalyptus deglupta, Ficus, Gliricidia sepium, Gmelina arborea, Gossypium anomalum, Guazuma ulmifolia, Guettarda macrosperma, Haematoxylum brasiletto, Hymenaea courbaril, Inga sapindoides, Lonchocarpus acuminatus, Lonchocarpus costaricensis, Lonchocarpus guatemalensis, Lonchocarpus minimiflorus, Mabea occidentalis, Malus »hillieri«, Mimosa xanti,*

Mosquitoxylum jamaicense, Myrospermum frutescens, Nectandra hihua, Ocotea veraguensis, Persea americana, Phoradendron quadrangulare, Prunus laurocerasus, Psidium guajava, Quercus oleoides, Rehdera trinervis, Roupala montana, Sambucus nigra, Spondias mombin, Swietenia macrophylla, Terminalia oblonga, Thouinidium decandrum, Trema micrantha, Trichospermum mexicanum, Virola koschnyi.

imperialis hallwachsae Brechlin & Meister, 2011
Foodplants: Unknown

imperialis imperialis (Drury, 1773)
didyma (Palisot de Beauvois, [1806])
imperatoria (Abbot & Smith, 1797)
nobilis Neumoegen, 1891
punctatissima Neumoegen, 1891
Foodplants: *Abies grandis, Acer negundo, Acer pseudoplatanus, Acer rubrum, Acer saccharinum, Acer saccharum, Aesculus hippocastanum, Alnus serrulata, Arbutus, Betula papyrifera, Betula pendula, Betula populifolia, Betula pubescens, Bursera tomentosa, Carpinus caroliniana, Carya, Castanea dentata, Castanea pumila, Castanea sativa, Cedrus, Cephalanthus angustifolius, Chamaecyparis thyoides, Cochlospermum vitifolium, Cornus, Crataegus, Cupressus, Diospyros guianensis* subsp. *akaraiensis, Fagus grandifolia, Fagus sylvatica, Ficus, Gleditsia triacanthos, Gossypium anomalum, Gossypium herbaceum, Juglans cinerea, Juniperus communis, Juniperus virginiana, Koelreuteria paniculata, Larix griffithii, Larix laricina, Ligustrum ovalifolium, Ligustrum vulgare Lindera benzoin, Liquidambar styraciflua, Malosma laurina, Malus »hillieri«, Myrica cerifera, Ostrya virginiana, Padus virginiana, Picea, Pinus banksiana, Pinus echinata, Pinus serotina, Pinus strobus, Pinus taeda, Platanus occidentalis, Platanus orientalis, Prunus laurocerasus, Prunus serotina, Pyracantha coccinea, Quercus alba, Quercus coccinea, Quercus laevis, Quercus macrocarpa, Quercus palustris, Quercus rubra, Rhus glabra, Robinia pseudoacacia, Rosa, Salix babylonica, Salix caprea, Sambucus canadensis, Sambucus nigra, Sassafras albidum, Taxodium distichum, Tilia americana, Tilia americana* var. *heterophylla, Tilia europaea, Tsuga canadensis, Ulmus alata, Ulmus americana, Ulmus procera, Ulmus rubra, Ulmus thomasii, Viburnum dentatum.*

imperialis magnifica Walker, 1855
Foodplants: *Anacardium occidentale, Apuleia ferrea, Arbutus, Betula pendula, Ficus, Gossypium anomalum, Ligustrum ovalifolium, Liquidambar styraciflua, Mangifera indica, Prunus laurocerasus, Prunus serotina, Quercus robur, Salix babylonica, Salix caprea, Sambucus nigra.*

imperialis opaca (Burmeister, 1878)
Foodplants: *Betula pendula, Erythrina crista-galli, Fraxinus americana, Ligustrum ovalifolium, Prunus laurocerasus, Prunus serotina, Quercus robur, Salix babylonica, Salix caprea.*

imperialis pini Michener, 1950
Foodplants: *Pinus banksiana, Picea glauca, Pinus resinosa, Pinus strobus, Pinus sylvestris, Quercus palustris, Sapindus saponaria.*

imperialis piurensis Brechlin & Meister, 2011
Foodplants: *Quercus robur*

imperialis quintanensis Lemaire, 1971
Foodplants: *Malus »hillieri«, Quercus robur.*

imperialis tucumana W. Rothschild, 1907
Foodplants: *Betula pendula, Celtis, Ligustrum ovalifolium, Prunus laurocerasus, Quercus × hispanica, Schinus polygama.*

imperialis typica Schüssler, 1936
Foodplants: *Astronium graveolens, Betula pendula, Billia hippocastanum, Bixa orellana, Bursera simaruba, Bursera tomentosa, Calycophyllum candidissimum, Calyptrogyne trichostachys, Cedrela odorata, Cochlospermum vitifolium, Cupressus lusitanica, Dalbergia retusa, Erythrina costaricensis, Eucalyptus deglupta, Ficus, Gliricidia sepium, Gmelina arborea, Gossypium anomalum, Guazuma ulmifolia, Guettarda macrosperma, Haematoxylum brasiletto, Hymenaea courbaril, Inga sapindoides, Lonchocarpus acuminatus, Lonchocarpus costaricensis, Lonchocarpus guatemalensis, Lonchocarpus minimiflorus, Mabea occidentalis, Malus »hillieri«, Mimosa xanti, Mosquitoxylum jamaicense, Myrospermum frutescens, Nectandra hihua, Ocotea veraguensis, Persea americana, Phoradendron quadrangulare, Prunus laurocerasus, Psidium guajava, Quercus oleoides, Rehdera trinervis, Roupala montana, Sambucus nigra, Spondias mombin, Swietenia macrophylla, Terminalia oblonga, Thouinidium decandrum, Trema micrantha, Trichospermum mexicanum, Virola koschnyi.*

kaechi Brechlin & Meister, 2011
Foodplants: Unknown

lauroi Oiticica Filho, 1938
Foodplants: Unknown

lemairei Rêgo-Barros & Tangerini, 1973
Foodplants: Unknown

manuelita Oiticica Filho, 1941
Foodplants: Unknown

masoni centralamericana Brechlin & Meister, 2011
Foodplants: Unknown

masoni machiapasiana Brechlin & Meister, 2011
Foodplants: Unknown

masoni masoni Schaus, 1896
Foodplants: *Liquidambar styraciflua, Malosma laurina, Schinus terebinthifolia.*

mayi Schaus, 1920
lombardi Bouvier, 1924
Foodplants: Unknown

ormondei janzeni Brechlin & Meister, 2011
Foodplants: Unknown

ormondei niepelti Draudt, 1930
Foodplants: *Liquidambar styraciflua, Malosma laurina, Quercus, Salix babylonica, Salix caprea.*

ormondei ormondei Schaus, 1889
Foodplants: *Guarea glabra, Hampea appendiculata, Liquidambar styraciflua, Malosma laurina, Meliosma glabrata, Panopsis suaveolens, Prunus annularis, Quercus, Tapirira mexicana.*

ormondei peruviana Bouvier, 1927
amazonica Schüssler, 1936
Foodplants: *Eucalyptus gunnii, Salix babylonica, Salix caprea.*

ormondei vanschaycki Brechlin & Meister, 2011
Foodplants: Unknown

ormondei violacea Lemaire, 1975
Foodplants: Unknown

ormondei yucatanensis Lemaire, 1988
Foodplants: *Malosma laurina, Quercus robur.*

oslari W. Rothschild, 1907
Foodplants: *Arctostaphylos patula, Juglans, Liquidambar styraciflua, Malosma laurina, Pinus ponderosa, Pinus strobus, Prunus spinosa, Pinus sylvestris, Quercus chihuahuensis, Quercus emoryi, Quercus gambelii, Quercus × hispanica, Quercus oblongifolia, Quercus robur, Quercus rubra, Quercus turbinella, Rhus glabra, Rhus typhina, Robinia neomexicana, Salix babylonica, Sapindus saponaria Sapindus drummondii, Schinus molle, Schinus terebinthifolia*

paraadoxa Brechlin & Meister, 2009
Foodplants: Unknown

penelope (Cramer, 1775)
cybele (Olivier, 1790)
majestalis Draudt, 1930
Foodplants: *Betula, Eucalyptus gunnii, Fagus grandifolia* subsp. *grandifolia, Fagus sylvatica, Psidium, Juglans regia, Laburnum, Ligustrum ovalifolium, Ligustrum vulgare, Malosma laurina, Malus »hillieri«, Melastoma, Psidium guajava, Quercus ilex, Salix babylonica, Salix caprea.*

silkae Brechlin & Meister, 2011
Foodplants: Unknown

tyrannus Draudt, 1930
Foodplants: *Liquidambar, Malus »hillieri«, Quercus × hispanica, Quercus robur, Prunus domestica.*

Giacomellia Bouvier, 1930

bilineata (Burmeister, 1878)
floresi (Giacomelli, 1915)
formosa (Dognin, 1916)
inversa (Giacomelli, 1911) **syn. nov.**
inversoatomasa (Giacomelli, 1911)
oberthueri (Bouvier, 1927)
Foodplants: *Prosopis nigra*

drechseli Lampe, 1995
Foodplants: Unknown

escobari Lampe, 1995
Foodplants: Unknown

vanschaycki Brechlin & Meister, 2011
Foodplants: Unknown

Jaiba Lemaire, Tangerini & Mielke, 1999

kesselringi Lemaire, Tangerini & Mielke, 1999
Foodplants: Unknown

Megaceresa Michener, 1949

pulchra (Bouvier, 1923)
Foodplants: *Anadenanthera peregrina* var. *falcata, Calliandra parviflora.*

Mielkesia Lemaire, 1988

paranaensis (Rêgo-Barros & Mielke, 1968)
Foodplants: Unknown

Neorcarnegia Draudt, 1930

basirei (Schaus, 1892)
bispinosa Naumann, 2006 **syn. nov.**
Foodplants: *Acacia baileyana, Caesalpinia ferrea, Cassia, Machaerium acutifolium, Salix pentandra.*

Oiticella Travassos & Noronha, 1965

brevis (Walker, 1855)
lanaris (W. Rothschild, 1907)
Foodplants: *Mimosa*

convergens (Herrich-Schäffer, [1854])
pallida (Bouvier, 1927)
Foodplants: Unknown

luteciae (Bouvier, 1924)
Foodplants: Unknown

Othorene Boisduval, 1872

bernardoespinozai Brechlin & Meister, 2011
Foodplants: Unknown

cadmus (Herrich-Schäffer, [1854])
Foodplants: *Gleditsia triacanthos*

hodeva (Druce, 1904)
rubra Schaus, 1904
Foodplants: *Acacia baileyana., Crataegus monogyna, Fagus sylvatica, Gleditsia triacanthos, Quercus × hispanica × pseudoturneri, Quercus robur, Sclerolobium paniculatum.*

intermedia Rothschild, 1907
Foodplants: *Chrysophyllum brenesii, Fagus sylvatica, Gleditsia triacanthos, Malus »hillieri«, Manilkara chicle, Quercus robur, Pouteria exfoliata, Psidium, Salix caprea, Terminalia catappa.*

lapaziana Brechlin & Meister, 2011
Foodplants: Unknown

peggyae Brechlin & Meister, 2011
Foodplants: Unknown

purpurascens (Schaus, 1905)
dollii (Packard, 1905)
intensiva (Draudt, 1930)
Foodplants: *Chrysophyllum brenesii, Fagus sylvatica, Gleditsia triacanthos, Malus »hillieri«, Manilkara chicle, Quercus robur, Pouteria exfoliata, Psidium, Salix caprea, Terminalia catappa.*

vanschayckorum Brechlin & Meister, 2011
Foodplants: Unknown

verana verana Schaus, 1900
castanea (Packard, 1905)
Foodplants: *Fagus sylvatica, Gleditsia triacanthos, Quercus gambelii, Quercus oleoides, Quercus robur.*

verana verapaziana Brechlin & Meister, 2011
Foodplants: Unknown

verana veroaxensis Brechlin & Meister, 2011
Foodplants: Unknown

winbrechlini Brechlin & Meister, 2011
Foodplants: Unknown

Procitheronia Michener, 1949

fenestrata (W. Rothschild, 1907)
Foodplants: *Prunus domestica, Prunus spinosa.*

principalis (Walker, 1855)
afenestrata (Draudt, 1930)
Foodplants: *Croton floribundus*

purpurea (Oiticica, 1942)
Foodplants: *Codiaeum variegatum, Psychotria.*

vanschaycki Brechlin & Meister, 2011
Foodplants: Unknown

Psilopygida Michener, 1949

Subgenus *Psilopygida* Michener, 1949

crispula (Dognin, 1905)
wagneri (Bouvier, 1923)
Foodplants: *Acacia farnesiana, Gleditsia triacanthos.*

Subgenus *Psigida* Oiticica Filho, 1959

apollinairei (Dognin, 1919)
Foodplants: Unknown

basalis Michener, 1952
Foodplants: *Acacia farnesiana, Gleditsia triacanthos.*

walkeri (Grote, 1867)
rosea (Mabilde, 1896)
rosea (Schüssler, 1936)
Foodplants: *Acacia baileyana, Acacia farnesiana, Calliandra parviflora, Gleditsia triacanthos, Mimosa diplotricha, Psidium.*

Psilopygoides Michener, 1949

oda (Schaus, 1905)
Foodplants: Unknown

Ptiloscola Michener, 1949

bipunctata Lemaire, 1972
Foodplants: *Acacia baileyana*

burmeisteri Meister & Brechlin, 2008
Foodplants: Unknown

cinerea (Schaus, 1900)
picklei (Schaus, 1933)
Foodplants: Unknown

cuscoensis Brechlin & Meister, 2011
Foodplants: Unknown

dargei Lemaire, 1971
Foodplants: *Acacia baileyana, Acacia collinsii, Acacia riparia, Acacia tenuifolia, Calliandra rubescens.*

descimoni Lemaire, 1971
Foodplants: Unknown

lilacina lilacina (Schaus, 1900)
Foodplants: Unknown

lilacina affinis (W. Rothschild, 1907)
Foodplants: *Salix*

paraguayensis Brechlin, Meister & Drechsel, 2008
Foodplants: Unknown

photophila (W. Rothschild, 1907)
lasiocampa (Bryk, 1953)
Foodplants: Unknown

rorerae (Schaus, 1928)
Foodplants: *Quercus* × *hispanica*

surrotunda (Dyar, 1925)
Foodplants: Unknown

wellingi Lemaire, 1971
Foodplants: Unknown

wolfei Brechlin & Meister, 2008
Foodplants: *Robinia pseudoacacia*

Rachesa Michener, 1949

adusta (W. Rothschild, 1907)
Foodplants: Unknown

alegrensis Brechlin & Meister, 2011
Foodplants: Unknown

breteuili (Bouvier, 1927)
caucensis Lemaire, 1969
Foodplants: Unknown

chrisbrechlinae Brechlin & Meister, 2011
Foodplants: Unknown

lampei Brechlin & Meister, 2011
Foodplants: Unknown

nisa (Druce, 1904)
Foodplants: Unknown

reventador Lemaire, 1975
Foodplants: Unknown

sinjaevorum Brechlin & Meister, 2011
Foodplants: Unknown

viksinjaevi Brechlin & Meister, 2011
Foodplants: Unknown

Schausiella Bouvier, 1930

arpi (Schaus, 1892)
Foodplants: *Embelia laeta, Gleditsia triacanthos, Hymenaea courbaril, Hymenaea courbaril* var. *altissima, Quercus, Robinia pseudoacacia.*

carabaya (W. Rothschild, 1907)
Foodplants: *Hymenaea courbaril* var. *altissima, Quercus.*

denhezorum Lemaire, 1969
Foodplants: *Hymenaea courbaril* var. *altissima*

janeira (Schaus, 1892)
Foodplants: *Hymenaea courbaril* var. *altissima, Quercus.*

longispina (W. Rothschild, 1907)
vellosoi Travassos, 1958
Foodplants: Unknown

moinieri Lemaire, 1969
Foodplants: Unknown

polybia (Stoll, 1781)
schausi Bouvier, 1931
Foodplants: *Gleditsia triacanthos, Robinia pseudoacacia.*

ronaldzunigi Brechlin & Meister, 2011
Foodplants: Unknown

sanjuensis Brechlin & Meister, 2011
Foodplants: Unknown

santarosensis Lemaire, 1982
Foodplants: *Hymenaea courbaril*

spitzi Travassos, 1958
Foodplants: *Hymenaea courbaril*

subochreata (Schaus, 1904)
adocima (Dyar, 1914)
bipuncta (Dognin, 1923)

klagesi (W. Rothschild, 1907)
polybioides (Bouvier, 1927)
Foodplants: Unknown

toulgoeti Lemaire, 1969
Foodplants: Unknown

Scolesa Michener, 1949

hypoxantha (W. Rothschild, 1907)
flavidorsata (Dognin, 1908)
Foodplants: Unknown

leucantha (Boisduval, 1872)
argyracantha (Boisduval, 1872)
pseudoargyracantha Travassos, 1957
Foodplants: Unknown

nebulosa Lemaire, 1971
Foodplants: Unknown

totoma (Schaus, 1900)
paranensis (Bouvier, 1923)
Foodplants: Unknown

viettei Travassos 1959
Foodplants: Unknown

vinacea (W. Rothschild, 1907)
Foodplants: Unknown

Syssphinx Hübner, [1819]

albolineata (Grote & Robinson, 1866)
Foodplants: *Acacia angustissima, Acacia baileyana, Gleditsia triacanthos, Leucaena pulverulenta, Mimosa, Robinia pseudoacacia.*

amena (Travassos, 1941)
Foodplants: Unknown

bicolor (Harris, 1841)
distigma (Walsh, 1864)
immaculata (Jewett, 1882)
suprema (Neumoegen, 1885)
Foodplants: *Acacia baileyana, Gleditsia triacanthos, Gymnocladus dioica, Hymenaea courbaril, Hymenaea courbaril* var. *altissima, Ligustrum japonicum, Quercus, Robinia pseudoacacia.*

bidens (W. Rothschild, 1907)
centrimacula (Strand, 1912)
Foodplants: *Acacia baileyana, Fagus sylvatica, Gleditsia triacanthos, Quercus × hispanica, Robinia pseudoacacia, Salix.*

bidentalis Brechlin & Meister, 2011
Foodplants: Unknown

bisecta (Lintner, 1879)
nebulosa (Neumoegen, 1890)
Foodplants: *Acacia baileyana, Gleditsia triacanthos, Gymnocladus dioica, Robinia pseudoacacia.*

blanchardi (Ferguson, 1971)
Foodplants: *Acacia angustissima, Acacia berlandieri, Acacia farnesiana, Acacia rigidula, Acacia wrightii, Cercidium floridum* subsp. *floridum, Ebenopsis ebano, Gleditsia triacanthos, Leucaena pulverulenta, Malpighia glabra, Mimosa malacophylla, Parkinsonia aculeata, Prosopis pubescens, Robinia pseudoacacia, Sideroxylon celastrinum.*

chocoensis Lemaire, 1988
Foodplants: Unknown

colla colla Dyar, 1907
jacca (Schaus, 1920)
pollens (Schaus, 1911)
talamanca (Schaus, 1911)
vilderi (Schaus, 1911)
Foodplants: *Albizia adinocephala, Albizia saman, Caesalpinia platyloba, Gleditsia triacanthos, Zapoteca formosa.*

colla grisea Lemaire, 1988
Foodplants: *Malus »hillieri«*

colloida (Dyar, 1925)
Foodplants: *Acacia baileyana, Acacia collinsii, Acacia dealbata, Albizia saman, Cercidium microphyllum, Ebenopsis ebano, Gleditsia triacanthos, Prosopis juliflora.*

digueti (Bouvier, 1929)
Foodplants: *Acacia baileyana, Gleditsia triacanthos.*

erubescens (Boisduval, 1872)
Foodplants: Unknown

gadouae (Lemaire, 1971)
Foodplants: Unknown

gomezi Lemaire, 1984
Foodplants: *Acacia baileyana, Robinia pseudoacacia.*

heiligbrodti (Harvey, 1877)
Foodplants: *Acacia farnesiana, Acacia rigidula, Cercidium floridum* subsp. *floridum, Cercidium microphyllum, Gleditsia triacanthos, Larrea divaricata, Prosopis glandulosa, Prosopis juliflora.*

hubbardi Dyar, 1903
Foodplants: *Acacia baileyana, Acacia greggii, Acacia wrightii, Cercidium floridum* subsp. *floridum, Cercidium microphyllum, Gleditsia triacanthos,*

Prosopis glandulosa, Prosopis juliflora, Prosopis velutina, Robinia pseudoacacia.

jasonoides (Lemaire, 1971)
Foodplants: Unknown

jennyphillipsae Brechlin & Meister, 2011
Foodplants: Unknown

lapazcellata Brechlin & Meister, 2011
Foodplants: Unknown

malinalcoensis (Lemaire, 1975)
Foodplants: Unknown

misionescellata Brechlin & Meister, 2011
Foodplants: Unknown

mexicana (Boiduval, 1872)
xanthochroia (Dyar, 1912)
Foodplants: *Acacia collinsii, Acacia cornigera, Gleditsia triacanthos, Robinia pseudoacacia.*

mexisinaloana Brechlin & Meister, 2011
Foodplants: Unknown

modena Dyar, 1913
Foodplants: Unknown

molina molina (Cramer, 1780)
agenor (Bouvier, 1923)
grandis (Grote & Robinson, 1867)
obtusa Strassberger, 1932
petersii von Bönninghausen, 1901
propinqua (Bouvier, 1923)
purpurpunctata Lampe, 1986
simulatilis (Grote & Robinson, 1867)
Foodplants: *Acacia baileyana, Acacia collinsii, Acosmium panamense, Albizia adinocephala, Albizia guachepele, Albizia lebbeck, Albizia saman, Carpinus betulus, Cassia fistula, Cassia grandis, Cassia javanica, Cotoneaster praecox, Dalbergia retusa, Diospyros kaki, Ficus microcarpa, Gleditsia triacanthos, Gliricidia sepium, Inga sapindoides, Mimosa xanti, Quercus oleoides, Quercus robur, Pyracantha coccinea, Robinia pseudoacacia, Salix caprea, Schinus polygama, Senna occidentalis, Senna reticulata, Senna uniflora.*

molina molpiurica Brechlin & Meister, 2011
Foodplants: Unknown

montana (Packard, 1905)
kuschei (Dyar, 1925)
phaedima (Dyar, 1925)
sinaloana (Schaus, 1920)
Foodplants: *Acacia angustissima, Acacia angustissima* var. *Texensis, Acacia baileyana, Acacia farnesiana, Cercidium floridum* subsp. *floridum, Gleditsia triacanthos, Haematoxylum brasiletto, Mimosa aculeaticarpa, Paraserianthes lophantha, Parkinsonia aculeata, Prosopis glandulosa, Prosopis pubescens, Robinia pseudoacacia, Senna candolleana.*

ocellata (W. Rothschild, 1907)
Foodplants: Unknown

paracolla Brechlin & Meister, 2011
Foodplants: Unknown

parocellata Brechlin & Meister, 2011
Foodplants: Unknown

pescadori Lemaire, 1988
Foodplants: Unknown

quadrilineata quadrilineata (Grote & Robinson, 1867)
hoegei (Druce, 1886)
isias (Boisduval, 1872)
Foodplants: *Acacia baileyana, Acacia collinsii, Acacia dealbata, Acacia tenuifolia, Albizia adinocephala, Albizia saman, Calliandra calothyrsus, Calliandra rubescens, Cassia grandis, Cercidium microphyllum, Ebenopsis ebano, Entada gigas, Gleditsia triacanthos, Haematoxylum brasiletto, Leucaena multicapitula, Lysiloma aurita, Mimosa pigra, Mimosa xanti, Prosopis juliflora, Robinia pseudoacacia, Salix caprea, Senna pallida, Senna papillosa, Senna reticulata, Sphinga platyloba.*

quadrilineata occlusa (Dognin, 1916)
babaulti (Bouvier, 1929)
cananche (Bouvier, 1927)
unimacula (Dognin 1916)
Foodplants: Unknown

raspa (Boisduval, 1871)
Foodplants: *Acacia angustissima, Acacia angustissima* var. *Texensis, Acacia baileyana, Acacia farnesiana, Cercidium floridum* subsp. *floridum, Gleditsia texana, Gleditsia triacanthos, Paraserianthes lophantha, Parkinsonia aculeata, Prosopis pubescens, Robinia pseudoacacia, Salix caprea.*

riekerti Brechlin & Meister, 2011
Foodplants: Unknown

smithi (Druce, 1904)
andrea (Dognin 1912)
Foodplants: *Salix caprea*

thiaucourti (Lemaire, 1975)
Foodplants: *Acacia*

xanthina Lemaire, 1984
Foodplants: Unknown

yucatana (Druce, 1904)
Foodplants: *Acacia collinsii*

Hemileucinae

Hemileucini

Adetomeris Michener, 1949

erythrops (Blanchard, 1852)
acharon (Butler, 1882)
annaehildegardae (Bryk, 1945)
betzholdi, (Ureta, 1942)
bullocki (Ureta, 1942)
contulma (Draudt 1929)
debilis (Butler, 1882)
erythraea (Philippi, 1859)
fusca (Ureta 1942)
gayi (Boisduval, 1875)
griseoflava (Philippi 1859)
herrerai (Ureta, 1942)
izquierdoi (Draudt, 1929)
jaffueli (Ureta 1942)
lucasii (Boisduval, 1875)
olivacea (Butler, 1882)
reedi (Ureta, 1942)
ruizi (Ureta, 1942)
Foodplants: *Acacia melanoxylon, Castanea pumila, Fraxinus, Juglans regia, Malosma laurina, Maytenus boaria, Nothofagus antarctica, Nothofagus dombeyi, Nothofagus pumilio, Pinus radiata, Populus, Prunus armeniaca, Puya chilensis, Rhamnus, Robinia pseudoacacia, Rubus ulmifolius, Rubus ulmifolius* var. *inermis.*

microphthalma (Izquierdo, 1895)
Foodplants: Unknown

Ancistrota Hübner, [1819]

plagia Hübner, [1819]
geayi (Bouvier, 1933)
Foodplants: Unknown

Arias Lemaire, 1995

inbio Lemaire, 1995
Foodplants: Unknown

Automerella Michener, 1949

aurora (Maassen & Weyding, 1885)
rubicunda (Schaus, 1892)
Foodplants: *Fagus, Laburnum, Ligustrum ovalifolium, Quercus, Robinia pseudoacacia.*

flexuosa (C. Felder & R. Felder, 1874)
bouvieri (Conte, 1906)
mendosa (Boisduval, 1875)
Foodplants: *Acacia, Fagus, Laburnum, Ligustrum ovalifolium, Quercus ilex, Quercus robur, Robinia pseudoacacia.*

miersi Lemaire & C. Mielke, 1998
Foodplants: Unknown

Automerina Michener, 1949

Subgenus *Automerina* Michener, 1949

carina Meister, Naumann & Brechlin, 2005
Foodplants: Unknown

caudatula (C. Felder & R. Felder, 1874)
Foodplants: *Acacia baileyana*

cypria (Gmelin, 1790)
jo (Cramer, [1780])
pericinctus (Conte, 1906)
stollii (Boisduval, 1875)
Foodplants: *Pyracantha coccinea, Theobroma cacao.*

vala (Kirby, 1871)
crameri (Boisduval, 1875)
Foodplants: *Theobroma cacao*

Subgenus *Automerula* Michener, 1949

auletes (Herrich-Schäffer, [1854])
Foodplants: *Anacardium occidentale, Eugenia, Gleditsia triacanthos, Malosma laurina, Myrcia, Protium heptaphyllum, Robinia pseudoacacia Salix caprea, Schinus terebinthifolia.*

beneluzi Lemaire, 2002
Foodplants: Unknown

Automeris Hübner, [1819]

abdomicajamarcensis Brechlin & Meister, 2011
Foodplants: Unknown

abdomimeridensis Brechlin & Meister, 2011
Foodplants: Unknown

abdominalis (C. Felder & R. Felder, 1874)
aristei Dognin, 1923
rufobrunnea Bouvier, 1936
Foodplants: *Robinia pseudoacacia, Salix caprea.*

abdominapoensis Brechlin & Meister, 2011
Foodplants: Unknown

abdomipichinchensis Brechlin & Meister, 2011
Foodplants: *Robinia pseudoacacia, Salix caprea.*

abdomipiurensis Brechlin & Meister, 2011
Foodplants: Unknown

adusta C. C. Hoffmann, 1942
Foodplants: Unknown

ahuitzolti Lemaire & Wolfe, 1993
Foodplants: *Acacia baileyana, Robinia pseudoacacia.*

altapazia Brechlin & Meister, 2011
Foodplants: Unknown

alticola Lemaire, 1975
Foodplants: Unknown

altotridens Brechlin & Meister, 2011
Foodplants: Unknown

amagabriellae Brechlin & Meister, 2011
Foodplants: Unknown

amaloretensis Brechlin & Meister, 2011
Foodplants: Unknown

amanda amanda Schaus, 1900
Foodplants: *Fagus sylvatica, Geoffroea decorticans, Piptadenia, Robinia pseudoacacia, Salix caprea, Schinus terebinthifolia, Selinum.*

amanda amandocuscoensis Brechlin & Meister, 2011
Foodplants: Unknown

amanda amandojunica Brechlin & Meister, 2011
Foodplants: Unknown

amanda lichyi Lemaire, 1966
Foodplants: *Fagus sylvatica, Quercus × hispanica.*

amanda limpida Lemaire, 1966
proximus peruvianus Bouvier, 1930
Foodplants: *Fagus sylvatica, Pyracantha coccinea, Schinus terebinthifolia.*

amanda subobscura Weymer, 1909
colombianus Bouvier, 1927
subpicta Dognin, 1923
Foodplants: *Fagus sylvatica*

amanda tucumana Bouvier, 1930
nigrolineata Eckerlein, 1935
conjuncta Köhler, 1935
Foodplants: *Geoffroea decorticans, Piptadenia, Schinus terebinthifolia.*

amasanata Brechlin & Meister, 2011
Foodplants: Unknown

amoena (Boisduval, 1875)
amoenoides Bouvier, 1930
ferrugineus Bouvier, 1929
parilis Schaus, 1906
pictus Conte, 1906
tucumana Bouvier, 1936
Foodplants: *Crataegus, Fagus sylvatica, Ligustrum ovalifolium, Quercus × hispanica, Quercus robur, Robinia pseudoacacia, Rubus fructicosus, Salix caprea, Schinus terebinthifolia.*

andicola Bouvier, 1930
Foodplants: *Rubus fructicosus*

angulatus Conte, 1906
Foodplants: *Cornus mas, Crataegus monogyna, Fagus sylvatica, Gleditsia triacanthos, Ixora floribunda, Laburnum, Ligustrum ovalifolium, Pyracantha coccinea, Quercus robur, Robinia pseudoacacia, Rubus fructicosus, Salix caprea.*

anika Brechlin, Meister & Van Schayck, 2011
Foodplants: Unknown

anikmeisterae Brechlin & Meister, 2011
Foodplants: Unknown

annulata Schaus, 1906
pallens Conte, 1906
Foodplants: *Acacia monticola, Inga sapindoides, Robinia pseudoacacia.*

argentifera Lemaire, 1966
Foodplants: *Ligustrum ovalifolium, Quercus robur, Robinia pseudoacacia, Salix caprea.*

arianae Brechlin & Meister, 2011
Foodplants: Unknown

arminia (Stoll, 1781)
altus Conte, 1906
falcifer Conte, 1906
surinamensis Kirby, 1892
Foodplants: *Fagus grandifolia* subsp. *grandifolia, Pyracantha coccinea.*

atrolimbata Lemaire, 1973
Foodplants: *Fagus sylvatica, Robinia pseudoacacia.*

averna Druce, 1886
Foodplants: *Fagus, Ligustrum ovalifolium, Quercus × hispanica.*

balachowskyi Lemaire, 1966
Foodplants: *Robinia pseudoacacia*

banus banumediata Brechlin & Meister, 2011
Foodplants: Unknown

banus banus (Boisduval, 1875)
proximus denudata Bouvier, 1930
Foodplants: *Abuta panamensis, Acalypha diversifolia, Acidoton nicaraguensis, Allophylus psilospermus, Alsophila firma, Andira inermis, Anemopaegma orbiculatum, Arachnothryx buddleioides, Attalea cohune, Augusta rivalis, Banisteriopsis muricata, Brosimum alicastrum, Calathea macrosepala, Calyptrogyne ghiesbreghtiana, Canna indica, Cecropia obtusifolia, Cecropia peltata, Cestrum racemosum, Cissampelos pareira, Citharexylum costaricense, Clethra lanata, Clethra mexicana, Combretum laxum, Connarus vulcanicus, Cordia alliodora, Cordia cymosa, Cordia eriostigma, Cordia polycephala, Coussapoa nymphaeifolia, Cupania glabra, Cynodon nlemfuensis, Desmodium incanum, Dichapetalum morenoi, Dioclea malacocarpa, Erythrina lanceolata, Fagus sylvatica, Gleditsia triacanthos, Gouania colombiana, Guarea glabra, Guazuma ulmifolia, Guettarda foliacea, Hampea appendiculata, Heliocarpus appendiculatus, Hiraea reclinata, Inga chocoensis, Inga herrerae, Inga oerstediana, Inga punctata, Inga sapindoides, Lacistema aggregatum, Lasiacis nigra, Lonchocarpus costaricensis, Lonchocarpus oliganthus, Machaerium arboreum, Machaerium cobanense, Malosma laurina, Malus pumila, Malvaviscus palmanus, Meliosma glabrata, Mucuna holtonii, Musa acuminata, Myrcia splendens, Myriocarpa longipes, Ochroma pyramidale, Paullinia bracteosa, Pavonia schiedeana, Pentagonia donnell-smithii, Pharus parvifolius, Pouteria exfoliata, Pouteria reticulata, Prunus annularis, Psidium guajava, Pterocarpus rohrii, Quercus × hispanica × pseudoturneri, Quercus robur, Robinia pseudoacacia, Rubus urticifolius, Salix babylonica, Salmea scandens, Sapium glandulosum, Senna papillosa, Sloanea faginea, Solanum rudepannum, Sorocea trophoides, Styrax argenteus, Swartzia cubensis, Tibouchina longifolia, Tournefortia bicolor, Triumfetta lappula, Urera caracasana, Urera elata, Vernonanthura patens, Weinmannia wercklei, Zexmenia virgulta.*

banus proxima Conte, 1906
coresus ecuadora Weymer, 1908
Foodplants: Unknown

barinasmargaritae Brechlin & Meister, 2011
Foodplants: Unknown

basalis (Walker, 1855)
griseorosea Bouvier, 1930
joiceyi Bouvier, 1930
macareis Schaus, 1892
penicillatus Bouvier, 1930
Foodplants: *Bambusa, Robinia pseudoacacia.*

beckeri (Herrich-Schäffer, [1856])
Foodplants: *Fagus, Prunus, Quercus × hispanica, Sorocea bonplandii.*

belti Druce, 1886
Foodplants: *Abuta panamensis, Acalypha diversifolia, Alsophila firma, Canna indica, Capulin, Cestrum racemosum, Chimarrhis parviflora, Clethra mexicana, Cordia alliodora, Coussapoa nymphaeifolia, Croton billbergianus, Dioclea malacocarpa, Fagus sylvatica, Ficus pertusa, Gmelina arborea, Guazuma ulmifolia, Hampea appendiculata, Heliocarpus appendiculatus, Hyeronima oblonga, Ilex skutchii, Inga oerstediana, Inga punctata, Ipomoea batatas, Ligustrum ovalifolium, Malus »hillieri«, Myrcia splendens, Myriocarpa longipes, Nectandra hihua, Ocotea insularis, Pavonia schiedeana, Pentagonia donnell-smithii, Perrottetia longistylis, Piper auritum, Piper fimbriulatum, Psidium guajava, Psychotria eurycarpa, Pterocarpus rohrii, Quercus robur, Rhynchosia pyramidalis, Robinia pseudoacacia, Salix, Senna papillosa, Sida rhombifolia, Swietenia macrophylla, Tabernaemontana alba, Talisia macrophylla, Terminalia catappa, Trema micrantha, Urena lobata, Vernonanthura patens, Vitex cooperi.*

beutelspacheri Lemaire, 2003
Foodplants: *Robinia pseudoacacia*

bilinea bilinea (Walker, 1855)
modesta (Ménétriés, 1863)
pallida Bouvier, 1930
vividior Draudt, 1929
Foodplants: *Fagus, Quercus robur, Robinia pseudoacacia, Rosa, Salix caprea.*

bilinea tamphilus Schaus, 1892
Foodplants: *Aspidosperma macrocarpon, Quercus robur.*

boops (C. Felder & R. Felder, 1874)
Foodplants: Unknown

boucardi Druce, 1886
violacea Conte, 1906
Foodplants: *Fagus sylvatica, Laburnum, Quercus robur, Robinia pseudoacacia, Salix caprea.*

boudinoti Lemaire, 1982
Foodplants: Unknown

boudinotiana Lemaire, 1986
Foodplants: *Quercus*

brenneri Brechlin, Meister & Van Schayck, 2011
Foodplants: Unknown

castrensis Schaus, 1898
Foodplants: Unknown

caucensis Lemaire, 1976
Foodplants: *Bambusa*

cecrojaliscensis Brechlin & Meister, 2011
Foodplants: Unknown

cecrops cecrops (Boisduval, 1875)
herse C. C. Hoffmann, 1942
mexicana Bouvier, 1929
Foodplants: *Cercocarpus betuloides, Liriodendron tulipifera, Prunus serotina, Malosma laurina, Quercus alba, Quercus × hispanica, Quercus robur, Robinia pseudoacacia, Rubus fructicosus, Salix caprea.*

cecrops pamina (Neumoegen, 1882)
aureosea (Neumoegen, 1882)
Foodplants: *Arctostaphylos patula, Arctostaphylos pungens, Ceanothus fendleri, Celtis ehrenbergiana, Cercis canadensis, Cercocarpus betuloides, Cercocarpus montanus* var. *glaber, Fagus sylvatica, Fouquieria splendens, Ligustrum ovalifolium, Malosma laurina, Mimosa aculeaticarpa, Prunus serotina, Quercus alba, Quercus emoryi, Quercus gambelii, Quercus grisea, Quercus marilandica, Quercus nigra, Quercus oblongifolia, Quercus palustris, Quercus phellus, Quercus robur, Quercus rubra, Quercus suber, Quercus turbinella, Quercus undulata, Opuntia, Robinia neomexicana, Robinia pseudoacacia, Rubus fructicosus, Salix caprea.*

celata Lemaire, 1969
rosea Bouvier, 1930
Foodplants: *Acalypha diversifolia, Adelia triloba, Cordia bicolor, Dioclea malacocarpa, Fagus sylvatica, Ligustrum ovalifolium, Ocotea cernua, Quercus, Robinia pseudoacacia, Salix caprea, Trichospermum mexicanum, Urera elata.*

chacona chacona Draudt, 1929
Foodplants: *Erythrina, Robinia pseudoacacia, Salix caprea.*

chacona cochabambae Lemaire, 1971
Foodplants: *Erythrina, Robinia pseudoacacia, Salix caprea.*

chaconoides Brechlin & Meister, 2008
Foodplants: Unknown

choco Brechlin & Meister, 2011
Foodplants: Unknown

chrisbrechlinianae Brechlin & Meister, 2011
Foodplants: Unknown

chrismeisterae Brechlin & Meister, 2011
Foodplants: Unknown

cinctinicaraguana Brechlin & Meister, 2011
Foodplants: Unknown

cinctistriga (C. Felder & R. Felder, 1874)
rubiginosa Bouvier, 1930
Foodplants: Unknown

claryi Naumann, Brosch & Wenczel, 2005
Foodplants: Unknown

colenon Dyar, 1912
thyreon Dyar, 1912
Foodplants: *Cynodon*

complicata (Walker, 1855)
Foodplants: *Bauhinia forficata, Citrus aurantium, Lagerstroemia indica, Passiflora, Prunus salicina* var. *salicina, Psidium guajava, Pyrus communis, Spartina patens, Tipuana tipu, Wisteria sinensis.*

coresus (Boisduval, 1859)
roseus Köhler, 1935
mirabilis Köhler, 1935
Foodplants: *Acer pseudoplatanus, Ligustrum ovalifolium, Platanus orientalis, Quercus robur, Salix caprea.*

cryptica Dognin, 1911
malvacea Bouvier, 1930
Foodplants: Unknown

cundinamarcensis Brechlin & Meister, 2011
Foodplants: Unknown

curvilinea Schaus, 1906
maculatus Conte, 1906
Foodplants: *Erythrina, Fagus sylvatica, Pyracantha coccinea, Quercus, Robinia pseudoacacia.*

cuscoata Brechlin & Meister, 2011
Foodplants: Unknown

cuscosylviae Brechlin & Meister, 2011
Foodplants: Unknown

dagmarae Brechlin & Meister, 2011
Foodplants: Unknown

dandemon Dyar, 1912
Foodplants: Unknown

daudiana Druce, 1894
Foodplants: Unknown

denhezorum Lemaire, 1966
Foodplants: Unknown

denticulata Conte, 1906
moerens Jordan, 1910
Foodplants: *Fagus sylvatica, Prunus domestica, Quercus robur, Quercus × hispanica, Robinia pseudoacacia, Salix babylonica, Salix caprea.*

descimoni Lemaire, 1972
Foodplants: Unknown

despicata Draudt, 1929
Foodplants: Unknown

dianae Brechlin & Meister, 2011
Foodplants: Unknown

diavolanda Naumann, Brosch & Wenczel, 2005
Foodplants: *Carpinus betulus, Fagus sylvatica, Quercus robur, Quercus × hispanica, Robinia pseudoacacia.*

doelfi Brechlin, Käch & Meister, 2011
Foodplants: Unknown

dognini Lemaire, 1967
Foodplants: Unknown

draudtiana Lemaire 1973
Foodplants: *Quercus alba, Quercus emoryi, Quercus lobata, Quercus nigra, Quercus palustris, Quercus robur, Quercus rubra.*

duchartrei Bouvier, 1936
Foodplants: *Malus »hillieri«, Robinia pseudoacacia.*

ecuata Brechlin & Meister, 2011
Foodplants: Unknown

egeus (Cramer, 1775)
draudti Röber, 1934
egeides Bouvier, 1936
umbracticus Eckerlein, 1935
zelleri (Grote & Robinson, 1868)
Foodplants: *Erythrina, Fagus sylvatica, Malus domestica, Quercus × hispanica × pseudoturneri, Quercus robur, Rosa, Rubus fructicosus, Salix caprea.*

elenensis Lemaire, 2002
Foodplants: Unknown

eogena (C. Felder & R. Felder, 1874)
adelon C. C.Hoffmann, 1942
erubescens Conte, 1906
Foodplants: Unknown

escalantei Lemaire, 1969
Foodplants: Unknown

excreta Draudt, 1929
grisea Bouvier, 1936
Foodplants: *Fagus sylvatica, Malus pumila, Prunus domestica, Quercus × hispanica, Quercus robur, Robinia pseudoacacia.*

exigua Lemaire, 1977
Foodplants: *Erythrina, Malus pumila, Ochroma pyramidale, Poa.*

fabiani Brechlin & Meister, 2011
Foodplants: Unknown

falco Jordan, 1910
Foodplants: Unknown

fieldi Lemaire, 1969
Foodplants: *Acacia baileyana, Inga edulis.*

fletcheri Lemaire, 1966
Foodplants: Unknown

frankae Brechlin & Meister, 2011
Foodplants: Unknown

gabriellae Lemaire, 1966
orneatoides Lemaire, 1966
Foodplants: *Fagus sylvatica, Pyracantha coccinea, Robinia pseudoacacia.*

gadouae Lemaire, 1966
Foodplants: *Cornus mas, Cornus sanguinea, Fagus, Quercus, Robinia pseudoacacia, Salix.*

godartii (Boisduval, 1875)
guyanensis Bouvier, 1936
Foodplants: *Gleditsia triacanthos, Salix babylonica, Salix purpurea.*

goiasensis Lemaire, 1977
Foodplants: Unknown

goodsoni Lemaire, 1966
Foodplants: Unknown

grammoboliviana Brechlin & Meister, 2011
Foodplants: Unknown

grammocajamarcensis Brechlin & Meister, 2011
Foodplants: Unknown

grammocuscoensis Brechlin & Meister, 2011
Foodplants: Unknown

grammodes Jordan, 1910
Foodplants: *Fargesia murielae, Phyllostachys nigra.*

grammopiurensis Brechlin & Meister, 2011
Foodplants: Unknown

granulosa Conte, 1906
Foodplants: *Bauhinia, Byrsonima verbascifolia, Erythroxylum anguifugum, Erythroxylum tortuosum, Rhododendron, Robinia pseudoacacia, Vatairea macrocarpa.*

hamata Schaus, 1906
columbiana Draudt, 1929
insolens Bouvier, 1927
panamensis Draudt, 1929
Foodplants: *Cornus mas, Crataegus monogyna, Fagus sylvatica, Gleditsia triacanthos, Ixora floribunda, Laburnum, Ligustrum ovalifolium, Pyracantha coccinea, Quercus robur, Robinia pseudoacacia, Rubus fructicosus, Salix caprea.*

harriamazonica Brechlin & Meister, 2011
Foodplants: *Fagus sylvatica,*

harricajamarcensis Brechlin & Meister, 2011
Foodplants: Unknown

harriyungasiana Brechlin & Meister, 2011
Foodplants: Unknown

harrisorum Lemaire, 1967
Foodplants: *Acacia baileyana, Fagus sylvatica, Robinia pseudoacacia.*

haxairei Herbin, 2003
Foodplants: Unknown

hebe (Walker, 1865)
Foodplants: *Quercus robur, Robinia pseudoacacia.*

hegena Brechlin & Meister, 2011
Foodplants: Unknown

heppneri Lemaire, 1982
Foodplants: Unknown

hesselorum Ferguson, 1972
Foodplants: *Acacia farnesiana, Chilopsis linearis, Eysenhardtia polystachya, Laburnum anagyroides, Mimosa, Prosopis velutina, Quercus emoryi, Quercus oblongifolia, Quercus palustris, Robinia pseudoacacia.*

iguaquensis Lemaire & Amarillo, 1992
Foodplants: Unknown

illustris (Walker, 1855)
coffeae (Boisduval, 1875)
gradli Sageder, 1939
pelotas Strand, 1920
phales (Boisduval, 1875)
pylades (Boisduval, 1875)
Foodplants: *Crataegus, Fagus grandifolia* subsp. *grandifolia, Ficus benjamina, Ligustrum ovalifolium, Ligustrum vulgare, Platanus orientalis, Prunus laurocerasus, Pyracantha coccinea, Robinia pseudoacacia, Rubus fructicosus, Syringa vulgaris.*

incarnata (Walker, 1865)
approximata (Walker, 1865)
meridana Schaus, 1906
vinosus Conte, 1906
Foodplants: *Acacia baileyana, Eucalyptus, Fagus sylvatica, Flemingia strobilifera, Laburnum, Ligustrum ovalifolium, Ligustrum vulgare, Malus »hillieri«, Prunus spinosa, Quercus robur, Robinia pseudoacacia, Salix babylonica, Salix caprea.*

innoxia Schaus, 1906
seclusa Bouvier, 1930
Foodplants: *Acacia baileyana, Fagus sylvatica, Prunus cerasifera, Pyracantha coccinea, Salix caprea.*

inornata (Walker, 1855)
fumosa (Boisduval, 1875)
obsoleta Niepelt, 1927
Foodplants: *Acacia baileyana, Prunus, Quercus × hispanica, Robinia pseudoacacia.*

io io (Fabricius, 1775)
argus Neumoegen & Dyar, 1893
caeca Igel, 1928
coloradensis Cockerell, 1914
corollaria (Perry, 1810)
fabricii (Boisduval, 1875)
fuscus Luther, 1907
lilith (Strecker, 1877)
lutheri Cockerell, 1914
mexicana Draudt, 1929
neomexicana Barnes & Benjamin, 1922
packardi Schüssler, 1934
texana Barnes & Benjamin, 1922
varia (Walker, 1855)
Foodplants: *Abies balsamea, Acer negundo, Acer rubrum, Acer saccharinum, Acosmium panamense, Aeschynomene brasiliana, Amorpha fruticosa, Ateleia herbert-smithii, Baptisia tinctoria, Betu-*

la pendula, Calopogonium caeruleum, Canavalia brasiliensis, Carpinus caroliniana, Carya, Cerasus vulgaris, Chamaecrista chamaecristoides, Chamaecrista fasciculata, Celtis ehrenbergiana, Celtis laevigata, Celtis laevigata var. *reticulata, Cercis canadensis, Cercocarpus betuloides, Chilopsis linearis, Comptonia peregrina, Conocarpus erectus, Cordia alliodora, Cornus florida, Corylus avellana, Crataegus, Crescentia alata, Croton morifolius, Dalbergia retusa, Dalea carthagenensis, Diospyros, Eysenhardtia polystachya, Fagus grandifolia, Fagus sylvatica, Fraxinus, Galactia striata, Gleditsia triacanthos, Gliricidia sepium, Glycyrrhiza, Gossypium herbaceum, Hibiscus lasiocarpos, Humulus lupulus, Hyparrhenia rufa, Ilex verticillata, Indigofera costaricensis, Inga sapindoides, Juglans nigra, Juglans regia, Laburnum anagyroides, Lespedeza, Leucaena pulverulenta, Ligustrum ovalifolium, Ligustrum vulgare, Liquidambar styraciflua, Liriodendron tulipifera, Litchi chinensis, Lonchocarpus costaricensis, Lonchocarpus eriocarinalis, Lonchocarpus minimiflorus, Magnolia virginiana, Malus pumila, Malvaviscus arboreus, Melilotus officinalis, Melilotus officinalis* subsp. *Alba, Mimosa camporum, Mimosa pigra, Morus, Myrica caroliniensis, Myrica pensylvanica, Litchi chinensis, Ostrya virginiana, Pachira quinata, Padus avium* var. *avium, Persea americana, Phaseolus lunatus, Phaseolus vulgaris, Platanus occidentalis, Populus balsamifera, Populus deltoides, Populus gileadensis, Populus tremuloides, Prunus domestica, Prunus emarginata* subsp. *mollis, Prunus serotina, Prunus spinosa, Padus virginiana, Psidium guajava, Pyrus communis, Quercus alba, Quercus lobata, Quercus emoryi, Quercus nigra, Quercus palustris, Quercus robur, Quercus rubra, Rhapis, Rhizophora mangle, Rhododendron, Rhynchosia calycosa, Rhynchosia reticulata, Ribes sativum, Robinia pseudoacacia, Rosa, Rubus allegheniensis, Rubus flagellaris, Rubus idaeus, Salix babylonica, Salix caprea, Salix interior, Sassafras albidum, Senna pallida, Serenoa repens, Sorbus coronaria* var. *angustifolia, Spiraea, Symphoricarpos albus, Syringa vulgaris, Tabebuia aurea, Tilia americana, Trifolium repens, Turnera ulmifolia, Ulmus, Viburnum dentatum, Wisteria frutescens, Wisteria sinensis, Zea mays, Zexmenia virgulta.*

io potosiana Brechlin, Meister & Van Schayck, 2011
Foodplants: Unknown

intertridens Brechlin & Meister, 2011
Foodplants: Unknown

iris (Walker, 1865)
ferruginea Draudt, 1929
occidentalis Draudt, 1929
Foodplants: *Acacia farnesiana, Chilopsis linearis, Eysenhardtia polystachya, Mimosa, Prosopis velutina, Quercus emoryi, Quercus oblongifolia, Quercus palustris, Robinia pseudoacacia, Salix caprea.*

jalishena Brechlin & Meister, 2011
Foodplants: Unknown

janrudloffi Brechlin & Meister, 2011
Foodplants: Unknown

janus (Cramer, 1775)
cruentus Niepelt, 1934
Foodplants: *Cocos nucifera, Erythrina, Fagus, Malus »hillieri«, Malus pumila, Malus sylvestris, Prunus laurocerasus, Malus pumila, Salix babylonica, Salix caprea, Salix pentandra, Salix viminalis.*

jinotegana Brechlin & Meister, 2011
Foodplants: Unknown

jivaros Dognin, 1898
Foodplants: *Acacia dealbata, Robinia pseudoacacia.*

jolantheae C. Mielke & Almeida, 2007
Foodplants: Unknown

juarezia Brechlin & Meister, 2011
Foodplants: Unknown

jucunda (Cramer, 1779)
divergens (Boisduval, 1875)
flavomarginatus Conte, 1906
junonia (Walker, 1855)
morescoides Bouvier, 1927
oblonga (Walker, 1855)
sinuatus Conte, 1906
titania (C. Felder & R. Felder, 1874)
Foodplants: *Acacia, Fagus sylvatica, Ligustrum ovalifolium, Prunus, Robinia pseudoacacia, Rubus fructicosus, Salix caprea.*

jucundoides Schaus, 1906
excentricus Bouvier, 1936
Foodplants: Unknown

junogabriellae Brechlin & Meister, 2011
Foodplants: Unknown

kaechi Brechlin & Meister, 2011
Foodplants: Unknown

kitchingi Brechlin & Meister, 2011
Foodplants: Unknown

knorkeorum Brechlin & Meister, 2011
Foodplants: Unknown

kopturae Lemaire, 1982
Foodplants: *Inga densiflora, Robinia pseudoacacia.*

labriquei Naumann, Brosch & Wenczel, 2005
Foodplants: Unknown

lachaumei Lemaire, 2002
Foodplants: Unknown

lapazchowskyi Brechlin & Meister, 2011
Foodplants: Unknown

larra (Walker, 1855)
palegon (Boisduval, 1875)
Foodplants: *Erythrina, Fagus sylvatica, Malus pumila, Quercus robur, Quercus × hispanica, Rubus fructicosus, Rubus idaeus.*

latenigra Lemaire, 1967
Foodplants: *Pyracantha coccinea, Rubus fructicosus.*

lauroia Oiticica, 1965
Foodplants: Unknown

lauta F. Johnson & Michener, 1948
Foodplants: *Erythrina, Robinia pseudoacacia.*

lecourti Decaëns & Herbin, 2002
Foodplants: Unknown

lemairei Beutelspacher, 1990
Foodplants: *Carpinus betulus, Robinia pseudoacacia.*

lemensis Lemaire, 1972
Foodplants: Unknown

lenarti Brechlin, Meister & Van Schayck, 2011
Foodplants: Unknown

liberia (Cramer, 1780)
erisichton (Boisduval, 1875)
geayi Bouvier, 1936
guyanensis Bouvier, 1929
megalops (Walker, 1865)
oweni Dognin, 1896
strandi Schüssler, 1934
Foodplants: *Acacia baileyana, Acacia farnesiana, Arbutus unedo, Arracacia xanthorrhiza, Citrus, Gleditsia triacanthos, Ligustrum ovalifolium, Malus pumila, Platanus, Pyracantha coccinea, Quercus, Robinia pseudoacacia, Rosa, Rubus fructicosus, Saccharum officinarum, Salix babylonica, Salix caprea, Sorghum vulgare, Zea mays.*

louisiana Ferguson & Brou, 1981
Foodplants: *Acer rubrum, Celtis laevigata, Cercis canadensis, Gleditsia triacanthos, Ligustrum ovalifolium, Liquidambar styraciflua, Padus virginiana, Populus deltoides, Prunus serotina, Quercus alba, Quercus falcata, Quercus nigra, Quercus virginiana, Robinia pseudoacacia, Spartina alterniflora, Salix babylonica, Salix caroliniana, Salix caprea.*

macphaili Schaus, 1921
Foodplants: *Robinia pseudoacacia*

maeonia (Druce, 1897)
Foodplants: *Acacia baileyana, Quercus × hispanica, Robinia pseudoacacia, Salix babylonica, Salix caprea.*

magdaleniana Brechlin & Meister, 2011
Foodplants: Unknown

mailinae Brechlin, Meister & Van Schayck, 2011
Foodplants: Unknown

managuana Brechlin & Meister, 2011
Foodplants: Unknown

manantlanensis Balcázar-Lara, 2000
Foodplants: *Quercus, Salix.*

margaritae Lemaire, 1967
Foodplants: *Salix caprea*

masti Lemaire, 1972
Foodplants: Unknown

melanops (Walker, 1865)
brasiliensis (Boisduval, 1875)
euryopa (Möschler, 1872)
Foodplants: *Hibiscus tiliaceus, Platanus orientalis, Robinia pseudoacacia, Schinus terebinthifolia, Tamarindus indica, Terminalia catappa, Tipuana tipu.*

melmon Dyar, 1912
Foodplants: *Quercus robur, Salix caprea.*

meridionalis Bouvier, 1936
Foodplants: *Fagus sylvatica, Ligustrum ovalifolium, Quercus robur, Robinia pseudoacacia, Salix babylonica, Salix caprea.*

metzli (Sallé, 1853)
Foodplants: *Cassia grandis, Cornus mas, Crataegus, Dalbergia retusa, Erythrina, Erythrina cos-*

taricensis, Fagus sylvatica, Hymenaea courbaril, Malus »hillieri«, Malus pumila, Melicoccus bijugatus, Ochroma pyramidale, Platanus occidentalis, Prunus domestica, Quercus alba, Quercus coccinea, Quercus × hispanica, Quercus oleoides, Quercus pubescens, Quercus robur, Robinia pseudoacacia, Tabebuia rosea, Trema micrantha, Zexmenia virgulta.

miamazonica Brechlin & Meister, 2011
Foodplants: Unknown

michneri Lemaire, 1966
Foodplants: Unknown

michoacana Balcázar-Lara, 1999
Foodplants: Unknown

midea (Maassen & Weyding, 1885)
aspersus Bouvier, 1927
sybilla Bouvier, 1929
Foodplants: *Fagus grandifolia* subsp. *grandifolia, Fagus sylvatica, Pyracantha coccinea, Robinia pseudoacacia, Salix caprea.*

mideloretensis Brechlin & Meister, 2011
Foodplants: Unknown

midenapoensis Brechlin & Meister, 2011
Foodplants: Unknown

mideperuensis Brechlin & Meister, 2011
Foodplants: Unknown

mixtus Bouvier, 1936
eitschbergeri Racheli & Racheli, 1999
Foodplants: *Quercus × hispanica, Quercus robur, Rubus fructicosus, Rubus idaeus.*

moenchorum Brechlin & Meister, 2011
Foodplants: Unknown

moloneyi Druce, 1897
Foodplants: *Ligustrum ovalifolium, Prunus domestica, Quercus, Robinia pseudoacacia, Saccharum officinarum.*

montegabriellae Brechlin & Meister, 2011
Foodplants: Unknown

montezuma (Boisduval, 1875)
crudelis (Maassen & Weyding, 1885)
Foodplants: *Robinia pseudoacacia*

moresca Schaus, 1906
crassus Conte, 1906
Foodplants: *Gleditsia triacanthos, Prunus laurocerasus, Pyracantha coccinea, Robinia pseudoacacia, Salix.*

muscula (Vuillot, 1893)
tristis (Bouvier, 1929)
Foodplants: *Prunus serotina, Rosa.*

nadezhdae Brechlin & Meister, 2011
Foodplants: Unknown

napoensis Lemaire, 2002
Foodplants: Unknown

naranja Schaus, 1898
aurantiaca Weymer 1907
lineatus Bouvier, 1930
pernambucensis Lemaire, 1973
umbrosa lampei Lemaire, 2002 **syn. nov.**
Foodplants: *Acacia longifolia, Acca sellowiana, Bauhinae forficata, Ilex paraguariensis, Ligustrum ovalifolium, Lonicera, Mimosa bimucronata, Psidium guajava, Prunus laurocerasus, Pyrus communis, Quercus × hispanica, Quercus robur, Robinia pseudoacacia, Salix caprea, Syringa vulgaris, Terminalia catappa.*

nebulosa Conte, 1906
elegans Conte, 1906
laticinctus Bouvier, 1930
levrati Bouvier, 1929
Foodplants: Unknown

niepelti Draudt, 1929
Foodplants: *Acacia Rubus fructicosus.*

nogueirai Herbin, 2009
Foodplants: Unknown

nubila (Walker, 1855)
intermedius (Bouvier, 1929)
obscura Schaus, 1900
violascens (Maassen & Weyding, 1885)
Foodplants: *Terminalia catappa*

oaxacensis Lemaire, 2002
Foodplants: *Acacia, Ligustrum ovalifolium, Prunus domestica, Robinia pseudoacacia.*

oaxhena Brechlin, Meister & Van Schayck, 2011
Foodplants: Unknown

oberthurii (Boisduval, 1875)
Foodplants: *Prosopis alba, Prosopis nigra, Pyracantha angustifolia, Pyracantha coccinea, Robinia pseudoacacia.*

occidentalis Draudt, 1929
Foodplants: *Chilopsis linearis, Eysenhardtia polystachya, Mimosa, Quercus emoryi, Quercus oblongifolia, Quercus palustris, Robinia pseudoacacia.*

occidentorestes Brechlin & Meister, 2011
Foodplants: Unknown

occiecuatoriana Brechlin & Meister, 2011
Foodplants: Unknown

oiticicai Lemaire, 1966
Foodplants: Unknown

orestes (Boisduval, 1875)
affinis Bouvier, 1929
nigrescens Bouvier, 1929
nigrocinctus Bouvier, 1930
Foodplants: Unknown

ovalina Conte, 1906
curitiba Schaus, 1921
Foodplants: Unknown

pallida Bouvier, 1930
Foodplants: Unknown

pallidior Draudt, 1929
Foodplants: *Acalypha apodanthes, Acalypha diversifolia, Calopogonium caeruleum, Cordia alliodora, Eugenia acapulcensis, Fagus, Guettarda macrosperma, Inga punctata, Lasiacis procerrima, Lonchocarpus oliganthus, Malosma laurina, Malvaviscus arboreus, Melanthera aspera, Paullinia cururu, Petrea volubilis, Poa, Prunus laurocerasus, Quercus, Rhus typhina, Robinia pseudoacacia, Salix, Schinus terebinthifolia, Tournefortia bicolor, Trigonia rugosa, Triumfetta bogotensis.*

paracelata Brechlin & Meister, 2011
Foodplants: Unknown

parachacona Brechlin & Meister, 2011
Foodplants: Unknown

paracreta Brechlin & Meister, 2011
Foodplants: Unknown

parageus Brechlin & Meister, 2011
Foodplants: Unknown

parainornata Brechlin & Meister, 2011
Foodplants: Unknown

paralarra Brechlin & Meister, 2011
Foodplants: Unknown

paramaculata Lemaire, 1966
Foodplants: Unknown

parapichinchensis Brechlin & Meister, 2011
Foodplants: Unknown

paramelanops Brechlin & Meister, 2011
Foodplants: Unknown

paraxigua Brechlin & Meister, 2011
Foodplants: Unknown

pastaziana Brechlin & Meister, 2011
Foodplants: Unknown

patagoniensis Lemaire, M. J. Smith & Wolfe, 1992
Foodplants: *Bambusa, Bouteloua, Cynadon dactylon, Poa.*

pelaezi Beutelspacher & Baigts, 1988
Foodplants: Unknown

peggyae Brechlin & Meister, 2011
Foodplants: Unknown

peigleri Lemaire, 1981
Foodplants: *Populus alba, Quercus grisea.*

peruviana Bouvier, 1929
Foodplants: Unknown

phrynon Druce, 1897
orneates Schaus, 1897
Foodplants: *Byrsonima crassifolia, Cecropia obtusifolia, Cestrum racemosum, Chimarrhis parviflora, Clethra mexicana, Coccoloba tuerckheimii, Coussapoa nymphaeifolia, Desmodium incanum, Dioclea malacocarpa, Gleditsia triacanthos, Hampea appendiculata, Heliocarpus appendiculatus, Inga marginata, Inga punctata, Inga sapindoides, Ligustrum, Litchi chinensis, Lonchocarpus heptaphyllus, Luehea seemannii, Malosma laurina, Myriocarpa longipes, Paullinia faginea, Pterocarpus rohrii, Terminalia catappa, Trichospermum galeottii, Trichospermum mexicanum, Uncaria tomentosa, Zexmenia virgulta.*

pichichensis Lemaire, 1976
Foodplants: Unknown

pomifera Schaus, 1906
Foodplants: *Malus pumila, Tilia americana.*

pomiferoides Brechlin & Meister, 2011
Foodplants: Unknown

postalbida Schaus, 1900
panamensis Draudt, 1929
Foodplants: *Abuta panamensis, Acalypha diversifolia, Acidoton costarricensis, Acidoton nicaraguensis, Adelia triloba, Alfaroa guanacastensis, Arachnothryx buddleioides, Bertiera bracteosa, Brosimum alicastrum, Brosimum guianense, Calathea leucostachys, Calophyllum brasiliense, Carapa guianensis, Casearia arborea, Casearia sylvestris, Casearia tacanensis, Cecropia obtusifolia, Cedrela odorata, Cespedesia spathulata,*

Chimarrhis parviflora, Chomelia microloba, Chrysobalanus, Chrysophyllum brenesii, Clethra mexicana, Coffea arabica, Colubrina spinosa, Cosmibuena valerii, Coussapoa nymphaeifolia, Croton schiedeanus, Cupania glabra, Cynometra retusa, Desmodium adscendens, Dichapetalum grayumii, Dichapetalum morenoi, Dioclea malacocarpa, Dioclea wilsonii, Drypetes standleyi, Erythrina, Erythroxylum macrophyllum, Eugenia basilaris, Fagus sylvatica, Ficus insipida, Ficus maxima, Ficus pertusa, Garcinia intermedia, Geonoma ferruginea, Gmelina arborea, Guarea glabra, Guarea rhopalocarpa, Guatteria amplifolia, Guatteria tonduzii, Guettarda foliacea, Guettarda macrosperma, Heisteria concinna, Heliconia latispatha, Heliocarpus appendiculatus, Hirtella racemosa, Hirtella triandra, Hirtella triandra subsp. *media, Hyeronima oblonga, Hylaeanthe hoffmannii, Inga chocoensis, Inga herrerae, Inga longispica, Inga oerstediana, Inga punctata, Inga samanensis, Inga sapindoides, Lacistema aggregatum, Licania hypoleuca, Licania riverae, Lonchocarpus guatemalensis, Lonchocarpus oliganthus, Lozania pittieri, Machaerium arboreum, Machaerium floribundum, Machaerium seemannii, Macrosamanea pubiramea* var. *pubiramea, Malpighia glabra, Maquira guianensis* subsp. *costaricana, Matudaea trinervia, Meliosma glabrata, Musa acuminata, Myrcia splendens, Myriocarpa longipes, Naucleopsis capirensis, Nectandra hihua, Nectandra salicifolia, Ocotea insularis, Ormosia cruenta, Ouratea lucens, Oxandra venezuelana, Paullinia grandifolia, Pavonia schiedeana, Pentagonia donnell-smithii, Perrottetia longistylis, Persea schiedeana, Pilea costaricensis, Piptocarpha poeppigiana, Pitcairnia arcuata, Posoqueria latifolia, Pourouma bicolor, Pourouma minor, Pouteria exfoliata, Pouteria reticulata, Prestoea decurrens, Protium confusum, Pseudolmedia mollis, Pseudolmedia quadrangularis, Psidium guajava, Psychotria berteroana, Psychotria eurycarpa, Pterocarpus officinalis, Quercus cortesii, Renealmia alpinia, Renistipulia costaricensis, Rhynchosia calycosa, Rochefortia lundellii, Rourea schippii, Salacia petenensis, Salix babylonica, Sapium glandulosum, Scleria, Senna hayesiana, Sorocea trophoides, Symphonia globulifera, Tabernaemontana alba, Tachigali costaricensis, Talisia macrophylla, Tapirira mexicana, Terminalia amazonia, Tetragastris panamensis, Tetrapterys discolor, Trichilia adolfi, Uncaria tomentosa, Urera elata, Urera simplex, Vaccinium poasanum, Vernonanthura patens, Vitex cooperi, Vochysia guatemalensis, Zanthoxylum juniperinum, Zanthoxylum melanostictum, Zygia longifolia.*

praemargaritae Lemaire, 2002
Foodplants: *Salix*

pueblensis Brechlin, Meister & Van Schayck, 2011
Foodplants: Unknown

punochacona Brechlin & Meister, 2011
Foodplants: Unknown

randa Druce, 1894
Foodplants: *Celtis ehrenbergiana, Cornus mas, Cornus sanguinea, Malus »hillieri«, Prunus serotina, Quercus emoryi, Quercus gambelii, Quercus palustris, Quercus pubescens, Quercus robur, Quercus rubra, Robinia pseudoacacia.*

rectilinea Bouvier, 1927
attenuata Bouvier, 1930
immodicus Bouvier, 1930
Foodplants: Unknown

rostralis Lemaire, 2002
Foodplants: Unknown

rotunda Lemaire, 1971
Foodplants: Unknown

rougeoti Lemaire, 1967
Foodplants: Unknown

rubrescens (Walker, 1855)
nopaltzin Schaus, 1892
unilateralis Bouvier, 1936
Foodplants: *Acacia tenuifolia, Amphilophium paniculatum, Aeschynomene brasiliana, Augusta rivalis, Bambusa, Banisteriopsis muricata, Bignonia diversifolia, Bignonia heterophylla, Billia hippocastanum, Byrsonima crassifolia, Byttneria catalpifolia, Caesalpinia bonduc, Calliandra calothyrsus, Calycophyllum candidissimum, Canavalia brasiliensis, Casearia corymbosa, Cassia grandis, Conocarpus erectus, Cordia alliodora, Cordia collococca, Cordia inermis, Cordia panamensis, Coursetia elliptica, Crescentia alata, Crotalaria pumila, Dalbergia retusa, Diphysa americana, Enterolobium cyclocarpum, Erythroxylum havanense, Fagus sylvatica, Genipa americana, Guazuma ulmifolia, Haematoxylum brasiletto, Handroanthus impetiginosus, Heliconia latispatha, Hirtella racemosa, Hymenaea courbaril, Inga oerstediana, Inga punctata, Inga samanensis, Inga sapindoides, Karwinskia calderonii, Laburnum anagyroides, Lantana camara, Licania arborea, Ligustrum ovalifolium, Lonchocarpus costaricen-*

sis, Lonchocarpus guatemalensis, Lonchocarpus heptaphyllus, Lonchocarpus minimiflorus, Lonchocarpus oliganthus, Lonchocarpus retiferus, Lysiloma aurita, Machaerium arboreum, Machaerium biovulatum, Machaerium kegelii, Machaerium salvadorense, Mimosa acantholoba, Mimosa camporum, Mimosa watsonii, Mimosa xanti, Muntingia calabura, Ouratea lucens, Paullinia cururu, Piscidia carthagenensis, Pithecellobium bipinnatum, Platymiscium parviflorum, Prunus cerasifera, Prunus domestica, Prunus laurocerasus, Psidium guajava, Quercus oleoides, Quercus robur, Rehdera trinervis, Rottboellia cochinchinensis, Rourea glabra, Salix viminalis, Senna alata, Senna pallida, Serjania schiedeana, Solanum hazenii, Stigmaphyllon lindenianum, Syringa vulgaris, Tabebuia ochracea, Tabebuia rosea, Tectona grandis, Trigonia rugosa, Triplaris melaenodendron, Uncaria tomentosa, Verbesina gigantea, Zuelania guidonia.

rudloffiani Brechlin & Meister, 2011
Foodplants: Unknown

salkorum Brechlin & Meister, 2011
Foodplants: Unknown

santaclariana Brechlin & Meister, 2011
Foodplants: Unknown

schwartzi Lemaire, 1967
Foodplants: Unknown

serpina Butler, 1878
Foodplants: Unknown

silkae Brechlin & Meister, 2011
Foodplants: Unknown

siri Brechlin, Meister & Van Schayck, 2011
Foodplants: Unknown

stacieae Lemaire & Wolfe, 1993
Foodplants: *Robinia pseudoacacia*

stumpei Brechlin & Meister, 2011
Foodplants: Unknown

styx Lemaire, 1982
Foodplants: *Fagus sylvatica, Ligustrum ovalifolium, Quercus × hispanica, Quercus robur, Prunus cerasifera, Prunus domestica, Prunus serotina, Pyracantha coccinea, Quercus × hispanica, Robinia pseudoacacia, Rosa, Salix caprea.*

submacula (Walker, 1855)
dioxippus (Boisduval, 1875)
Foodplants: *Bambusa, Robinia pseudoacacia.*

suteri Naumann, Brosch & Wenczel, 2005
Foodplants: Unknown

sylviae Decaëns, [2005]
Foodplants: Unknown

tamaulipasiana Brechlin & Meister, 2011
Foodplants: Unknown

tamsi Lemaire, 1966
Foodplants: Unknown

tatiae Naumann, Brosch & Wenczel, 2005
Foodplants: *Erythrina, Fagus sylvatica, Pyracantha coccinea, Robinia pseudoacacia, Rubus fructicosus, Salix caprea.*

themis Dognin, 1919
Foodplants: *Gleditsia triacanthos*

tolimaiensis Brechlin & Meister, 2011
Foodplants: Unknown

tridens Herrich-Schäffer, [1855]
Foodplants: *Acacia tenuifolia, Aeschynomene brasiliana, Amphilophium paniculatum, Augusta rivalis, Bambusa, Banisteriopsis muricata, Bignonia diversifolia, Bignonia heterophylla, Billia hippocastanum, Byrsonima crassifolia, Byttneria catalpifolia, Caesalpinia bonduc, Calliandra calothyrsus, Calycophyllum candidissimum, Canavalia brasiliensis, Casearia corymbosa, Cassia grandis, Conocarpus erectus, Cordia alliodora, Cordia collococca, Cordia inermis, Cordia panamensis, Coursetia elliptica, Crescentia alata, Crotalaria pumila, Dalbergia retusa, Diphysa americana, Enterolobium cyclocarpum, Erythroxylum havanense, Fagus sylvatica, Genipa americana, Guazuma ulmifolia, Haematoxylum brasiletto, Handroanthus impetiginosus, Heliconia latispatha, Hirtella racemosa, Hymenaea courbaril, Inga oerstediana, Inga punctata, Inga samanensis, Inga sapindoides, Karwinskia calderonii, Laburnum anagyroides, Lantana camara, Licania arborea, Ligustrum ovalifolium, Lonchocarpus costaricensis, Lonchocarpus guatemalensis, Lonchocarpus heptaphyllus, Lonchocarpus minimiflorus, Lonchocarpus oliganthus, Lonchocarpus retiferus, Lysiloma aurita, Machaerium arboreum, Machaerium biovulatum, Machaerium kegelii, Machaerium salvadorense, Mimosa acantholoba, Mimosa camporum, Mimosa watsonii, Mimosa xanti, Muntingia calabura, Ouratea lucens, Paullinia cururu, Piscidia carthagenensis, Pithecellobium bipinnatum, Platymiscium parviflorum, Prunus cerasifera, Prunus domestica, Prunus lau-*

rocerasus, Psidium guajava, Quercus oleoides, Quercus robur, Rehdera trinervis, Rottboellia cochinchinensis, Rourea glabra, Salix viminalis, Senna alata, Senna pallida, Serjania schiedeana, Solanum hazenii, Stigmaphyllon lindenianum, Syringa vulgaris, Tabebuia ochracea, Tabebuia rosea, Tectona grandis, Trigonia rugosa, Triplaris melaenodendron, Uncaria tomentosa, Verbesina gigantea, Zuelania guidonia.

tristis (Boisduval, 1875)
hagar Niepelt, 1927
Foodplants: *Bambusa, Fargesia murielae.*

umbrosa Weymer, 1906
ater Conte, 1906
Foodplants: *Acacia baileyana, Fagus sylvatica, Ligustrum ovalifolium, Ocotea porphyria, Phyllostylon rhamnoides, Prunus laurocerasus, Pyracantha coccinea, Quercus robur, Robinia pseudoacacia, Salix babylonica, Tecoma stans, Tipuana tipu.*

unifasciatus Bouvier, 1927
Foodplants: Unknown

vanschaycki Brechlin & Meister, 2011
Foodplants: Unknown

vergnei Bouvier, 1936
Foodplants: Unknown

viksinjaevi Brechlin & Meister, 2011
Foodplants: Unknown

vomona Schaus, 1906
pupilla Draudt, 1929
Foodplants: *Fagus sylvatica, Robinia pseudoacacia.*

watsoni Lemaire, 1966
Foodplants: Unknown

wayampi Lemaire & Bénelúz, 2002
Foodplants: Unknown

wenczeli Brechlin, Meister & Van Schayck, 2011
Foodplants: Unknown

wernermeisteri Brechlin & Meister, 2011
Foodplants: Unknown

winbrechlini Brechlin, Käch & Meister, 2011
Foodplants: Unknown

winbrechliniani Brechlin & Meister, 2011
Foodplants: Unknown

windiana Lemaire, 1972
Foodplants: Unknown

witti Brechlin & Meister, 2011
Foodplants: Unknown

yungaiana Brechlin & Meister, 2011
Foodplants: Unknown

yungasmargaritae Brechlin & Meister, 2011
Foodplants: Unknown

zaruma Schaus, 1898
equatorialis Bouvier, 1936
foucheri Bouvier, 1927
Foodplants: *Fagus sylvatica, Robinia pseudoacacia, Salix caprea.*

zephyria (Grote, 1882)
zephyriata Barnes & Benjamin, 1924
Foodplants: *Cercis canadensis, Cercocarpus betuloides, Malosma laurina, Malus pumila, Padus virginiana, Prunus domestica, Prunus serotina, Pyracantha coccinea, Quercus alba, Quercus emoryi, Quercus gambelii, Quercus nigra, Quercus oblongifolia, Quercus palustris, Quercus petraea, Quercus phellos, Quercus robur, Quercus rubra, Robinia pseudoacacia, Rubus fructicosus, Salix babylonica, Salix caprea, Tilia.*

zoziboucardi Brechlin & Meister, 2011
Foodplants: *Fagus sylvatica, Laburnum anagyroides, Salix caprea.*

zozimanaguana Brechlin & Meister, 2011
Foodplants: Unknown

zozinicaraguana Brechlin & Meister, 2011
Foodplants: *Fagus sylvatica, Salix caprea.*

zozine Druce, 1886
Foodplants: *Fagus sylvatica, Laburnum anagyroides, Ligustrum ovalifolium, Malosma laurina, Malus »hillieri«, Padus avium* var. *avium, Quercus robur, Robinia pseudoacacia, Salix caprea.*

zugana Druce, 1886
Foodplants: *Acacia tenuifolia, Acalypha diversifolia, Acosmium panamense, Adelia triloba, Albizia adinocephala, Alibertia edulis, Allophylus racemosus, Andira inermis, Annona purpurea, Annona rensoniana, Apeiba tibourbou, Ateleia herbert-smithii, Augusta rivalis, Banisteriopsis muricata, Bauhinia glabra, Bauhinia ungulata, Billia hippocastanum, Briquetia spicata, Bronwenia cornifolia, Bursera simaruba, Byrsonima crassifolia, Byttneria catalpifolia, Calathea lutea, Calliandra calothyrsus, Calliandra rubescens, Calliandra tergemina, Callichlamys latifolia, Calopogonium caeruleum, Calycophyllum candidis-*

simum, Casearia arborea, Casearia corymbosa, Cassia fistula, Cecropia obtusifolia, Centrosema macrocarpum, Centrosema pubescens, Cespedesia spathulata, Chomelia spinosa, Chrysophyllum brenesii, Citharexylum costaricense, Clethra mexicana, Coccoloba tuerckheimii, Coffea arabica, Cordia alliodora, Cordia eriostigma, Cordia guanacastensis, Cordia panamensis, Coussapoa nymphaeifolia, Crescentia alata, Croton schiedeanus, Cupania glabra, Cupania guatemalensis, Cupania macropoda, Bignonia aequinoctialis, Bignonia diversifolia, Bignonia heterophylla, Dalbergia retusa, Desmodium incanum, Dioclea malacocarpa, Dioclea violacea, Diospyros salicifolia, Diphysa americana, Enterolobium cyclocarpum, Erythrina lanceolata, Erythroxylum havanense, Exostema mexicanum, Ficus crocata, Fridericia mollissima, Fridericia patellifera, Gleditsia triacanthos, Gliricidia sepium, Gmelina arborea, Guarea glabra, Guazuma ulmifolia, Guettarda macrosperma, Heisteria concinna, Hirtella racemosa, Hyeronima oblonga, Hymenaea courbaril, Hyptis obtusiflora, Inga chocoensis, Inga densiflora, Inga oerstediana, Inga punctata, Inga sapindoides, Laburnum anagyroides, Lacistema aggregatum, Lantana camara, Licania arborea, Ligustrum ovalifolium, Lonchocarpus costaricensis, Lonchocarpus eriocarinalis, Lonchocarpus guatemalensis, Lonchocarpus minimiflorus, Lonchocarpus oliganthus, Lonchocarpus retiferus, Lonchocarpus rugosus, Lygodium venustum, Mabea occidentalis, Machaerium arboreum, Machaerium biovulatum, Machaerium cobanense, Machaerium kegelii, Malvaviscus arboreus, Manilkara chicle, Melanthera aspera, Melicoccus bijugatus, Miconia argentea, Miconia dolichopoda, Mimosa pigra, Mimosa skinneri, Mucuna urens, Muntingia calabura, Musa acuminata, Myrcia splendens, Myriocarpa longipes, Myrospermum frutescens, Nectandra hihua, Ochroma pyramidale, Ocotea veraguensis, Oryza latifolia, Padus avium var. *avium, Panicum maximum, Paullinia bracteosa, Paullinia cururu, Pentagonia donnell-smithii, Piper tuberculatum, Piptadenia flava, Piscidia carthagenensis, Pithecellobium furcatum, Platymiscium parviflorum, Pleiostachya leiostachya, Poa, Pouteria juruana, Prestoea decurrens, Pterocarpus officinalis, Pterocarpus orbiculatus, Pterocarpus rohrii, Quercus oleoides, Quercus robur, Rhamnus, Robinia pseudoacacia, Rottboellia cochinchinensis, Roupala montana, Rourea glabra, Securidaca sylvestris, Senna hayesiana, Senna pallida, Senna papillosa, Senna undulata, Serjania atrolineata, Serjania mexicana, Serjania schiedeana, Sesbania emerus, Sloanea terniflora, Solanum hazenii, Sphinga platyloba, Spondias mombin, Spondias purpurea, Styrax argenteus, Swartzia simplex, Syzygium malaccense, Tabebuia ochracea, Tabebuia rosea, Tephrosia multifolia, Thunbergia alata, Tibouchina longifolia, Trema micrantha, Trichospermum mexicanum, Trigonia rugosa, Triplaris melaenodendron, Triumfetta lappula, Ulmus, Uncaria tomentosa, Vigna vexillata, Vitex cooperi, Xylosma flexuosa, Zanthoxylum setulosum, Zygia longifolia.*

zurobara Druce, 1886
simplex Bouvier, 1936
Foodplants: *Cornus mas, Cornus sanguinea, Fagus sylvatica, Gleditsia triacanthos, Inga sapindoides, Prunus laurocerasus, Pyracantha, Quercus robur, Robinia pseudoacacia, Salix caprea.*

zurouae Brechlin & Meister, 2011
Foodplants: Unknown

Automeropsis Lemaire, 1969

umbrata (Boisduval, 1875)
accuminata (Maassen, 1885)
Foodplants: Unknown

Callodirphia Michener, 1949

arpi (Schaus, 1908)
satanas (Jones, 1908)
seitzi (Draudt, 1930)
zikani (Draudt, 1930)
Foodplants: *Eucalyptus, Myrsine coriacea.*

Catacantha Bouvier, 1930

evitae Brechlin, Meister & Van Schayck, 2010
Foodplants: Unknown

ferruginea (Draudt, 1929)
Foodplants: Unknown

juliae Brechlin, Meister & Van Schayck, 2010
Foodplants: Unknown

latifasciata Bouvier, 1930
Foodplants: Unknown

nataliae Brechlin, Meister & Van Schayck, 2010
Foodplants: Unknown

obliqua Bouvier, 1930
Foodplants: Unknown

oculata (Schaus, 1921)
Foodplants: Unknown

siriae Brechlin, Meister & Van Schayck, 2010
Foodplants: Unknown

sofiae Brechlin, Meister & Van Schayck, 2010
Foodplants: Unknown

stramentalis (Draudt, 1929)
Foodplants: *Phyllostachis*

tabeae Brechlin, Meister & Van Schayck, 2010
Foodplants: Unknown

Catharisa Jordan, 1911

cerina Jordan, 1911
Foodplants: Unknown

Cerodirphia Michener, 1949

apunctata Dias & Lemaire, 1991
Foodplants: *Campomanesia phaea, Nectandra, Persea americana, Pittosporum undulatum, Psidium guajava, Quercus robur.*

araguensis Lemaire, 1971
Foodplants: Unknown

avenata (Draudt, 1930)
Foodplants: Unknown

bahiana Lemaire, 2002
Foodplants: Unknown

barbuti Rougerie & Herbin, 2004
Foodplants: Unknown

brunnea brunnea (Draudt, 1930)
Foodplants: Unknown

brunnea interrupta (Bouvier, 1930)
Foodplants: Unknown

candida Lemaire, 1969
Foodplants: Unknown

cutteri (Schaus, 1927)
upanona (Draudt, 1929)
Foodplants: Unknown

flammans Lemaire, 1973
Foodplants: Unknown

flavoscripta (Dognin, 1901)
Foodplants: Unknown

flavosignata (F. Johnson & Michener, 1948)
Foodplants: Unknown

gualaceensis Lemaire, 2002
Foodplants: Unknown

harrisae Lemaire, 1975
Foodplants: Unknown

inopinata Lemaire, 1982
Foodplants: Unknown

lojensis Lemaire, 1988
Foodplants: Unknown

marahuaca Lemaire, 1971
Foodplants: Unknown

mielkei Lemaire, 2002
Foodplants: Unknown

mota mota (Druce, 1909)
Foodplants: Unknown

mota napoensis Lemaire, 1981
Foodplants: *Erythrina edulis, Inga sapindoides, Nectandra, Quercus.*

mota pachona (Draudt, 1929)
fuliginosa (Bouvier, 1930)
Foodplants: Unknown

nadiana Lemaire, 1975
Foodplants: *Inga sapindoides, Quercus.*

ockendeni Lemaire, 1985
Foodplants: Unknown

opis (Schaus, 1892)
albicincta (Bouvier, 1929)
arpi (Draudt, 1930)
castanea (Bouvier, 1930)
colax (Draudt, 1930)
infuscata (Bouvier, 1929)
Foodplants: *Croton urucurana*

peigleri Naumann, Brosch & Wenczel, 2005
Foodplants: *Quercus × hispanica, Quercus robur.*

porioni Lemaire, 1982
Foodplants: Unknown

radama radama (Druce, 1904)
Foodplants: Unknown

radama consaepta (Draudt, 1930)
Foodplants: Unknown

rosacordis (Walker, 1855)
schausi (Bouvier, 1929)
Foodplants: Unknown

rubripes (Draudt, 1930)
brasiliensis (Bouvier, 1930)
Foodplants: Unknown

sanctimartinensis Lemaire, 1982
Foodplants: Unknown

speciosa (Cramer, 1777)
quadricolor (Walker, 1865)
sexfasciata (Walker, 1869)
Foodplants: *Ligustrum ovalifolium, Pyracantha coccinea, Quercus robur, Rubus fructicosus, Salix caprea.*

vagans (Walker, 1855)
minor (Bouvier, 1929)
simplex (Dognin, 1924)
Foodplants: Unknown

wellingi roseissima Lemaire, 2002
Foodplants: *Inga sapindoides, Quercus.*

wellingi wellingi Lemaire, 1973
Foodplants: *Quercus*

zikani (Schaus, 1921)
Foodplants: Unknown

zulemae Decaëns & Rougerie, 2008
Foodplants: Unknown

Cinommata Butler, 1882

bistrigata Butler, 1882
izquierdoi Ureta, 1944
Foodplants: Unknown

Coloradia Blake, 1863

casanovai Beutelspacher, 1993
Foodplants: *Pinus montezumae, Pinus oocarpa, Pinus pseudostrobus, Pinus teocote.*

doris Barnes, 1900
duffneri Barnes & Benjamin, 1926
lois Dyar, 1911
Foodplants: *Pinus contorta latifolia, Pinus monophylla, Pinus ponderosa, Pinus strobus.*

euphrosyne Dyar, 1912
Foodplants: *Pinus*

guerreroiana Brechlin & Meister, 2010
Foodplants: Unknown

hoffmanni Beutelspacher, 1978
Foodplants: *Pinus ponderosa, Pinus strobus.*

jaliscensis Brechlin & Meister, 2010
Foodplants: Unknown

luski Barnes & Benjamin, 1926
Foodplants: *Pinus ponderosa, Pinus strobus.*

oaxacensis Brechlin & Meister, 2010
Foodplants: Unknown

pandora davisi Barnes & Benjamin, 1926
boniwelli Barnes & Benjamin, 1926
chiricahua Barnes & Benjamin, 1926
Foodplants: *Pinus contorta, Pinus edulis, Pinus ponderosa scopulorum, Pinus strobus, Pinus sylvestris.*

pandora pandora Blake, 1863
lindseyi Barnes & Benjamin, 1926
loiperda Dyar, 1912
Foodplants: *Pinus contorta, Pinus contorta* var. *murrayana, Pinus coulteri, Pinus echinata, Pinus jeffreyi, Pinus lambertiana, Pinus ponderosa ponderosa, Pinus ponderosa scopulorum, Pinus strobus, Pinus sylvestris, Pinus thunbergii, Populus tremuloides.*

paraguerreroiana Brechlin & Meister, 2010
Foodplants: Unknown

prchali Lemaire & M. J. Smith, 1992
Foodplants: *Juniperus deppeana, Pinus engelmannii, Pinus leilophylla.*

smithi Lemaire, 2002
Foodplants: *Pinus ponderosa, Pinus strobus.*

vazquezae Beutelspacher, 1978
Foodplants: *Pinus*

velda Johnson & Walter, 1979
Foodplants: *Pinus edulis, Pinus jeffreyi, Pinus monophylla, Pinus strobus, Pinus sylvestris, Pinus thunbergii.*

Dirphia Hübner, [1819]

abhorca Lemaire, 1969
Foodplants: Unknown

acidalia Hübner, [1819]
stolli Bouvier, 1929
Foodplants: Unknown

aculea Vuillot, 1892
Foodplants: *Quercus × hispanica*

albescens Brechlin & Meister, 2008
Foodplants: *Quercus × hispanica, Quercus robur, Salix caprea.*

araucariae Jones, 1908
divisa Jordan, 1910
Foodplants: *Araucaria angustifolia*

avia (Stoll, 1780)
concolor Walker, 1855
nora Druce, 1897
orasia (Stoll, 1792)

triangulum Walker, 1855
Foodplants: *Cedrela odorata, Crataegus curvisepala, Fagus sylvatica, Gleditsia triacanthos, Guarea glabra, Hymenaea courbaril, Malus »hillieri«, Prunus laurocerasus, Quercus ilex, Quercus oleoides, Quercus robur, Malosma laurina, Salix caprea, Spondias mombin, Swietenia macrophylla.*

avrilae Lemaire, 1980
Foodplants: *Mangifera indica*

barinasensis Meister & Wenczel, 2002
Foodplants: *Mangifera indica, Prunus spinosa, Salix caprea.*

baroma (Schaus, 1906)
fasciata Bouvier, 1930
glauca Mabilde, 1896
walkeri Bouvier, 1929
Foodplants: *Betula, Fagus sylvatica, Quercus ilex, Quercus × hispanica.*

brevifurca Strand, 1911
Foodplants: Unknown

cadioui Lemaire, 1980
Foodplants: *Malosma laurina*

carimaguensis Decaëns, Bonilla & Naumann, 2004
Foodplants: Unknown

centralis F. Johnson & Michener, 1948
Foodplants: *Fagus sylvatica*

centrifurca Naumann, Brosch & Wenczel, 2005
Foodplants: Unknown

crassifurca Lemaire, 1971
Foodplants: *Rhus*

curitiba Draudt, 1930
Foodplants: *Fagus, Prunus persica, Quercus.*

demarmelsi Naumann, Brosch, Wenczel & Clavijo, 2005
Foodplants: Unknown

dentimaculata (Schaus, 1921)
Foodplants: Unknown

diasi Lemaire, 1994
Foodplants: Unknown

docquinae Lemaire, 1993
Foodplants: Unknown

dolosa Bouvier, 1929
Foodplants: Unknown

fernandezi Lemaire, 1972
Foodplants: Unknown

fornax (Druce, 1903)
rothschildi (Dognin, 1916)
Foodplants: Unknown

fraterna fraterna (C. Felder & R. Felder, 1874)
brunneoalba Bouvier, 1929
callosa Draudt, 1930
javarina Butler, 1878
theobromana Draudt, 1930
Foodplants: Unknown

fraterna seraphini Bouvier, 1929
Foodplants: Unknown

horca Dognin, 1894
Foodplants: Unknown

horcana Schaus, 1911
Foodplants: *Ardisia crassiramea, Blakea chlorantha, Cavendishia bracteata, Chrysochlamys psychotriifolia, Conostegia montana, Eugenia glandulosopunctata, Gonocalyx pterocarpus, Miconia brenesii, Miconia lonchophylla, Myrcia citrifolia, Panopsis suaveolens, Schwartzia costaricensis, Vaccinium poasanum, Weinmannia alta, Weinmannia wercklei.*

inexpectata Racheli & Racheli, 2005
Foodplants: Unknown

irradians Lemaire, 1972
Foodplants: Unknown

lemoulti Bouvier, 1930
Foodplants: *Melastoma, Salix caprea.*

lichyi Lemaire, 1971
Foodplants: *Anacardium excelsum, Ligustrum ovalifolium, Malosma laurina, Mangifera indica, Prunus laurocerasus, Prunus spinosa, Pyracantha coccinea. Salix babylonica, Salix caprea, Salix fragilis, Salix viminalis, Syzygium jambos, Syzygium malaccense.*

ludmillae Lemaire, 1974
Foodplants: Unknown

mielkeorum Naumann, Brosch & Wenczel, 2005
Foodplants: Unknown

moderata Bouvier, 1929
avicula Draudt, 1930
Foodplants: *Anacardium occidentale, Myrsine umbellata, Schinus gracilipes, Schinus terebinthifolia, Salix.*

monticola Zerny, 1923
fallax Bouvier, 1930
Foodplants: *Myrsine coriacea*

muscosa Schaus, 1898
colorata Bouvier, 1930
hortensia (Schaus, 1913)
sinuosa Bouvier, 1929
Foodplants: *Juglans regia, Quercus ilex, Quercus robur.*

napoensis Racheli & Racheli, 2005
Foodplants: Unknown

panamensis fassli (Dognin, 1923)
areolata Bouvier, 1930
Foodplants: *Quercus × hispanica, Quercus robur, Robinia pseudoacacia, Salix babylonica, Salix caprea.*

panamensis panamensis (Schaus, 1921)
decolor Bouvier, 1929
Foodplants: *Fagus sylvatica, Gleditsia triacanthos, Malosma laurina, Quercus × hispanica, Quercus robur, Robinia pseudoacacia, Rubus fructicosus, Salix caprea, Schinus terebinthifolia.*

proserpina Lemaire, 1982
Foodplants: Unknown

purpurascens Walker, 1855
Foodplants: Unknown

radiata Dognin, 1916
Foodplants: *Melastoma, Salix caprea.*

rengei Naumann, Lindt & Viidalepp, 2010
Foodplants: Unknown

riograndensis C. Mielke & A. Moser, 2007
Foodplants: Unknown

rubricauda Bouvier, 1929
Foodplants: Unknown

rufescens F. Johnson & Michener, 1948
Foodplants: Unknown

rustica (Stoll, 1780)
Foodplants: Unknown

satellitia Walker, 1855
Foodplants: Unknown

sombrero Le Cerf, 1934
dalcyra Oiticica Filho, 1953
Foodplants: Unknown

somniculosa confluens Bouvier, 1930
Foodplants: *Castanea, Miconia, Prunus, Quercus.*

somniculosa eximia Lemaire, 1977
Foodplants: Unknown

somniculosa juninensis Lemaire, 1977
Foodplants: *Salix caprea*

somniculosa somniculosa (Cramer, 1777)
rochereaui Bouvier, 1929
Foodplants: *Arbutus unedo, Salix caprea.*

subhorca Dognin, 1901
Foodplants: *Quercus*

tarquinia (Cramer, 1775)
bouvieri (Schüssler, 1934)
tarpejus (Hübner, [1811])
tarquinius (Cramer, 1775)
Foodplants: *Cerasus lusitanica, Crataegus curvisepala, Malosma laurina, Melastoma, Prunus armeniaca, Salix caprea, Schinus terebinthifolia.*

thliptophana pacifica Lemaire, 1981
Foodplants: Unknown

thliptophana thliptophana (C. Felder & R. Felder, 1874)
saturata Bouvier, 1930
Foodplants: Unknown

undulosa Walker, 1855
Foodplants: Unknown

ursina Walker, 1855
jorgenseni (Schaus, 1921)
melanosoma Niepelt, 1934
Foodplants: *Malosma laurina, Schinus gracilipes, Wisteria.*

Dirphiella Michener, 1949

albofasciata (Johnson & Michener, 1948)
Foodplants: Unknown

niobe (Lemaire, 1978)
Foodplants: Unknown

taylori (Donahue & Lemaire, 1975)
Foodplants: Unknown

Dirphiopsis Bouvier, 1928

ayuruoca (Foetterle, 1902)
Foodplants: Unknown

brockmanni C. Mielke, S. Naumann & U. Brosch, 2008
Foodplants: Unknown

cochabambensis (Lemaire, 1977)
Foodplants: *Malosma laurina, Pyracantha coccinea, Rubus.*

curvilineata Decaëns, Wolfe & Herbin, 2003
Foodplants: *Malosma laurina*

delta (Foetterle, 1902)
aphrodite (Schaus, 1915)
Foodplants: *Prunus*

epiolina (C. Felder & R. Felder, 1874)
spumosa (Vuillot, 1893)
Foodplants: *Fagus sylvatica, Psidium guajava, Rhus typhina, Salix caprea.*

flora (Schaus, 1911)
orientalis (Lemaire, 1976)
Foodplants: *Fagus sylvatica, Malosma laurina, Quercus seemannii, Robinia pseudoacacia, Schinus gracilipes, Schinus molle, Wisteria.*

herbini Wolfe, 2003
Foodplants: *Malosma laurina*

janzeni Lemaire, 2002
Foodplants: *Malosma laurina*

lombardi (Bouvier, 1924)
Foodplants: Unknown

multicolor Walker, 1855
brasiliensis (Bouvier, 1924)
ramigera (C. Felder & R. Felder, 1874)
Foodplants: *Erythrina speciosa, Gossypium herbaceum, Prunus, Sapium glandulosum, Terminalia catappa.*

oridocea (Schaus, 1924)
Foodplants: Unknown

picturata (Schaus, 1913)
tripicata (F. Johnson, 1937)
Foodplants: Unknown

pulchricornis Walker, 1855
pomona (C. Felder & R. Felder, 1874)
Foodplants: *Fagus sylvatica, Ligustrum ovalifolium, Quercus robur, Salix caprea.*

schreiteri (Schaus, 1925)
Foodplants: *Malosma laurina, Pyracantha coccinea, Rubus.*

trisignata (C. Felder & R. Felder, 1874)
irregularis (Foetterle, 1902)
pulchra (Bouvier, 1930)
Foodplants: *Eucalyptus saligna, Quercus robur, Prunus laurocerasus.*

undulinea (F. Johnson, 1937)
Foodplants: Unknown

unicolor Lemaire, 1982
Foodplants: Unknown

wanderbilti Pearson, 1958
Foodplants: Unknown

wolfei Lemaire, 1992
Foodplants: *Malosma laurina, Quercus, Schinus terebinthifolia.*

Erythromeris Lemaire, 1969

flexilineata (Dognin, 1911)
Foodplants: *Acer, Ligustrum ovalifolium, Platanus orientalis, Robinia pseudoacacia, Solanum tuberosum, Solanum variabile.*

obscurior Lemaire, 1975
Foodplants: *Acer, Ligustrum, Platanus orientalis, Robinia pseudoacacia, Solanum tuberosum, Solanum variabile.*

saturniata (Walker, 1865)
coesa (Boisduval, 1875)
Foodplants: *Acer, Erica, Ligustrum ovalifolium, Pieris, Platanus orientalis, Robinia pseudoacacia, Solanum tuberosum, Solanum variabile.*

Eubergia Bouvier, 1929

altoparanensis Brechlin & Meister, 2011
Foodplants: Unknown

argyrea (Weymer, 1908)
Foodplants: Unknown

boetifica (Druce, 1899) **rev. stat.**
Foodplants: *Mimosa lawranceana*

caisa (Berg, 1883)
bedoci (Bouvier, 1925)
rufa Bouvier, 1930
Foodplants: *Mimosa lawranceana*

chuquisaciana Brechlin & Meister, 2011
Foodplants: Unknown

radians (Dognin, 1911)
Foodplants: Unknown

santacruziana Brechlin & Meister, 2011
Foodplants: Unknown

sinjaevorum Brechlin & Meister, 2011
Foodplants: Unknown

Eubergioides Michener, 1949

bertha (Schaus, 1896)
Foodplants: Unknown

Eudyaria Grote, 1896

sigridae Mielke & Drechsel, 2009
Foodplants: Unknown

venata *(Butler, 1871)*
consularis (Burmeister, 1878)
tribunalis (Burmeister, 1878)
Foodplants: *Conopodium majus, Eryngium agavifolium, Eryngium campestre, Pimpinella saxifraga, Poterium, Sanguisorba.*

zeta (Berg, 1885)
boliviana (Bouvier, 1930)
carilapha (Schaus, 1924)
mansosotoi (Orfila, 1951)
Foodplants: *Conopodium majus, Eryngium agavifolium, Pimpinella saxifraga, Poterium, Sanguisorba.*

Gamelia Hübner, [1819]

abas (Cramer, 1775)
banghaasi (Draudt, 1929)
barii (Boisduval, 1875)
episcopus (Boisduval, 1875)
Foodplants: *Psidium guajava*

abasia (Stoll, 1781)
irmina (Stoll, 1781)
Foodplants: *Cerasus fruticosa, Fagus sylvatica, Psidium, Psidium guajava, Quercus × hispanica, Robinia pseudoacacia, Rubus fructicosus.*

abasiella Lemaire, 1973
Foodplants: *Psidium guajava*

anableps (C. Felder & R. Felder, 1874)
arguta (Boisduval, 1875)
pandarus (Boisduval, 1875)
theseus (Boisduval, 1875)
Foodplants: *Psidium guajava*

berliozi Lemaire, 1967
Foodplants: *Psidium guajava*

catharina (Draudt, 1929)
Foodplants: *Psidium guajava*

centralis Naumann, 2011
Foodplants: unknown

cimarrones Decaéns, Bonilla & Ramirez, 2005
Foodplants: Unknown

dargei Naumann, Brosch &Wenczel, 2005
Foodplants: Unknown

denhezi Lemaire, 1967
Foodplants: *Psidium guajava*

herbini Naumann, 2011
Foodplants: Unknown

kiefferi Lemaire, 1967
Foodplants: Unknown

lampei Naumann, 2011
Foodplants: Unknown

lichyi Lemaire, 1973
Foodplants: *Fagus sylvatica, Psidium guajava, Quercus × hispanica, Robinia pseudoacacia, Salix caprea, Salix viminalis.*

longispina Naumann, Brosch & Wenczel, 2005
Foodplants: Unknown

musta Schaus, 1912
Foodplants: *Acalypha diversifolia, Acidoton nicaraguensis, Adelia triloba, Alchornea latifolia, Alfaroa guanacastensis, Arachnothryx buddleioides, Banisteriopsis muricata, Billia hippocastanum, Calathea crotalifera, Citharexylum costaricense, Clethra mexicana, Cnestidium rufescens, Coccoloba tuerckheimii, Cojoba valerioi, Combretum laxum, Cordia panamensis, Croton billbergianus, Cupania glabra, Erythrina lanceolata, Guarea glabra, Hampea appendiculata, Heisteria macrophylla, Inga densiflora, Inga herrerae, Inga longispica, Inga mortoniana, Inga oerstediana, Inga punctata, Inga sapindoides, Lacistema aggregatum, Lonchocarpus oliganthus, Lonchocarpus retiferus, Machaerium arboreum, Malvaviscus arboreus, Meliosma glabrata, Myrcia splendens, Myriocarpa longipes, Ormosia cruenta, Pachyrhizus erosus, Pausandra trianae, Pithecellobium furcatum, Posoqueria latifolia, Pouteria exfoliata, Pouteria reticulata, Pseudolmedia spuria, Pterocarpus officinalis, Pterocarpus rohrii, Rourea schippii, Senna papillosa, Serjania mexicana, Serjania valerii, Sloanea faginea, Sorocea trophoides, Spermacoce ovalifolia, Tachigali costaricensis, Trema micrantha, Trophis involucrata, Urera elata, Vernonanthura patens, Xylophragma seemannianum, Zygia longifolia.*

neidhoeferi *Lemaire, 1967*
Foodplants: *Cassia, Fagus sylvatica, Inga sapindoides, Prunus serotina, Psidium guajava, Quercus robur, Robinia pseudoacacia, Rubus.*

paraensis Lemaire, 1973
Foodplants: Unknown

pygmaea (Schaus, 1904)
nigropunctata (Bouvier, 1930)
Foodplants: Unknown

pyrrhomelas (Walker, 1855)
Foodplants: Unknown

remissa (Weymer, 1907)
punctatus (Bouvier, 1929)
Foodplants: Unknown

remissoides Lemaire, 1967
Foodplants: Unknown

rindgei Lemaire, 1967
Foodplants: *Fagus sylvatica, Padus avium* var. *avium, Psidium guajava, Quercus robur, Robinia pseudoacacia, Salix caprea.*

rubriluna (Walker, 1862)
occidentalis (Bouvier, 1936)
Foodplants: *Psidium guajava, Quercus robur, Robinia pseudoacacia.*

septentrionalis (Bouvier, 1936)
Foodplants: *Clethra uleana, Dipteryx oleifera, Fagus sylvatica, Guatteria amplifolia, Inga, Ligustrum ovalifolium, Myriocarpa longipes, Quercus robur, Robinia pseudoacacia, Senna fruticosa.*

vanschaycki Naumann, Brosch & Wenczel, 2005
Foodplants: Unknown

viettei Lemaire, 1967
Foodplants: *Inga sapindoides, Psidium guajava, Quercus Salix caprea.*

Gamelioides Lemaire, 1988

anneliesae Brechlin, Käch & Meister, 2011
Foodplants: Unknown

denisae Naumann, Brosch & Wenczel, 2005
Foodplants: Unknown

elainae (Lemaire, 1967)
Foodplants: Unknown

ifei Brechlin, Käch & Meister, 2011
Foodplants: Unknown

kattyae Brechlin, Käch & Meister, 2010
Foodplants: Unknown

meisteri Brechlin, 2010
Foodplants: Unknown

sachai Brechlin, Käch & Meister, 2011
Foodplants: Unknown

seitzi (Draudt, 1929)
Foodplants: Unknown

Heliconisa Walker, 1855

pagenstecheri (Geyer, [1835])
costosa (Herrich-Schäffer, [1855])
immaculata Testout, 1939
impar Walker, 1855
lancea (Butler, 1871)
Foodplants: *Eryngium floribundum, Eucalyptus, Actinocephalus polyanthus, Rapanea.*

Hemileuca Walker, 1855

annulata Ferguson, 1971
Foodplants: *Amelanchier alnifolia* var. *utahensis, Ceanothus velutinus, Cercocarpus betuloides, Prunus emarginata, Purshia tridentata, Symphoricarpos vaccinoides.*

artemis Packard, 1893
Foodplants: *Corylus, Populus fremontii, Populus tremuloides, Prunus, Quercus agrifolia, Salix babylonica, Salix eriocephala* var. *eriocephala, Salix interior, Salix lasiopes, Salix longifolia, Salix lucida* var. *lasiandra.*

burnsi J. H. Watson, 1910
ilmae J. H. Watson, 1910
nigrovenosa J. H. Watson, 1910
conjuncta J. H. Watson, 1912
paradoxa J. H. Watson, 1913
bifasciata (Schüssler, 1911)
Foodplants: *Acacia greggii, Eriogonum fasciculatum, Gleditsia triacanthos, Padus virginiana, Prunus fasciculata, Psorothamnus arborescens* var. *simplifolius, Psorothamnus fremontii, Tetradymia axillaris, Tetradymia glabrata, Tetradymia spinosa.*

chinatiensis (Tinkham, 1943)
conwayae Peigler, 1985
Foodplants: *Acacia, Berberis trifoliolata, Calliandra eriophylla, Cercocarpus, Condalia ericoides, Ephedra aspera, Forestiera angustifolia, Fremontodendron californicum, Krameria erecta, Mimosa aculeaticarpa, Mimosa borealis, Mimosa lindheimeri, Rhus aromatica flabelliformis, Rhus microphylla.*

dyari (Draudt, 1930)
Foodplants: Unknown

eglanterina eglanterina (Boisduval, 1852)
boisduvali (Oberthür, 1914)

denudata (Neumoegen, 1891)
harrisi (Oberthür, 1914)
Foodplants: *Acer circinatum, Adenostoma fasciculatum, Amelanchier sanguinea* var. *alnifolia, Arctostaphylos glauca, Arctostaphylos uva-ursi, Artemisia tridentata, Betula, Cascaronia, Ceanothus spinosus, Ceanothus thyrsiflorus, Ceanothus velutinus, Cerasus vulgaris, Cercocarpus betuloides, Cercocarpus montanus, Eriogonum fasciculatum, Frangula californica, Frangula purshiana, Holodiscus discolor, Malus, Malus pumila, Populus tremuloides, Populus, Potentilla, Prunus avium, Prunus domestica, Prunus emarginata, Prunus serotina, Padus virginiana, Purshia tridentata, Pyracantha, Quercus lobata, Ribes, Rosa, Rubus allegheniensis, Salix babylonica, Salix exigua, Salix eriocephala* var. *eriocephala, Spiraea douglasii, Spiraea douglasii* subsp. *menziesii, Symphoricarpos albus, Symphoricarpos mollis, Symphoricarpos vaccinoides, Tamarix, Vitis californica, Vitis vulpina.*

eglanterina shastaensis (Grote, 1880)
normalis (Dyar & Cockerell, 1914)
Foodplants: *Arctostaphylos, Ceanothus spinosus, Cercocarpus betuloides, Prunus avium, Purshia tridentata.*

electra clio Barnes & McDunnough, 1918
Foodplants: *Bursera microphylla, Eriogonum fasciculatum, Eriogonum giganteum, Eriogonum latefolium, Fouquieria columnaris, Simmondsia chinensis, Tamarix gallica.*

electra electra W. G. Wright, 1884
ricksеckeri J. H. Watson, 1912
watsoni Schüssler, 1934
Foodplants: *Adenostoma fasciculatum, Bursera microphylla, Eriogonum fasciculatum, Eriogonum giganteum, Eriogonum latefolium, Fouquieria columnaris, Simmondsia chinensis, Tamarix gallica.*

electra mojavensis Tuskes & McElfresh, 1995
Foodplants: *Bursera microphylla, Eriogonum fasciculatum, Fouquieria columnaris, Simmondsia chinensis, Tamarix gallica.*

electra rubra McElfresh & Tuskes, 1995
Foodplants: *Fouquieria columnaris, Fouquieria splendens, Pachycormus discolor, Tamarix.*

griffini Tuskes, 1978
Foodplants: *Amelanchier, Cercocarpus betuloides, Ephedra.*

grotei diana Packard, 1874
Foodplants: *Quercus emoryi, Quercus gambelii, Quercus lobata, Quercus oblongifolia, Quercus palustris, Quercus robur, Quercus turbinella.*

grotei grotei Grote & Robinson, 1868
Foodplants: *Quercus fusiformis, Quercus emoryi, Quercus gambelii, Quercus havardii, Quercus marilandica, Quercus stellata, Quercus texana, Quercus turbinella, Quercus virginiana.*

hera hera (Harris, 1841)
chrysocarena (Harris, 1869)
gunderi (Hill, 1924)
pica (Walker, 1855)
Foodplants: *Artemisia cana, Artemisia filifolia, Artemisia tridentata, Eriogonum, Lupinus, Purshia tridentata.*

hera marcata (Neumoegen, 1891)
Foodplants: *Artemisia filifolia, Artemisia tridentata.*

hualapai (Neumoegen, 1882)
Foodplants: *Bouteloua curtipendula, Cynodon.*

juno Packard, 1872
yavapai Neumoegen, 1881
Foodplants: *Acacia, Eriogonum, Gleditsia triacanthos, Parkinsonia, Populus deltoides, Populus fremontii, Prosopis glandulosa, Prosopis juliflora, Prosopis pubescens, Salix.*

lares (Druce, 1897)
Foodplants: Unknown

lex (Druce, 1897)
Foodplants: *Acacia, Eriogonum wrightii, Fagus, Quercus arizonica, Robinia pseudoacacia.*

lucina Henry Edwards, 1887
lutea Reiff, 1910
obsoleta Reiff, 1910
Foodplants: *Betula populifolia, Menyanthes trifoliata, Quercus, Prunus serotina, Salix, Spiraea alba* var. *latifolia, Spiraea mongolica* var. *tomentulosa, Spiraea salicifolia, Vaccinium.*

magnifica (Rotger, 1948)
Foodplants: *Artemisia filifolia, Artemisia tridentata, Eriogonum effusum.*

maia (Drury, 1773)
lintneri Cockerell, 1914
proserpina (Fabricius, 1775)
Foodplants: *Betula pumila, Cerasus vulgaris, Corylus, Populus fremontii, Populus tremuloides, Prosopis glandulosa, Prunus serotina, Quercus*

coccinea, Quercus falcata, Quercus ilicifolia, Quercus laevis, Quercus prinoides, Quercus rubra, Quercus shumardii, Quercus stellata, Quercus velutina, Quercus virginiana, Salix alba, Salix babylonica, Salix caprea, Salix cinerea, Salix exigua var. *hindsiana, Salix interior, Spiraea.*

mania (Druce, 1897)
macellaria (Draudt, 1930)
Foodplants: *Zea mays*

marillia Dyar, 1911
Foodplants: Unknown

mexicana (Druce, 1887)
Foodplants: Unknown

neumoegeni (H. Edwards, 1881)
Foodplants: *Eriogonum fasciculatum, Fallugia paradoxa, Forestiera pubescens* var. *parvifolia, Padus virginiana, Prunus fasciculata, Rhus trilobata, Schinus terebinthifolia.*

nevadensis Stretch, 1872
californica W. G. Wright, 1888
lucina latifascia Barnes & McDunnough, 1916
Foodplants: *Acer rubrum, Corylus, Crataegus, Populus fremontii, Populus tremuloides, Prunus, Quercus agrifolia, Salix alba, Salix babylonica, Salix caprea, Salix cinerea, Salix eriocephala* var. *eriocephala, Salix interior, Salix lasiopes, Salix longifolia, Salix lucida* var. *lasiandra.*

numa (Druce, 1897)
nitria (Druce, 1897)
packardi Cockerell, 1914
Foodplants: Unknown

nuttalli nuttalli (Strecker, 1875)
arizonensis (Strecker, [1878])
washingtonensis (Medlar, 1944)
Foodplants: *Artemisia tridentata, Basilima millefolium, Cercocarpus betuloides, Purshia tridentata, Pyracantha, Ribes cereum, Symphoricarpos albus, Symphoricarpos rotundifolius* var. *parishii.*

nuttalli uniformis (Cockerell, 1914)
Foodplants: *Cercocarpus betuloides, Purshia tridentata, Pyracantha, Symphoricarpos albus, Symphoricarpos rotundifolius* var. *parishii, Symphoricarpos vaccinoides.*

oliviae Cockerell, 1898
grisea Cockerell, 1914
suffusa Cockerell, 1914
Foodplants: *Andropogon hallii, Bouteloua curtipendula, Chondrosum gracile, Chondrosum hirsutum, Bromus polyanthus, Buchloe dactyloides, Distichlis spicata, Elymus smithii, Hilaria cenchroides, Hordeum jubatum, Lycurus phleoides, Muhlenbergia texana, Sitanion jubatum, Sporobolus airoides, Sorghum.*

peigleri, Lemaire, 1981
Foodplants: *Quercus fusiformis, Quercus havardii, Quercus robur, Quercus shumardii, Quercus texana.*

peninsularis Lemaire, 1993
Foodplants: *Acacia baileyana, Cercidium microphyllum, Fouquieria columnaris, Prosopis juliflora.*

rubridorsa C. Felder & R. Felder, 1874
minette Dyar, 1911
norba (Druce, 1897)
Foodplants: *Acacia baileyana, Mimosa dealbata, Quercus, Robinia pseudoacacia, Salix.*

slossleri Peigler & Stone, 1989
Foodplants: *Quercus fusiformis, Quercus macrocarpa, Quercus havardii, Quercus palustris, Quercus robur.*

sororia (H. Edwards, 1881)
Foodplants: Unknown

stonei Lemaire, 1993
Foodplants: *Quercus emoryi, Quercus oblongifolia, Quercus turbinella.*

tricolor (Packard, 1872)
Foodplants: *Acacia baileyana, Acacia constricta, Acacia greggii, Cercidium microphyllum, Gleditsia triacanthos, Fouquieria columnaris, Mimosa, Prosopis juliflora, Quercus rubra, Robinia neomexicana, Robinia pseudoacacia, Sarcobatus vermiculatus.*

Hidripa Draudt, 1930

albipellis Draudt, 1930
Foodplants: Unknown

gschwandneri Draudt, 1930
Foodplants: Unknown

paranensis (Bouvier, 1929)
weymeri Draudt, 1930
Foodplants: Unknown

perdix (Maassen & Weyding, 1885)
Foodplants: Unknown

ruscheweyhi Draudt, 1930
Foodplants: Unknown

taglia (Schaus, 1896)
Foodplants: Unknown

Hirpida Draudt, 1930

amazonensis Brechlin & Meister, 2011
Foodplants: Unknown

boliviochoba Brechlin & Meister, 2010
Foodplants: Unknown

choba (Druce, 1904)
rubella (Dognin, 1910)
Foodplants: *Quercus*

chuquisaciana Brechlin & Meister, 2010
Foodplants: Unknown

cuscolinea Brechlin & Meister, 2011
Foodplants: Unknown

gaujoni (Dognin, 1894)
Foodplants: Unknown

junopascoensis Brechlin & Meister, 2010
Foodplants: Unknown

jujuylinea Brechlin & Meister, 2010
Foodplants: Unknown

levis (Johnson & Michener, 1948)
Foodplants: Unknown

levocuscoensis Brechlin & Meister, 2010
Foodplants: Unknown

levopascoensis Brechlin & Meister, 2010
Foodplants: Unknown

mavanschaycki Brechlin & Meister, 2010
Foodplants: Unknown

nigrolinea (Druce, 1904)
Foodplants: Unknown

olgae Brechlin & Meister, 2010
Foodplants: Unknown

pomacochasensis Brechlin & Meister, 2010
Foodplants: Unknown

santacruziana Brechlin & Meister, 2010
Foodplants: Unknown

sinjaevorum Brechlin & Meister, 2010
Foodplants: Unknown

Hispaniodirphia Lemaire, 1999

lemaireiana Rougerie & Herbin, 2006
Foodplants: Unknown

plana (Walker, 1855)
Foodplants: Unknown

Hylesia Hübner, [1820]

acuta Druce, 1886
petena Schaus, 1927
Foodplants: *Bursera simaruba*, *Quercus*, *Schinus terebinthifolia*.

aeneides (Druce, 1897)
Foodplants: *Croton schiedeanus*, *Laburnum*, *Prunus*, *Robinia pseudoacacia*.

alticola Lemaire, 1988
Foodplants: Unknown

anchises Lemaire, 1988
Foodplants: Unknown

andensis Lemaire, 1988
Foodplants: Unknown

andrei Dognin, 1923
Foodplants: *Cedrela odorata*

angulex Draudt, 1929
Foodplants: Unknown

annulata Schaus, 1911
ochrifex Dyar, 1913
scalex Draudt, 1929
Foodplants: Unknown

ascodex Dyar, 1913
Foodplants: Unknown

athlia Dyar, 1913
Foodplants: Unknown

beneluzi Lemaire, 1988
Foodplants: Unknown

bertrandi Lemaire, 1982
Foodplants: Unknown

biolleya Schaus, 1927
Foodplants: Unknown

bouvereti Dognin, 1889
Foodplants: *Prunus serotina*

canitia (Cramer, 1780)
amazonica Draudt, 1929
Foodplants: Unknown

cedomnibus Dyar, 1913
Foodplants: Unknown

coex Dyar, 1913
caucanex Draudt, 1929

Foodplants: *Annona purpurea, Mimosa arenosa, Senna alata.*

coinopus Dyar, 1913
Foodplants: Unknown

colimatifex Dyar, 1926
dognini Bouvier, 1930
Foodplants: *Guazuma ulmifolia*

colombex Dognin, 1923
Foodplants: Unknown

composita Dognin, 1912
Foodplants: Unknown

continua alinda Druce, 1886
Foodplants: *Annona spraguei, Desmopsis panamensis, Persea americana, Psidium friedrichsthalianum.*

continua colombiana Dognin, 1922
pernex Draudt, 1929
Foodplants: *Salix caprea*

continua continua (Walker, 1865)
chirex Schaus, 1921
euphemia Dyar, 1913
rosex Bouvier, 1930
Foodplants: *Acalypha diversifolia, Adelia triloba, Alfaroa guanacastensis, Allophylus psilospermus, Alsophila firma, Andira inermis, Annona purpurea, Ardisia compresa, Ardisia revoluta, Calliandra calothyrsus, Calyptranthes chytraculia, Casearia arborea, Casearia corymbosa, Casearia sylvestris, Casearia tacanensis, Cecropia peltata, Chimarrhis parviflora, Clethra mexicana, Coccoloba guanacastensis, Coccoloba porphyrostachys, Conceveiba pleiostemona, Cordia alliodora, Coussapoa nymphaeifolia, Croton billbergianus, Croton megistocarpus, Croton schiedeanus, Cupania glabra, Davilla nitida, Desmodium adscendens, Dioclea malacocarpa, Dioclea wilsonii, Erythrina costaricensis, Eugenia acapulcensis, Eugenia basilaris, Eugenia costaricensis, Eugenia glandulosopunctata, Eugenia siggersii, Gmelina arborea, Guarea glabra, Guazuma ulmifolia, Guettarda foliacea, Guettarda macrosperma, Hampea appendiculata, Hasseltia floribunda, Hedychium coronarium, Heliocarpus americanus, Heliocarpus appendiculatus, Hiraea fagifolia, Hirtella racemosa, Hirtella triandra, Inga chocoensis, Inga oerstediana, Inga punctata, Inga samanensis, Lonchocarpus oliganthus, Machaerium cobanense, Maripa nicaraguensis, Muntingia calabura, Myrcia splendens, Myriocarpa longipes, Nectandra hihua, Nectandra salicina, Ocotea insularis, Otoba novogranatensis, Pentagonia donnell-smithii, Persea americana, Piper arboreum, Pithecellobium furcatum, Prunus annularis, Prunus domestica, Psidium friedrichsthalianum, Psidium guajava, Quassia amara, Quercus, Randia genipoides, Renealmia aromatica, Renealmia cernua, Senna papillosa, Senna reticulata, Serjania mexicana, Serjania rhombea, Siparuna thecaphora, Solanum hayesii, Spondias mombin, Symphonia globulifera, Tabebuia rosea, Tachigali costaricensis, Terminalia amazonia, Tetrapterys tinifolia, Tibouchina longifolia, Trema micrantha, Trichospermum mexicanum, Uncaria tomentosa, Urera elata, Vernonanthura patens, Virola koschnyi, Vismia billbergiana, Vitex cooperi, Witheringia solanacea, Zexmenia virgulta, Zuelania guidonia, Zygia longifolia.*

cottica Schaus, 1932
Foodplants: Unknown

cressida Dyar, 1913
Foodplants: *Prunus domestica, Quercus.*

dalina Schaus, 1911
dalifex Dognin, 1916
schausi Dyar, 1913
Foodplants: *Casearia corymbosa, Casearia aculeata, Croton billbergianus, Croton schiedeanus, Gmelina arborea, Hampea appendiculata, Hyeronima oblonga, Inga punctata, Malvaviscus arboreus, Psidium guajava, Uncaria tomentosa.*

daryae Decaëns, Bonilla & Wolfe, 2003
Foodplants: Unknown

discifex Draudt, 1929
Foodplants: Unknown

dyarex Schaus, 1921
Foodplants: Unknown

ebalus (Cramer, 1775)
bolivex Dognin, 1922
ernestonis Strand, 1920
lilex Dognin, 1923
macellex Draudt, 1929
margarita Dognin, 1901
minasia Schaus, 1921
polyploca Draudt, 1929
purpurex Bouvier, 1930
wagneri Bouvier, 1923
Foodplants: *Arachis, Psidium guayava, Robinia pseudoacacia, Salix caprea, Tibouchina holosericea.*

egrex Draudt, 1929
violacea Bouvier, 1930
Foodplants: Unknown

extremex Naumann, Brosch, & Wenczel, 2005
Foodplants: Unknown

falcifera (Hubner, [1825])
Foodplants: Unknown

fallaciosa Lemaire, 2002
Foodplants: Unknown

frederici Lemaire, 1993
Foodplants: Unknown

frigida Schaus, 1911
vittex Draudt, 1929
Foodplants: *Arbutus, Pinus maximinoi, Quercus.*

gamelioides Michener, 1952
Foodplants: Unknown

gigantex gigantex Draudt, 1929
Foodplants: *Quercus*

gigantex orbana Schaus, 1932
Foodplants: *Quercus agrifolia*

gyrex Dyar, 1913
novex Dognin, 1922
Foodplants: Unknown

hamata Schaus, 1911
Foodplants: Unknown

haxairei Lemaire, 1988
Foodplants: Unknown

hawksi Lemaire, Wolfe & Monzon S., 2001
Foodplants: *Quercus*

hubbelli Lemaire, 1982
Foodplants: Unknown

humilis Dognin, 1923
Foodplants: Unknown

ileana Schaus, 1932
Foodplants: Unknown

index Dyar,1913
Foodplants: Unknown

indurata Dyar, 1910
lilacina Dognin, 1912
Foodplants: Unknown

inficita (Walker, 1865)
Foodplants: Unknown

invidiosa Dyar, 1914
Foodplants: *Byrsonima crassifolia, Hampea appendiculata.*

iola Dyar, 1913
omeva Dyar, 1916
Foodplants: *Acacia*

lamis (Cramer, [1780])
Foodplants: Unknown

lebeboides (Walker, 1869)
Foodplants: Unknown

leilex Dyar, 1913
mixtiplex Dognin, 1919
multiplex Draudt, 1916
peruvex Draudt, 1929
Foodplants: Unknown

lineata Druce, 1886
funebris (Conte, 1906)
ondulatus (Conte, 1906)
Foodplants: *Acacia baileyana, Acacia tenuifolia, Acosmium panamense, Aeschynomene brasiliana, Allophylus racemosus, Allosidastrum pyramidatum, Andira inermis, Annona purpurea, Annona rensoniana, Banisteriopsis, Bronwenia cornifolia, Banisteriopsis muricata, Bauhinia glabra, Bauhinia ungulata, Bunchosia nitida, Byrsonima crassifolia, Byttneria aculeata, Byttneria catalpifolia, Calliandra rubescens, Calliandra tergemina, Calliandra tergemina* var. *emarginata, Calycophyllum candidissimum, Casearia corymbosa, Casearia sylvestris, Casearia tremula, Cassia fistula, Cassia grandis, Cecropia peltata, Chloroleucon mangense, Chomelia spinosa, Chrysophyllum brenesii, Coccoloba guanacastensis, Combretum farinosum, Cordia alliodora, Cordia panamensis, Cordia polycephala, Croton glabellus, Croton jutiapensis, Croton niveus, Croton yucatanensis, Cupania guatemalensis, Dalbergia retusa, Diphysa americana, Enterolobium cyclocarpum, Erythroxylum havanense, Erythroxylum rotundifolium, Exostema mexicanum, Gliricidia sepium, Gmelina arborea, Gouania polygama, Guazuma ulmifolia, Guettarda macrosperma, Haematoxylum brasiletto, Helicteres baruensis, Helicteres guazumifolia, Heteropterys, Hiraea reclinata, Hirtella racemosa, Hymenaea courbaril, Hyptis pectinata, Inga punctata, Inga sapindoides, Lantana camara, Licania arborea, Lonchocarpus costaricensis, Lonchocarpus guatemalensis, Lonchocarpus minimiflorus, Luehea candida, Luehea speciosa, Lysiloma aurita, Machaerium*

biovulatum, Machaerium kegelii, Machaerium salvadorense, Malpighia glabra, Malvaviscus arboreus, Mimosa acantholoba, Mimosa pigra, Mimosa xanthocentra, Mimosa xanti, Muntingia calabura, Myrospermum frutescens, Ocotea veraguensis, Ouratea lucens, Pachira quinata, Passiflora pulchella, Paullinia cururu, Persea americana, Piscidia carthagenensis, Pithecellobium lanceolatum, Platymiscium parviflorum, Psidium guajava, Psidium guineense, Pterocarpus rohrii, Quercus oleoides, Rehdera trinervis, Robinia pseudoacacia, Securidaca sylvestris, Senna hayesiana, Senna pallida, Serjania schiedeana, Sida acuta, Sphinga platyloba, Sterculia apetala, Stigmaphyllon ellipticum, Tabebuia ochracea, Tabebuia rosea, Trema micrantha, Trigonia rugosa, Triplaris melaenodendron, Triumfetta bogotensis, Triumfetta lappula, Urvillea ulmacea, Xanthoxylum seulosum, Zuelania guidonia.

maurex Draudt, 1929
Foodplants: Unknown

medifex Dognin, 1916
pachobex Dognin, 1920
Foodplants: Unknown

melanops Lemaire, 2002
Foodplants: Unknown

melanostigma (Herrich-Schäffer, [1855])
propex Draudt, 1929
vespex Dognin, 1923
Foodplants: Unknown

metabus (Cramer, 1775)
caripitox Orfila, 1952
fusifascia Walker, 1862
grisoli Bouvier, 1923
obsoleta (Cramer, 1780)
urticans Floch & Abonnenc, 1944
Foodplants: *Avicennia germinans, Psidium guajava.*

metapyrrha (Walker, 1855)
corevia Schaus, 1900
Foodplants: *Psidium guayava*

moronensis Lemaire, 1976
Foodplants: Unknown

mortifex Dyar, 1913
lolamex Dyar, 1913
Foodplants: Unknown

munonia Schaus, 1927
violex Draudt, 1929
Foodplants: Unknown

murex Dyar, 1913
mystica Dyar, 1913
Foodplants: Unknown

mymex Dyar, 1913
Foodplants: Unknown

nanus (Walker, 1855)
dispar (Burmeister 1878)
dissimilis (Herrich-Schäffer, [1856])
myops (Walker, 1855)
Foodplants: *Croton*

natex Draudt, 1929
Foodplants: Unknown

nigricans (Berg, 1875)
fulviventris (Berg, 1883)
Foodplants: *Carpinus betulus, Ilex paraguariensis, Malus »hillieri«, Malus pumila, Ocotea laxa, Populus, Quercus × hispanica, Quercus robur, Salix babylonica, Salix caprea.*

nigridorsata Dognin, 1912
murmur Dyar, 1913
Foodplants: Unknown

nigripes Draudt, 1929
Foodplants: Unknown

oblonga Lemaire, 2002
Foodplants: Unknown

obtusa Dognin, 1923
Foodplants: Unknown

olivenca Schaus, 1927
Foodplants: *Populus, Pyrus, Quercus, Salix.*

oratex Dyar, 1913
sorana Schaus, 1927
Foodplants: Unknown

orbifex Dyar, 1913
Foodplants: Unknown

oroyex Dognin, 1922
Foodplants: Unknown

pallidex Dognin, 1923
coarya Schaus, 1932
gracilex Dognin, 1923
Foodplants: Unknown

paraguayensis Lemaire, 2002
Foodplants: Unknown

paulex Dognin, 1922
Foodplants: *Eucalyptus cloeziana*

pauper Dyar, 1913
Foodplants: Unknown

pearsoni Lemaire, 2002
Foodplants: *Padus virginiana, Prunus caroliniana, Prunus serotina, Spiraea bumalda.*

peigleri Lemaire, 2002
Foodplants: *Acacia drummondii, Padus virginiana, Prunus caroliniana, Prunus serotina, Spiraea bumalda.*

penai Lemaire, 1988
Foodplants: Unknown

praeda nigra Lemaire, 2002
Foodplants: Unknown

praeda praeda Dognin, 1901
rubriprocta Bouvier, 1930
Foodplants: Unknown

pseudomoronensis Camargo, 2007
Foodplants: Unknown

remex Dyar, 1913
livex Dyar, 1913
Foodplants: *Canna, Ilex paraguariensis, Laurus nobilis, Musa acuminata, Porlieria microphylla.*

rex Dyar, 1913
discutex Draudt, 1929
liturex Dyar, 1913
molpex Dyar, 1913
Foodplants: Unknown

rosacea rosacea Schaus, 1911
Foodplants: Unknown

rosacea thaumex Draudt, 1929
Foodplants: Unknown

roseata Dognin, 1914
Foodplants: Unknown

rubrifons muzoensis Draudt, 1929
Foodplants: Unknown

rubrifons rubrifons Schaus, 1911
multiplex Dyar, 1913
Foodplants: *Annona rensoniana, Anthurium obtusilobum, Hampea appendiculata, Myriocarpa longipes, Piptocarpha poeppigiana, Senna obtusifolia, Senna papillosa, Urera elata, Vismia baccifera.*

rufex Draudt, 1929
Foodplants: Unknown

rufipes Schaus, 1911
Foodplants: *Alchornea latifolia, Canavalia oxyphylla, Coccoloba, Dioclea wilsonii.*

santaelensis Lemaire, 1988
Foodplants: Unknown

schuessleri Strand, 1934
rufex Bouvier, 1930
Foodplants: *Aspidosperma macrocarpon, Aspidosperma tomentosum, Kielmeyera abdita.*

scortina Draudt, 1929,
Foodplants: *Croton floribundus*

sesostris (Vuillot, 1893)
Foodplants: Unknown

subaurea Schaus, 1900
coadjutor Dyar, 1907
Foodplants: *Quercus, Rhus, Robinia pseudoacacia.*

subcana (Walker, 1855)
Foodplants: Unknown

subcottica Lemaire, 2002
Foodplants: Unknown

subfasciata Dognin, 1916
Foodplants: Unknown

tapabex Dyar, 1913
Foodplants: Unknown

tapareba Dyar, 1913
derica Schaus, 1940
Foodplants: Unknown

teratex Draudt, 1929
Foodplants: Unknown

terranea Schaus, 1906
Foodplants: Unknown

terrosex Dognin, 1916
pluto Dognin, 1922
Foodplants: Unknown

tinturex Schaus, 1921
croex Schaus, 1921
Foodplants: *Pinus oocarpa*

tiphys Dognin, 1916
Foodplants: Unknown

travassosi Lemaire, 1988
Foodplants: Unknown

umbrata Schaus, 1911
darlingi Dyar, 1914
guayanensis Draudt, 1929
latex Draudt, 1929
lilacinex Draudt, 1929
palcazua Schaus, 1927
pollex Dyar, 1913

tendex Draudt, 1929
valvex Dyar, 1913
Foodplants: *Anacardium occidentale, Annona purpurea, Byrsonima crassifolia, Hura crepitans, Ochroma pyramidale, Ocotea, Sapium glandulosum, Zuelania guidonia.*

umbratula Dyar, 1915
Foodplants: Unknown

vassali Lemaire, 1988
Foodplants: Unknown

venezuelensis Lemaire, 2002
Foodplants: Unknown

vialactea Draudt, 1929
Foodplants: Unknown

vindex Dyar, 1913
erebus Bouvier, 1930
solvex Dyar, 1913
Foodplants: Unknown

wiebex Bouvier, 1923
Foodplants: Unknown

zonex Draudt, 1929
Foodplants: Unknown

Hylesiopsis Bouvier, 1929

festiva Bouvier, 1929
flavilineata (Draudt, 1930)
Foodplants: Unknown

Hyperchiria Hübner, [1819]

acuta (Conte, 1906)
flavus (Conte, 1906)
Foodplants: Unknown

acutapex Brechlin & Meister, 2010
Foodplants: Unknown

aniris (Jordan, 1910)
Foodplants: *Robinia pseudoacacia*

australoacuta Brechlin & Meister, 2010
Foodplants: Unknown

azteca (Draudt, 1929)
Foodplants: Unknown

bicolor (Bouvier, 1930)
Foodplants: *Berberis laurina, Celtis spinosa, Fagus, Ficus benjamina, Laburnum, Platanus orientalis, Prosopis alba, Pyracantha coccinea, Quercus ilex, Quercus robur, Rubus fructicosus, Senna corymbosa, Serjania.*

carabobensis Brechlin & Meister, 2010
Foodplants: Unknown

columbiana Brechlin & Meister, 2010
Foodplants: Unknown

cuscoincisoides Brechlin & Meister, 2010
Foodplants: Unknown

extremapex Brechlin & Meister, 2010
Foodplants: Unknown

gadouae Lemaire, 1967
Foodplants: *Berberis laurina, Celtis spinosa, Fagus, Ficus benjamina, Laburnum, Platanus orientalis, Prosopis alba, Quercus ilex, Quercus robur, Senna corymbosa, Serjania.*

guatemalensis Brechlin & Meister, 2010
Foodplants: Unknown

incisa Walker, 1855
orodes (Boisduval, 1875)
Foodplants: *Acacia baileyana, Acer negundo, Berberis laurina, Celtis spinosa, Fagus, Ficus benjamina, Laburnum, Myrocarpus frondosus, Platanus orientalis, Prunus laurocerasus, Quercus × hispanica, Quercus ilex, Robinia pseudoacacia, Rubus fructicosus, Salix, Senna corymbosa, Serjania, Trema micrantha, Wisteria.*

incisoides Brechlin & Meister, 2010
Foodplants: Unknown

jinotegaensis Brechlin & Meister, 2010
Foodplants: Unknown

meridaensis Brechlin & Meister, 2010
Foodplants: Unknown

misionincisoides Brechlin & Meister, 2010
Foodplants: Unknown

nausica (Cramer, [1779])
Foodplants: *Acacia baileyana, Anacardium, Fagus, Inga oerstediana, Laburnum, Mucuna holtonii, Rhus, Robinia pseudoacacia, Senna papillosa.*

nausioccidentalis Brechlin & Meister, 2010
Foodplants: Unknown

orodina (Schaus, 1906)
Foodplants: *Acacia baileyana, Fagus sylvatica, Ipomoea orizabensis, Malus »hillieri«, Pyracantha coccinea, Quercus robur, Robinia pseudoacacia, Rubus fructicosus, Salix viminalis, Sapindus, Serjania*

plicata (Herrich-Schäffer, [1855])
lacerata (Herrich-Schäffer, [1858])

Foodplants: *Berberis laurina, Celtis spinosa, Fagus, Ficus benjamina, Laburnum, Platanus orientalis, Quercus × hispanica, Quercus ilex, Senna corymbosa, Serjania.*

sanjuensis Brechlin & Meister, 2010
Foodplants: Unknown

schmiti Meister & Knorke, 2004
Foodplants: Unknown

Hyperchirioides Lemaire, 1981

bulaea (Maassen & Weyding, 1885)
Foodplants: Unknown

Hypermerina Lemaire, 1969

kasyi Lemaire, 1969
Foodplants: Unknown

Ithomisa Oberthür, 1881

catherina (Schaus, 1896)
impuncta Draudt, 1930
Foodplants: *Actinocephalus polyanthus*

kinkelini Oberthür, 1881
Foodplants: *Actinocephalus polyanthus, Eryngium penicillatum.*

lepta (Druce, 1890)
Foodplants: Unknown

umbrata Oiticica Filho, 1958
Foodplants: Unknown

Kentroleuca Draudt, 1929

albilinea (Schaus, 1908)
Foodplants: Unknown

boliviensis Brechlin & Meister, 2002
Foodplants: Unknown

brunneategulata Mielke & Furtado, 2006
Foodplants: Unknown

dukinfieldi (Schaus, 1894)
Foodplants: Unknown

griseoalbata Mielke & Furtado, 2006
Foodplants: Unknown

lineosa (Walker, 1855)
Foodplants: Unknown

novaholandensis Lemaire & Mielke, 2001
Foodplants: Unknown

spitzi Lemaire, 1971
Foodplants: Unknown

Leucanella Lemaire 1969

acutissima (Walker, 1865)
Foodplants: *Malosma laurina*

anikae Meister & Brechlin 2002
conlani Lemaire, 2002
Foodplants: *Ligustrum ovalifolium*

apollinairei (Dognin, 1923)
Foodplants: *Ligustrum ovalifolium, Prunus spinosa, Salix caprea, Salix helvetica.*

aspera aspera (C. Felder & R. Felder, 1874)
burmeisteri (Weyenbergh, 1886)
distincta (Köhler, 1935)
rosacea (Draudt, 1929)
tristis (Köhler, 1935)
Foodplants: *Acacia, Aspidosperma quebracho-blanco, Betula pendula, Celtis ehrenbergiana, Celtis spinosa, Fagus, Laburnum, Ligustrum ovalifolium, Malus »hillieri«, Mimosa hexandra, Mimosa lawranceana, Prosopis affinis, Prosopis alba, Prosopis nigra, Pyracantha coccinea, Robinia pseudoacacia, Salix caprea, Salix cinerea, Tamarix.*

aspera patagonica (Breyer, 1957)
Foodplants: *Ligustrum ovalifolium*

atahualpa Meister & Naumann, 2006
Foodplants: Unknown

bivius (Bouvier, 1927)
Foodplants: Unknown

bonillensis Decaëns & Rougerie, 2008
Foodplants: Unknown

christbrechlinae Brechlin & Meister, 2011
Foodplants: Unknown

contei (Lemaire, 1967)
Foodplants: *Ligustrum ovalifolium, Salix caprea.*

contempta contempta (Lemaire, 1967)
Foodplants: *Fagus, Juglans regia, Ligustrum ovalifolium, Prunus capollin, Prunus domestica, Prunus serotina, Prunus spinosa, Pyracantha, Rubus fructicosus, Salix caprea.*

contempta inti Käch & Racheli, 2010
Foodplants: Unknown

contempta flavissima (Lemaire, 1967)
Foodplants: *Ligustrum ovalifolium*

contempta nyctimenoides (Lemaire, 1967)
Foodplants: *Ligustrum ovalifolium*

contempta sacha Käch & Racheli, 2010
Foodplants: Unknown

contempta windi Lemaire, 1973
Foodplants: *Ligustrum ovalifolium, Salix caprea.*

evanschaycki Brechlin & Meister, 2011
Foodplants: Unknown

flammans (Schaus, 1900)
Foodplants: *Inga sapindoides, Ligustrum ovalifolium.*

fusca (Walker, 1855)
combusta (Walker, 1865)
oaxaca (F. Johnson & Michener, 1948)
Foodplants: Unknown

gibbosa (Conte, 1906)
hoffmanni (Sageder, 1935)
Foodplants: *Ligustrum ovalifolium*

hausmanni Brechlin & Meister, 2011
Foodplants: Unknown

heisleri (Jones, 1908)
miris (Sageder, 1935)
Foodplants: Unknown

hosmera (Schaus, 1941)
Foodplants: *Gonocalyx pterocarpus, Ligustrum ovalifolium, Schwartzia costaricensis, Ternstroemia tepezapote, Trema micrantha, Weinmannia wercklei.*

janeira (Westwood, [1854])
cinerea (Walker, 1855)
fumata (C. Felder & R. Felder, 1874)
norcestes (Boisduval, [1875])
Foodplants: *Ligustrum ovalifolium, Rosa.*

julianae Brechlin & Meister, 2011
Foodplants: Unknown

lama (Berg, 1883)
joannisi (Bouvier, 1927)
Foodplants: *Ligustrum ovalifolium, Prunus spinosa, Rubus fructicosus.*

leucane (Geyer, [1837])
budleyi (Fiedler, 1930)
cinnamomea (C. C. Hoffmann, 1942)
Foodplants: *Betula, Ligustrum ovalifolium, Malus pumila, Malus sylvestris, Prunus, Quercus, Robinia pseudoacacia, Syringa vulgaris.*

lynx (Bouvier, 1930)
Foodplants: *Juglans regia, Ligustrum ovalifolium, Malus »hillieri«, Padus avium* var. *avium, Prunus serotina, Prunus spinosa.*

maandensis Brechlin & Meister, 2011
Foodplants: Unknown

maasseni (Möschler, 1872)
falcata (Boisduval, [1875])
Foodplants: *Ligustrum ovalifolium, Ligustrum vulgare.*

memusae gardineri Lemaire, 1973
Foodplants: *Inga sapindoides, Laburnum anagyroides, Ligustrum ovalifolium, Prunus domestica, Trema micrantha.*

memusae memusae (Walker, 1855)
mimusops (Boisduval, [1875])
Foodplants: *Acacia, Acer, Cephalanthus glabratus, Erythrina, Erythrina crista-galli, Ilex paraguariensis, Laburnum anagyroides, Ligustrum ovalifolium, Lonicera caprifolia, Malus, Olea europaea, Platanus orientalis, Prunus domestica, Pyrus communis, Robinia pseudoacacia, Salix caprea, Salix cinerea, Solanum lycocarpum, Solanum melongena, Solanum paniculatum Solanum tuberosum, Solanum variabile, Trema micrantha.*

memusoides Lemaire, 1973
piura Lampe, 2004 **syn. nov.**
Foodplants: *Laburnum anagyroides, Ligustrum ovalifolium.*

muelleri (Draudt, 1829)
Foodplants: *Ligustrum ovalifolium*

newmani (Lemaire, 1967)
Foodplants: *Ligustrum ovalifolium*

nyctimene (Latreille, 1832)
Foodplants: *Acer, Alnus acuminata, Ligustrum ovalifolium, Platanus orientalis, Pyracantha, Robinia pseudoacacia, Solanum tuberosum, Solanum variabile.*

rufoboliviana Brechlin & Meister, 2011
Foodplants: Unknown

siriae Brechlin & Meister, 2011
Foodplants: Unknown

saturata (Walker, 1855)
schausii (H. Edwards, 1884)
Foodplants: *Ligustrum ovalifolium, Platanus mexicana.*

stuarti koehleri (Gemignani, 1931)
jordani (Bouvier, 1933)
Foodplants: *Adesmia, Berberis flexuosa, Ligustrum ovalifolium, Prunus, Senna birostris* var. *hookeriana.*

stuarti rothschildi (Bouvier, 1933)
Foodplants: Unknown

stuarti stuarti (W. Rothschild & Jordan, 1901)
Foodplants: *Adesmia, Berberis flexuosa, Prunus, Rosa, Senna birostris, Senna birostris* var. *hookeriana.*

viettei (Lemaire, 1967)
Foodplants: *Cornus, Ligustrum ovalifolium, Juglans regia, Morus, Prunus serotina, Prunus spinosa, Pyracantha coccinea, Rubus fructicosus, Rubus nemorosus, Salix.*

viridescens paraviridior Brechlin & Meister, 2011
Foodplants: Unknown

viridescens viridescens (Walker, 1855)
damoeus (Boisduval, [1875])
semicaeca (Schaus, 1932)
Foodplants: *Acer, Erythrina crista-galli, Laburnum anagyroides, Ligustrum japonicum, Ligustrum ovalifolium, Malus, Morus nigra, Platanus, Trema micrantha.*

viridescens viridior Lemaire, 1973
Foodplants: *Erythrina crista-galli, Laburnum anagyroides, Ligustrum japonicum, Ligustrum ovalifolium, Morus nigra.*

winfbrechlini Brechlin & Meister, 2011
Foodplants: Unknown

yungasensis Meister & Naumann, 2006
Foodplants: *Ligustrum ovalifolium, Pyracantha, Rubus fructicosus.*

Lonomia Walker, 1855

achelous achelous (Cramer, 1777)
acheloe (Hübner, [1819])
Foodplants: *Erythrina crista-galli, Malosma laurina, Malus »hillieri«, Pyrus communis, Robinia pseudoacacia, Schinus terebinthifolia, Tapirira guianensis.*

achelous diabolus Draudt, 1929
brunnea (Bouvier, 1930)
Foodplants: *Hevea brasiliensis*

belizonensis Brechlin, Meister & Van Schayck, 2011
Foodplants: Unknown

beneluzi Lemaire, 2002
Foodplants: Unknown

camox Lemaire, 1972
Foodplants: *Carpinus betulus, Fagus grandifolia* subsp. *grandifolia.*

canescens Brechlin & Meister, 2011
Foodplants: Unknown

columbiana Lemaire, 1972
Foodplants: Unknown

concordia Druce, 1886
Foodplants: *Arachnothryx buddleioides, Billia hippocastanum, Carpinus betulus, Casearia tacanensis, Chionanthus panamensis, Citharexylum costaricense, Clethra mexicana, Coffea arabica, Croton schiedeanus, Citharexylum costaricense, Geonoma interrupta, Gymnanthes riparia, Hampea appendiculata, Hasseltia floribunda, Ligustrum ovalifolium, Malosma laurina, Malvaviscus palmanus, Mosquitoxylum jamaicense, Parathesis glabra, Pentagonia donnell-smithii, Pouteria exfoliata, Prunus annularis, Psidium guajava, Quassia amara, Robinia pseudoacacia, Xylosma chlorantha.*

descimoni Lemaire, 1972
Foodplants: Unknown

electra Druce, 1886
Foodplants: *Arachnothryx buddleioides, Billia hippocastanum, Carpinus betulus, Casearia tacanensis, Chionanthus panamensis, Citharexylum costaricense, Clethra mexicana, Coffea arabica, Croton schiedeanus, Citharexylum costaricense, Geonoma interrupta, Gymnanthes riparia, Hampea appendiculata, Hasseltia floribunda, Ligustrum ovalifolium, Malosma laurina, Malvaviscus palmanus, Mosquitoxylum jamaicense, Parathesis glabra, Pentagonia donnell-smithii, Pouteria exfoliata, Prunus annularis, Psidium guajava, Quassia amara, Robina pseudoacacia, Xylosma chlorantha.*

francescae Racheli, 2005
Foodplants: Unknown

frankae Meister, Naumann, Brosch & Wenczel, 2005
Foodplants: *Fagus sylvatica, Malosma laurina.*

luteomexicana Brechlin & Meister, 2011
Foodplants: Unknown

madrediosiana Brechlin & Meister, 2011
Foodplants: Unknown

maranhensis Brechlin, Meister & Mielke, 2011
Foodplants: Unknown

mexilectra Brechlin & Meister, 2011
Foodplants: Unknown

obliqua Walker, 1855
albigutta Walker, 1855
carnica Schaus, 1892
vulpina Draudt, 1929
Foodplants: *Eucalyptus, Ficus subtrinervia, Handroanthus pulcherrimus, Rollinia emarginata.*

orientoandensis Brechlin & Meister, 2011
Foodplants: Unknown

paralectra Brechlin & Meister, 2011
Foodplants: Unknown

parobliqua Brechlin, Meister & Mielke, 2011
Foodplants: Unknown

parubrescens Brechlin & Meister, 2011
Foodplants: Unknown

pseudoobliqua Lemaire, [1973]
Foodplants: Unknown

puntarenasiana Brechlin & Meister, 2011
Foodplants: Unknown

rubrescens Brechlin & Meister, 2011
Foodplants: Unknown

rufescens Lemaire, 1972
Foodplants: *Cedrela odorata, Hampea appendiculata, Parathesis, Pentagonia donnell-smithii, Senna papillosa, Tetragastris panamensis.*

serranoi Lemaire, 2002
Foodplants: Unknown

sinjaevorum Brechlin & Meister, 2011
Foodplants: Unknown

venezuelensis Lemaire, 1972
Foodplants: *Salix caprea*

viksinyaevi Brechlin & Meister, 2011
Foodplants: Unknown

yucatensis Brechlin & Meister, 2011
Foodplants: Unknown

Meroleuca Packard, 1904

Subgenus *Dihirpa*, Draudt, 1929

lamasi Lemaire, 1982
Foodplants: Unknown

litura (Walker, 1855)
marginella (Walker, 1865)
Foodplants: Unknown

Subgenus *Meroleuca*, Packard 1904

decaensi Lemaire, 1995
Foodplants: Unknown

lituroides (Bouvier, 1929)
Foodplants: Unknown

mossi Lemaire, 1995
Foodplants: Unknown

nigra (Dognin, 1913)
bogotana (Bouvier, 1930)
Foodplants: Unknown

venosa (Walker, 1855)
Foodplants: Unknown

Subgenus *Meroleucoides*, Michener, 1949

albomaculata (Dognin, 1916)
Foodplants: Unknown

amarillae Lemaire & Wolfe, 1995
Foodplants: Unknown

bipectinata Lemaire, 2002
Foodplants: Unknown

bipunctata Lemaire, 1982
Foodplants: Unknown

bravera Lemaire, 2002
Foodplants: *Bambusa*

catamarcensis Meister & Brechlin, 2008
Foodplants: Unknown

dargei Lemaire, 1982
Foodplants: Unknown

diazmaurini Decaéns, Bonilla & Ramirez, 2004
Foodplants: Unknown

erythropus (Maassen, 1890)
sinuata Lemaire, 1982
Foodplants: Unknown

famula (Maassen, 1890)
Foodplants: Unknown

fassli Lemaire, 1995
Foodplants: *Brachychiton acerifolius, Quercus.*

flavodiscata (Dognin, 1916)
Foodplants: Unknown

laverna (Druce, 1890)
Foodplants: Unknown

microstyx Lemaire, 2002
Foodplants: Unknown

modesta Lemaire, 2002
Foodplants: Unknown

nadiana Lemaire, 2002
Foodplants: Unknown

naias (Bouvier, 1929)
Foodplants: Unknown

nata (Maassen, 1890)
Foodplants: Unknown

penai Lemaire, 1982
Foodplants: Unknown

ramicosa (Lemaire, 1975)
Foodplants: Unknown

rectilineata Lemaire & Venedictoff, 1989
Foodplants: Unknown

raineri Brechlin & Meister, 2008
Foodplants: Unknown

riveti Lemaire, 2002
Foodplants: Unknown

thibaudi Naumann, Meister & Brosch, 2006
Foodplants: Unknown

verae Van Schayck, 2000
Foodplants: *Bambusa, Phyllostachis.*

Molippa Walker, 1855

azuelensis Lemaire, 1976
Foodplants: Unknown

basina Maassen & Weyding, 1885
Foodplants: *Acacia baileyana, Ligustrum ovalifolium, Salix caprea, Salix viminalis.*

basinoides Bouvier, 1926
Foodplants: Unknown

bertrandi Lemaire, 1982
Foodplants: Unknown

bertrandoides Brechlin & Meister, 2008
Foodplants: *Prunus, Salix caprea, Salix helvetica.*

binasa (Schaus, 1924)
Foodplants: *Acacia baileyana, Pyracantha coccinea, Quercus robur, Rubus fructicosus, Salix caprea.*

convergens (Walker, 1855)
praecruenta (Maassen & Weyding, 1885)
Foodplants: Unknown

coracoralinae Lemaire & Tangerini, 2002
Foodplants: Unknown

cruenta (Walker, 1855)
semirosea (Weymer, 1907)
Foodplants: *Acacia baileyana, Robinia pseudoacacia.*

eophila (Dognin, 1919)
Foodplants: *Acacia caven*

kohlli Racheli, 2008
wenczeli Brechlin & Meister, 2008
Foodplants: *Cornus mas, Cornus sanguinea, Prunus, Salix caprea, Salix helvetica.*

larensis Lemaire, 1972
Foodplants: Unknown

latemedia (Druce, 1890)
boliviana (Bouvier, 1930)
Foodplants: *Acacia baileyana, Pyracantha coccinea, Quercus × hispanica, Quercus robur, Robinia pseudoacacia, Rubus fructicosus, Salix caprea.*

luzalessarum Naumann, Brosch & Wenczel, 2005
Foodplants: Unknown

nibasa Maassen & Weyding, 1885
centralis C. C. Hoffmann, 1942
Foodplants: *Acacia baileyana, Amphilophium paniculatum, Ateleia herbert-smithii, Bauhinia ungulata, Cassia grandis, Cassia javanica, Enterolobium cyclocarpum, Fridericia mollissima, Fridericia patellifera, Gliricidia sepium, Lonchocarpus minimiflorus, Pyracantha coccinea, Robinia pseudoacacia, Salix caprea, Senna pallida, Tamarindus indica.*

ninfa (Schaus, 1921)
Foodplants: *Robinia pseudoacacia*

pearsoni Lemaire, 1982
Foodplants: Unknown

pilarae Naumann, Brosch & Wenczel, 2005
Foodplants: Unknown

placida (Schaus, 1921)
amazonica (Bouvier, 1930)
Foodplants: Unknown

rosea (Druce, 1886)
Foodplants: *Platanus, Robinia pseudoacacia.*

sabina Walker, 1855
Foodplants: *Acacia mearnsii, Bauhinia, Carpinus betulus, Erythrina, Mimosa, Peltophorum dubium, Quercus × hispanica, Salix babylonica, Salix caprea.*

simillima Jones, 1907
flavocrinata Mabilde, 1896
Foodplants: *Carpinus betulus, Corylus avellana, Malus »hillieri«, Mimosa, Pyracantha coccinea, Quercus robur, Robinia pseudoacacia, Salix caprea, Trema micrantha.*

strigosa (Maassen & Weyding, 1885)
Foodplants: Unknown

superba superba (Burmeister, 1878)
lelusa (Schaus, 1925)
Foodplants: *Pyracantha coccinea, Quercus × hispanica, Robinia pseudoacacia, Rubus fructicosus, Salix caprea.*

superba tangerinii Lemaire, 1977
Foodplants: Unknown

tusina (Schaus, 1921)
denhezi Lemaire, 1969
Foodplants: *Dioclea malacocarpa, Hampea appendiculata, Hasseltia floribunda, Inga oerstediana, Lonchocarpus ferrugineus, Pterocarpus rohrii, Senna papillosa.*

wittmeri Lemaire, 1976
Foodplants: *Acacia baileyana, Mimosa, Robinia.*

Ormiscodes Blanchard, 1852

amphinome amphinome (Fabricius, 1775)
latifasciata Bouvier, 1935
humeralis (Walker, 1855)
hyadesi (Mabille, 1885)
marginata (Philippi, 1859)
vulpina (Philippi, 1859)
Foodplants: *Cupressus, Juglans regia, Nothofagus glauca, Nothofagus obliqua, Pinus radiata, Populus nigra, Pseudotsuga menziesii, Salix, Schinus latifolius, Schinus molle.*

amphinome lauta (Berg, 1881)
Foodplants: *Malus, Pyrus, Salix babylonica, Salix humboldtiana.*

amphinome nigrolutea (Bouvier, 1924)
patagonica (Bouvier, 1926)
Foodplants: Unknown

bruchi (Köhler, 1930)
Foodplants: Unknown

caxambua (Schaus, 1921)
Foodplants: Unknown

cinnamomea (Feisthamel, 1839)
claudia Bryk, 1945
crinita Blanchard, 1852
plana (Walker, 1855)
telifera (Bouvier, 1930)
Foodplants: *Cupressus, Juglans regia, Muhlenbeckia sagittaefolia, Nothofagus glauca, Nothofagus obliqua, Populus nigra, Pinus radiata, Piper, Pseudotsuga menziesii, Schinus latifolius, Schinus molle.*

cognata Philippi, 1859
alanus (Packard, 1914)
Foodplants: Unknown

eugeniae Brechlin & Meister, 2010
Foodplants: Unknown

joiceyi (Draudt, 1930)
requelae Béeche, 1999
Foodplants: Unknown

nigrosignata (Philippi, 1859)
luperina (Packard, 1914)
murina (Draudt, 1930)
Foodplants: Unknown

penai Lemaire & Parra, 1995
Foodplants: Unknown

rufosignata (Blanchard, 1852)
lupina (Draudt, 1930)
Foodplants: Unknown

schmidtnielseni Lemaire, 1985
Foodplants: *Araucaria araucana*

shapiroi Lemaire, 1978
Foodplants: Unknown

socialis grisea Ureta, 1957
Foodplants: Unknown

socialis socialis (Feisthamel, 1839)
angulifera (Walker, 1855)
ribesii (Edwards, 1874)
Foodplants: *Cryptocarya alba, Cryptocarya rubra.*

Paradirphia Michener, 1949

andicola Lemaire, 2002
Foodplants: *Cassia, Inga sapindoides, Malus, Prunus, Robinia pseudoacacia.*

andrei Brechlin & Meister, 2010
Foodplants: Unknown

anikae Brechlin & Meister, 2010
Foodplants: *Fagus sylvatica*

antonia (Dognin, 1911)
apollinairei (Bouvier, 1930)
Foodplants: *Malus, Prunus, Robinia pseudoacacia.*

boudinoti Lemaire & Wolfe, 1990
Foodplants: *Malus »hillieri«, Prunus serotina, Robinia pseudoacacia.*

citrina (Druce, 1886)
angrandi (Bouvier, 1923)
Foodplants: *Quercus robur, Robinia pseudoacacia.*

coprea (Draudt, 1930)
Foodplants: *Robinia pseudoacacia*

estivalisae Guererro & Passola, 2003
Foodplants: Unknown

frankae Brechlin & Meister, 2010
Foodplants: *Fagus sylvatica*

fumosa (C. Felder & R. Felder, 1874)
michoakana Beutelspacher, 1984
Foodplants: Unknown

gabrielae Brechlin & Meister, 2010
Foodplants: Unknown

geneforti (Bouvier, 1923)
gemeli (Bouvier, 1925)
Foodplants: *Inga sapindoides, Juglans neotropica, Pinus, Quercus × hispanica.*

hectori Brechlin & Meister, 2010
Foodplants: *Fagus sylvatica*

herediana Brechlin & Meister, 2010
Foodplants: Unknown

hoegei (Druce, 1886)
rufa (Bouvier, 1930)
Foodplants: *Fagus sylvatica, Malus pumila, Platanus mexicana, Prunus, Robinia pseudoacacia, Salix babylonica, Salix humboldtiana.*

ibarrai Balcázar-Lara, 1999
inornata (Bouvier, 1925)
Foodplants: Unknown

latipunctata (Lemaire, 1976)
wenczeli Racheli, 2008 **syn. nov.**
Foodplants: *Robinia pseudoacacia*

leoni Brechlin & Meister, 2010
Foodplants: Unknown

lieseorum Brechlin & Meister, 2010
Foodplants: Unknown

lasiocampina (C. Felder & R. Felder, 1874)
Foodplants: *Eriobotrya japonica, Fagus sylvatica, Malosma laurina, Malus pumila, Quercus robur, Robinia pseudoacacia, Salix caprea.*

manes (Druce, 1897)
Foodplants: Unknown

oblita (Lemaire, 1976)
Foodplants: Unknown

pararudloffi Brechlin & Meister, 2010
Foodplants: Unknown

peggyae Brechlin & Meister, 2010
Foodplants: Unknown

rectilineata Wolfe, 1994
Foodplants: *Prunus*

rudloffi Brechlin & Meister, 2008
Foodplants: Unknown

semirosea (Walker, 1855)
roseana (Draudt, 1930)
Foodplants: *Fagus sylvatica, Robinia pseudoacacia, Salix caprea.*

talamancaia Brechlin & Meister, 2010
Foodplants: Unknown

torra Weymer, 1907
Foodplants: Unknown

valverdei Lemaire & Wolfe, 1990
Foodplants: *Fagus sylvatica, Prunus, Robinia pseudoacacia.*

winifredae Lemaire & Wolfe, 1990
Foodplants: *Blakea chlorantha, Gonocalyx pterocarpus, Miconia lonchophylla, Vaccinium poasanum, Weinmannia pinnata, Weinmannia wercklei.*

Periga Walker, 1855

abnegata (Walker 1855)
Foodplants: Unknown

angulosa (Lemaire, 1972)
Foodplants: *Fagus grandifolia* subsp. *grandifolia*

anitae Naumann, Brosch & Wenczel, 2005
Foodplants: *Cornus mas, Cornus sanguinea.*

armata (Lemaire, [1973])
Foodplants: Unknown

aurantiaca Lemaire, 1972
Foodplants: Unknown

bispiniosa Lemaire, 1972
Foodplants: Unknown

boettgerorum Naumann, Brosch & Wenczel, 2005
Foodplants: Unknown

brechlini Naumann, Brosch & Wenczel, 2005
Foodplants: Unknown

circumstans Walker, 1855
Foodplants: *Coffea, Eriobotrya japonica, Persea americana.*

cluacina (Druce, 1886)
Foodplants: *Carapa guianensis, Melastomataceae.*

cynira (Cramer, 1777)
Foodplants: *Acacia, Astronium graveolens, Carpinus betulus, Fagus sylvatica, Ligustrum ovalifolium, Lithraea brasiliensis, Quercus × hispanica, Robinia pseudoacacia, Urera alceifolia, Urtica urens.*

elsa (Lemaire, [1973])
Foodplants: Unknown

extensiva Lemaire, 2002
Foodplants: Unknown

falcata (Walker, 1855)
apicalis (Walker, 1855)
subcynira (Bouvier, 1933)
submacula (Walker, 1855)
Foodplants: *Diospyros kaki, Joannesia princeps.*

galbimaculata (Lemaire, 1972)
Foodplants: Unknown

gueneei (Lemaire, [1973])
Foodplants: Unknown

herbini Lemaire, 2002
Foodplants: Unknown

inexpectata (Lemaire, 1972)
Foodplants: Unknown

insidiosa (Lemaire, 1972)
Foodplants: Unknown

intensiva (Lemaire, [1973])
Foodplants: Unknown

kindli Lemaire, 1993
Foodplants: Unknown

kishidae Naumann, Brosch & Wenczel, 2005
Foodplants: Unknown

lichyi (Lemaire, 1972)
Foodplants: Unknown

lobulata Lemaire, 2002
Foodplants: Unknown

occidentalis (Lemaire, 1972)
Foodplants: Unknown

parvibulbacea (Lemaire, 1972)
Foodplants: *Malosma laurina, Prunus serotina.*

prattorum (Lemaire, 1972)
Foodplants: Unknown

rasplusi (Lemaire, 1985)
Foodplants: Unknown

sanguinea Lemaire, 2002
Foodplants: Unknown

spatulata (Lemaire, [1973])
Foodplants: Unknown

squamosa (Lemaire, 1972)
Foodplants: Unknown

Periphoba Hübner, [1820]

albata (Draudt, 1930)
angelica (Bouvier, 1930)
Foodplants: Unknown

arcaei (Druce, 1886)
arctus (Bouvier, 1929)
Foodplants: *Acacia collinsii, Acalypha diversifolia, Aeschynomene brasiliana, Alibertia edulis, Anacardium excelsum, Anacardium occidentale, Andira inermis, Annona purpurea, Annona rensoniana, Ardisia compressa, Ardisia revoluta, Ardisia standleyana, Astronium graveolens, Augusta rivalis, Bignonia heterophylla, Bursera simaruba, Byrsonima crassifolia, Calophyllum brasiliense, Calycophyllum candidissimum, Calyptranthes chytraculia, Casearia corymbosa, Cassia grandis, Cespedesia spathulata, Chrysophyllum brenesii, Cnestidium rufescens, Combretum farinosum, Conostegia xalapensis, Coussapoa nymphaeifolia, Croton schiedeanus, Cupania guatemalensis, Dilodendron costaricense, Dioclea malacocarpa, Erythrina costaricensis, Erythrina lanceolata, Eugenia acapulcensis, Eugenia basilaris, Fagus sylvatica, Ficus benjamina, Garcinia intermedia, Gouania polygama, Guarea, Guarea rhopalocarpa, Guatteria amplifolia, Guazuma ulmifolia, Guettarda macrosperma, Hampea appendiculata, Hedychium coronarium, Hirtella racemosa, Hymenaea courbaril, Inga chocoensis, Inga oerstediana, Inga punctata, Inga sapindoides, Jatropha curcas, Licania arborea, Lonchocarpus retiferus, Luehea speciosa, Lysiloma aurita, Mabea occidentalis, Machaerium arboreum, Machaerium floribundum, Machaerium salvadorense, Mangi-*

fera indica, Manilkara chicle, Matudaea trinervia, Maytenus segoviarum, Melicoccus bijugatus, Miconia argentea, Mimosa xanti, Mortoniodendron guatemalense, Mouriri myrtilloides, Myrcia splendens, Myriocarpa longipes, Ochroma pyramidale, Ocotea mollifolia, Ocotea veraguensis, Ouratea lucens, Pachira quinata, Paullinia cururu, Pavonia schiedeana, Persea americana, Piper, Pithecellobium furcatum, Psidium guajava, Psidium guineense, Quassia amara, Quercus oleoides, Quercus pubescens, Quercus robur, Rhus, Robinia pseudoacacia, Rourea glabra, Salix viminalis, Sapium macrocarpum, Securidaca sylvestris, Senna alata, Senna papillosa, Simarouba amara, Sloanea terniflora, Spondias mombin, Spondias purpurea, Symphonia globulifera, Syzygium jambos, Syzygium malaccense, Syzygium salomonense, Talisia macrophylla, Tapirira mexicana, Tetracera volubilis, Thouinidium decandrum, Trema micrantha, Triplaris melaenodendron, Uncaria tomentosa, Vismia baccifera, Xylopia frutescens, Zygia longifolia

attali Lemaire & Terral, 1994
Foodplants: Unknown

augur (Bouvier, 1929)
Foodplants: *Carpinus betulus, Salix caprea.*

aurata Lemaire, 1994
Foodplants: Unknown

courtini Lemaire, 1994
Foodplants: Unknown

galmeidai C. Mielke & Furtado, 2006
Foodplants: Unknown

hidalgensis Brechlin & Meister, 2010
Foodplants: Unknown

hircia (Cramer, 1775)
afflata (Strand, 1911)
amalia (Stoll, 1782)
calchas (Cramer, 1780)
obtusa (Walker, 1855)
romani (Bryk, 1953)
roseigrisea (Bouvier, 1930)
Foodplants: *Carpinus, Crataegus, Elaeis guineensis, Fagus sylvatica, Malosma laurina, Malus pumila, Populus nigra, Prunus, Quercus ilex, Quercus robur, Robinia pseudoacacia, Rosa, Salix babylonica, Salix caprea.*

moseri C. Mielke & Furtado, 2006
Foodplants: Unknown

nigra (Dognin, 1901)
quaesita (Draudt, 1930)
Foodplants: *Malosma laurina, Prunus, Psidium guajava.*

ockendeni Lemaire, 1995
Foodplants: Unknown

parallela (Schaus, 1921)
Foodplants: Unknown

pascoensis Brechlin & Meister, 2010
Foodplants: Unknown

pessoai C. Mielke & Furtado, 2006
Foodplants: Unknown

porioni Lemaire, 1982
Foodplants: *Malosma laurina*

punoensis Brechlin & Meister, 2010
Foodplants: Unknown

rosacea (Draudt, 1930)
Foodplants: Unknown

rudloffi Brechlin & Meister, 2010
Foodplants: Unknown

tangerini C. Mielke & Furtado, 2006
Foodplants: Unknown

tarapoto Lemaire, 2002
Foodplants: *Fagus sylvatica, Malosma laurina, Pyracantha rogersiana.*

tolimaiana Brechlin & Meister, 2010
Foodplants: Unknown

unicolor (Lemaire, 1977)
Foodplants: Unknown

yungasiana Brechlin & Meister, 2010
Foodplants: Unknown

Prohylesia Draudt, 1929

friburgensis (Schaus, 1915)
Foodplants: Unknown

peruviana Lemaire, 1982
Foodplants: *Phyllostachys nigra*

rosalinda Draudt, 1929
Foodplants: Unknown

zikani Draudt, 1929
Foodplants: Unknown

Pseudautomeris Lemaire, 1967

antioquia (Schaus, 1921)
Foodplants: Unknown

arminicuscoensis Brechlin & Meister, 2010
Foodplants: Unknown

arminiyungasensis Brechlin & Meister, 2010
Foodplants: Unknown

boettgeri Naumann, Brosch & Wenczel, 2005
Foodplants: Unknown

brasiliensis (Walker, 1855)
scapularis (Boisduval, 1875)
Foodplants: *Bambusa, Ricinus communis, Senna bicapsularis.*

chinchipensis Racheli & Racheli, 2006
Foodplants: Unknown

chrisbrechlinae Brechlin & Meister, 2010
Foodplants: Unknown

coronis (Schaus, 1913)
coronoides (Bouvier, 1936)
parametea (Bouvier, 1936)
Foodplants: *Tilia*

erubescens (Boisduval, 1875)
eximius (Bouvier, 1929)
Foodplants: Unknown

fimbridentata (Dognin, 1916)
Foodplants: Unknown

grammivora (Jones, 1908)
deletum (Bouvier, 1929)
Foodplants: *Fagus, Hemarthria compressa, Ligustrum ovalifolium, Quercus robur, Robinia pseudoacacia, Salix caprea.*

hubneri (Boisduval, 1875)
amphirene (Boisduval, 1875)
atra (Bouvier, 1930)
porus (Boisduval, 1875)
quadridentata (Kirby, 1892)
Foodplants: *Crotalaria, Senna corymbosa.*

irene arminirene (Strand, 1920)
paucidentata (Bouvier, 1936)
rectangularis (Bouvier, 1929)
Foodplants: *Robinia pseudoacacia, Salix caprea.*

irene irene (Cramer, 1779)
obliqua (Bouvier, 1929)
roseus (Conte, 1906)
Foodplants: *Coffea, Carpotroche brasiliensis, Erythrina poepiggiana, Passiflora quadrangularis, Saccharum officinarum, Wisteria sinensis.*

lata (Conte, 1906)
brutus (Strand, 1911)
caesar (Bouvier, 1927)
sinuosa (Dognin, 1911)
Foodplants: *Inga, Ligustrum ovalifolium, Malosma laurina, Robinia pseudoacacia, Salix babylonica.*

luteata (Walker, 1865)
conjuncta (Bouvier, 1936)
lucidus (Bouvier, 1929)
nigrosignata (Köhler, 1935)
Foodplants: *Ligustrum ovalifolium, Prunus domestica, Prunus laurocerasus, Quercus × hispanica, Quercus robur, Robinia pseudoacacia.*

ophthalmica (Moore, 1883)
Foodplants: Unknown

pohli Lemaire, 1967
Foodplants: *Rubus fructicosus*

porifera (Strand, 1920)
Foodplants: Unknown

rudloffi Brechlin & Meister, 2010
Foodplants: Unknown

salmonea (Cramer, 1777)
armidia (Hübner, [1819])
arminia (Stoll, 1892)
hersilia (Boisduval, 1875)
metea (Cramer, 1780)
Foodplants: *Citrus, Cornus mas, Cornus sanguinea, Fagus sylvatica, Ligustrum ovalifolium, Malosma laurina, Passiflora quadrangularis, Quercus × hispanica, Robinia pseudoacacia, Salix caprea, Salix viminalis, Schinus terebinthifolia.*

sanmartini Racheli & Racheli, 2006
Foodplants: Unknown

stawiarskii (Gagarin, 1936)
suavina (F. Johnson, 1937)
Foodplants: Unknown

subcoronis Lemaire, 1967
Foodplants: Unknown

toulgoeti Lemaire, 2002
Foodplants: Unknown

yourii Lemaire, 1985
Foodplants: *Bambusa, Erythrina, Inga sapindoides.*

zamora Racheli & Racheli, 2006
Foodplants: Unknown

Pseudodirphia Bouvier, 1928

agis agis (Cramer, 1775)
angulata Bouvier, 1929
aphlebia Bouvier, 1935
Foodplants: *Robinia pseudoacacia, Salix caprea.*

agis inflexa Bouvier, 1930
Foodplants: Unknown

alba (Druce, 1911)
Foodplants: Unknown

albosignatus (Bouvier, 1924)
Foodplants: *Eucalyptus*

alticola Lemaire, 2002
Foodplants: Unknown

andicola Bouvier, 1930
Foodplants: *Inga sapindoides*

biremis (Draudt, 1930)
Foodplants: *Quercus robur, Robinia pseudoacacia.*

boliviana Lampe, 2004
Foodplants: Unknown

catarinensis (Lemaire, 1975)
Foodplants: Unknown

choroniensis (Lemaire, 1975)
Foodplants: Unknown

conjucta Lemaire, 2002
Foodplants: Unknown

cupripuncta Lemaire, 1982
Foodplants: Unknown

ducalis Lemaire, 2002
Foodplants: Unknown

eumedide (Stoll, 1782)
peruviana concava Bouvier, 1929
convexa Bouvier, 1929
griseoalba Bouvier, 1929
Foodplants: *Crataegus curvisepala, Erythrina, Fagus sylvatica, Laburnum anagyroides, Planera aquatica, Quercus ilex, Robinia pseudoacacia, Salix caprea.*

eumedidoides (Vuillot, 1892)
gregatus Bouvier, 1923
Foodplants: *Carpinus betulus, Fagus, Pyracantha coccinea, Quercus × hispanica, Robinia pseudoacacia, Salix babylonica, Salix caprea.*

frickei Meister & Brechlin, 2008
Foodplants: Unknown

guyanensis (Lemaire, 1975)
Foodplants: Unknown

herbuloti (Lemaire, 1975)
Foodplants: Unknown

imperialis (Draudt, 1930)
Foodplants: Unknown

infuscata (Bouvier, 1924)
Foodplants: *Carpinus betulus*

knorkei Meister & Schmidt, 2004
Foodplants: *Rubus fructicosus*

lasca Lemaire, 1996
Foodplants: Unknown

lesieuri Lemaire, 2002
Foodplants: Unknown

marxi Brechlin & Meister, 2008
Foodplants: Unknown

medinensis (Draudt, 1930)
Foodplants: Unknown

menander (Druce, 1886)
reducta (Hering, 1925)
Foodplants: *Acalypha diversifolia, Anthurium, Anthurium consobrinum, Asplundia utilis, Inga sapindoides, Monstera tenuis, Philodendron rhodoaxis, Philodendron sulcatum, Rhodospatha, Spathiphyllum laeve.*

mexicana (Bouvier, 1924)
Foodplants: *Robinia pseudoacacia, Stigmaphyllon lindenianum.*

niceros (Dognin, 1911)
Foodplants: Unknown

obliqua (Bouvier, 1924)
Foodplants: *Canna indica, Robinia pseudoacacia, Salix caprea.*

pallida (Walker, 1865)
apollinairei Bouvier, 1929
aurora (Vuillot, 1892)
Foodplants: Unknown

peruviana (Bouvier, 1924)
obtusa Bouvier, 1935
Foodplants: Unknown

regia (Draudt, 1930)
Foodplants: *Fagus silvatica, Robinia, Tetracera hydrophila, Vitis tiliifolia.*

sanctimartinensis Lemaire, 2002
Foodplants: Unknown

singeri Meister & Brechlin, 2008
Foodplants: Unknown

sinuosa Lemaire, 2002
Foodplants: Unknown

theodorici Lemaire, 1982
Foodplants: Unknown

thiaucourti Lemaire, 1982
Foodplants: *Robinia pseudoacacia*

undulata Lemaire, 2002
Foodplants: Unknown

uniformis (Lemaire, 1975)
Foodplants: *Robinia*, *Salix caprea*.

varia (Walker, 1855)
draudti Bouvier, 1935
Foodplants: *Erythrina*, *Robinia pseudoacacia*, *Rubus fructicosus*.

weritzi Brechlin & Meister, 2008
Foodplants: Unknown

Rhodirphia Michener, 1949

carminata (Schaus, 1902)
niepelti (Draudt, 1930)
Foodplants: Unknown

Travassosula Michener, 1949

mulierata Lemaire, 1971
Foodplants: Unknown

subfumata (Schaus, 1921)
Foodplants: Unknown

Xanthodirphia Michener, 1949

abbreviata Becker & Chacon, 2001
Foodplants: Unknown

amarilla (Schaus, 1908)
Foodplants: Unknown

Polythysanini

Polythysana Walker, 1855

apollina C. Felder & R. Felder, 1874
albescens Bartlett & Calvert, 1886
Foodplants: *Aristotelia chilensis*, *Crinodendron patagua*, *Cryptocarya alba*, *Cryptocarya rubra*, *Eucalyptus*, *Kageneckia oblonga*, *Liquidambar*, *Maytenus boaria*, *Nothofagus antarctica*, *Persea americana*.

cinerascens (Philippi, 1859)
andromeda (Philippi, 1859)
edmondsii Butler, 1882
latchami Ureta, 1943
Foodplants: *Aristotelia chilensis*, *Beilschmiedia miersii*, *Cryptocarya alba*, *Cryptocarya rubra*, *Cupressus macrocarpa*, *Cydonia oblonga*, *Eucalyptus*, *Fagus sylvatica*, *Maytenus boaria*, *Nothofagus antarctica*, *Nothofagus dombeyi*, *Nothofagus obliqua*, *Pinus halepensis*, *Pinus radiata*, *Pseudotsuga menziesii*, *Quercus* × *hispanica*, *Quercus robur*, *Rosa*, *Tsuga canadensis*.

rubrescens (Blanchard, 1852)
rhodoessa (Prittwitz, 1868)
maasseni Schüssler, 1934
watsoni Schüssler, 1934
Foodplants: *Aristotelia chilensis*, *Beilschmiedia miersii*, *Cryptocarya alba*, *Cryptocarya rubra*, *Cupressus macrocarpa*, *Cydonia oblonga*, *Eucalyptus*, *Fagus sylvatica*, *Maytenus boaria*, *Nothofagus antarctica*, *Nothofagus dombeyi*, *Nothofagus obliqua*, *Pinus halepensis*, *Pinus radiata*, *Pseudotsuga menziesii*, *Quercus* × *hispanica*, *Quercus robur*, *Rosa*, *Tsuga canadensis*.

Agliinae

Aglia Ochsenheimer, 1810

homora Jordan, 1911
Foodplants: Unknown

ingens Naumann, Brosch & Nässig, 2003
Foodplants: Unknown

japonica japonica (Leech, [1889])
Foodplants: *Crataegus monogyna*, *Fagus grandifolia* subsp. *grandifolia*, *Padus avium* var. *avium*.

japonica microtau Inoue, 1958
Foodplants: *Padus avium* var. *avium*

tau (Linnaeus, 1758)
amurensis Jordan, 1911
cerretanica Rougeot, 1965
eumense Gómez-Bustillo & Fernández-Vidal, 1980
spaniolissima Gómez-Bustillo & Fernández-Vidal, 1980
tanus (Haworth, 1802)
Foodplants: *Acer pseudoplatanus*, *Alnus cremastogyne*, *Alnus japonica*, *Alnus orientalis*, *Betula pendula*, *Betula platyphylla*, *Betula pubescens*, *Carpinus betulus*, *Castanea crenata*, *Cornus san-*

guinea, Corylus avellana, Crataegus monogyna, Daphne mezereum, Fagus crenata, Fagus longipetiolata, Fagus sylvatica, Ilex verticillata, Juglans regia, Malus domestica, Malus pumila, Padus avium var. *avium, Platanus, Populus nigra, Prunus avium, Prunus serotina, Pyrus communis, Quercus dentata, Quercus robur, Quercus serrata, Salix aurita, Salix babylonica, Salix caprea, Salix fragilis, Sorbus aucuparia, Tilia cordata, Tilia platyphyllos, Ulmus glabra, Ulmus minor.*

Salassinae

Salassa Moore, 1859

albocirculata Naumann, Löffler & Kohll, 2010
Foodplants: Unknown

arianae Brechlin & Kitching, 2010
Foodplants: Unknown

belinda aeos Witt & Pugaev, 2007
Foodplants: Unknown

belinda belinda Witt & Pugaev, 2007
Foodplants: *Padus avium* var. *avium*

bhutanensis Brechlin, 2009
Foodplants: Unknown

cottoni Naumann, Löffler & Kohll, 2010
Foodplants: *Padus avium* var. *avium*

extremorientalis Naumann, Löffler & Kohll, 2010
Foodplants: Unknown

fansipana Brechlin, 1997
Foodplants: *Salix pentandra*

hauensteini Naumann, Löffler & Kohll, 2010
Foodplants: *Alnus glutinosa, Padus avium* var. *avium.*

htayaungi Naumann, Löffler & Kohll, 2010
Foodplants: Unknown

inversa Naumann, Löffler & Kohll, 2010
Foodplants: Unknown

iris Jordan, 1910
Foodplants: *Padus avium* var. *avium, Pyracantha coccinea.*

katschinica Bryk, 1944
Foodplants: Unknown

kitchingi Brechlin, 2010
Foodplants: Unknown

lemaii chiangmaiensis Brechlin & Meister, 2009
Foodplants: Unknown

lemaii lemaii Le Moult, 1933
Foodplants: Unknown

lola (Westwood, 1847)
Foodplants: Unknown

megastica Swinhoe, 1894
Foodplants: Unknown

meisteri Brechlin, 2010
Foodplants: Unknown

mesosa excellens Bryk, 1944
melanops Bryk, 1944
Foodplants: Unknown

mesosa mesosa Jordan, 1910
episcopalis Kaiser, 1917
Foodplants: *Juglans regia, Padus avium* var. *avium, Pyracantha coccinea, Tilia platyphyllos.*

olivacea Oberthür, 1890
Foodplants: Unknown

parakatschinica Brechlin, 2009
Foodplants: Unknown

pararoyi Brechlin, 2009
Foodplants: Unknown

paratonkiniana Brechlin, 2009
Foodplants: *Salix babylonica, Salix caprea.*

royi Elwes, 1887
Foodplants: Unknown

shuyiae Zhang & Kohll, 2008
Foodplants: Unknown

thespis (Leech, 1890)
Foodplants: *Liquidambar styraciflua, Prunus, Rhamnus, Salix caprea.*

tibaliva Zhu & Wang, 1993
Foodplants: Unknown

tonkiniana Le Moult, 1933
Foodplants: *Salix babylonica, Salix caprea.*

viridis Naumann, Löffler & Kohll, 2010
Foodplants: *Frangula dodonei, Liquidambar styraciflua, Padus avium* var. *avium, Pyracantha coccinea.*

Saturniinae

Micragonini

Decachorda Aurivillius, 1898

aspersa Bouvier, 1927
Foodplants: *Chloris virgata, Cyperus alternifolius, Poa arvensis.*

bouvieri Bouvier, 1930
kitalina Bouvier, 1930
Foodplants: *Hyparrhenia*

congolana Bouvier, 1930
Foodplants: Unknown

fletcheri Rougeot, 1970
Foodplants: Unknown

fulvia (Druce, 1886)
mombasana Stoneham, 1962
pomona (Weymer, 1892)
Foodplants: Unknown

inspersa Hampson, 1910
orientalis Bouvier, 1930
talboti Bouvier, 1930
Foodplants: Unknown

pomona (Weymer, 1892)
Foodplants: Unknown

rosea Aurivillius, 1898
aurivillii Bouvier, 1930
gaedei Dufran, 1953
Foodplants: *Hyparrhenia*

seydeli Rougeot, 1970
Foodplants: Unknown

Eosia Le Cerf, 1911

digennaroi Bouyer, 2008
Foodplants: Unknown

insignis Le Cerf, 1911
crenulata (Fawcett, 1915)
Foodplants: Unknown

minettii Bouyer, 2008
Foodplants: Unknown

Goodia Holland, 1893

Subgenus *Campimoptilum* Karsch, 1896

hollandi (Butler, 1898)
boulardi (Rougeot, 1974)
Foodplants: *Acacia dealbata, Aframomum, Amomum.*

kuntzei (Dewitz, 1881)
ansorgei (Kirby, 1896)
hollandi Butler, 1898
ochraceum Aurivillius, 1901
Foodplants: *Acacia baileyana, Acacia dealbata, Acacia sieberiana, Bauhinia galpinii, Bauhinia petersiana, Brachystegia randii, Brachystegia spiciformis, Brachystegia venosa, Deinbollia, Dichrostachys cinerea, Julbernardia globiflora, Julbernardia magnistipulata, Julbernardia paniculata, Millettia, Quercus × hispanica, Pseudoberlinia globiflora, Pseudoberlinia paniculata.*

pareensis Darge, 2008
Foodplants: Unknown

sparsum Darge, 2008
Foodplants: Unknown

Subgenus *Goodia* Holland, 1893

addita Darge, 2008
Foodplants: Unknown

astrica Darge, 1977
Foodplants: Unknown

canui Bouyer, 2004
Foodplants: Unknown

dimonica Darge, 2008
Foodplants: Unknown

egeri Bouyer, 2009
Foodplants: Unknown

falcata (Aurivillius, 1893)
Foodplants: Unknown

hierax Jordan, 1922
Foodplants: Unknown

leonardi Bouyer, 2009
Foodplants: Unknown

lunata Holland, 1893
fulvescens Sonthonnax, 1898
unguiculata Bouvier, 1936
Foodplants: *Aframomum, Amomum, Ficus.*

nyungwensis Bouyer, 2009
Foodplants: Unknown

nubilata Holland, 1893
nodulifera (Karsch, 1893)
sentosa Jordan, 1922
Foodplants: *Aframomum, Amomum.*

obscuripennis Strand, 1912
Foodplants: Unknown

oxytela Jordan, 1922
Foodplants: *Aframomum melegueta*

perfulvastra perfulvastra Darge, 1994
Foodplants: Unknown

perfulvastra veneris Darge, 1994
Foodplants: Unknown

praedicta Bouyer, 2009
Foodplants: Unknown

smithii arabica Rougeot, 1977
Foodplants: Unknown

smithii smithii Holland, 1897
decolor Le Cerf, 1911
heptapora Fawcett, 1915
oriens Hampson, 1909
uniformis Joannis 1913
Foodplants: Unknown

stellata Darge, 1994
Foodplants: Unknown

thia Jordan, 1922
Foodplants: Unknown

Holocerina Pinhey, 1956

Subgenus *Eoholocerina* Cooper & Cooper, 2002

agomensis (Karsch, 1893)
Foodplants: *Bauhinia reticulata, Brachystegia, Laburnum, Monotes, Protea, Uapaca kirkiana, Uapaca nitida.*

Subgenus *Holocerina* Pinhey, 1956

angulata angulata (Aurivillius, 1893)
guineensis (Strand, 1912)
smilax (Schaus, 1893)
Foodplants: *Bauhinia reticulata, Combretum, Ficus afzelii* var. *caillei, Ficus sur, Ligustrum ovalifolium, Maesa lanceolata, Psorospermum, Piliostigma, Shirakiopsis elliptica, Spathodea campanulata.*

angulata istsariensis Stoneham, 1962
Foodplants: *Bauhinia reticulata, Brachystegia, Laburnum, Monotes, Protea, Uapaca kirkiana, Uapaca nitida.*

digennariana Darge 2008
Foodplants: Unknown

intermedia Rougeot, 1978
Foodplants: *Bauhinia reticulata, Brachystegia, Laburnum, Monotes, Protea, Uapaca kirkiana, Uapaca nitida.*

menieri Rougeot, 1973
Foodplants: Unknown

micropteryx (Hering, 1949)
Foodplants: Unknown

nilotica (Jordan, 1922)
Foodplants: Unknown

occidentalis Bouyer, 2008
Foodplants: Unknown

orientalis Bouyer, 2001
Foodplants: *Bauhinia reticulata, Psorospermum, Piliostigma.*

prosti Rougeot, 1978
Foodplants: Unknown

rhodesiensis (Janse, 1918)
Foodplants: *Cussonia spicata*

rougeoti Bouyer, 1997
Foodplants: Unknown

smilax basquini Darge, 1986
Foodplants: *Liquidambar styraciflua, Quercus ilex, Quercus robur.*

smilax smilax (Westwood, 1849)
Foodplants: *Ekebergia capensis, Halleria lucida, Harpephyllum caffrum, Jasminum humile* var. *humile, Jasminum officinale, Ligustrum ovalifolium, Ligustrum vulgare, Malosma laurina, Malus pumila, Nuxia floribunda, Piliostigma, Pinus patula, Protorhus longifolia, Psidium guajava, Quercus gambelii, Quercus robur, Rhus glabra, Rhus laevigata, Rhus trilobata, Rhus typhina, Rhus villosa, Syringa vulgaris.*

wensis Rougerie & Bouyer, 2005
Foodplants: Unknown

Ludia Wallengren, 1865

arguta Jordan, 1922
hansali Butler, 1898
russa Jordan, 1922
Foodplants: *Vernonia*

arida Jordan, 1938
Foodplants: *Asteraceae*

corticea Jordan, 1922
Foodplants: *Asteraceae*

delegorguei (Boisduval, 1847)
Foodplants: *Acacia, Brachylaena discolor, Brachylaena rotundata, Microglossa mespiloides, Microglossa mespilifolia, Robinia pseudoacacia, Salvia runcinata, Senecio deltoideus, Syringa vulgaris, Tarchonanthus camphoratus, Tarchonanthus minor, Uapaca kirkiana, Vernonia natalensis, Vernonia oligocephala.*

dentata (Hampson, 1891)
luciphila Strand, 1911
nyassana Strand, 1911
Foodplants: *Vernonia*

goniata Rothschild, 1907
Foodplants: *Vernonia*

hansalii eximia Rothschild, 1907
Foodplants: *Vernonia*

hansalii festiva Jordan, 1922
Foodplants: *Vernonia*

hansalii hansalii (R. Felder, [1874])
Foodplants: *Tarchonanthus camphoratus*

hansalii tanganyikae Strand, 1911
Foodplants: *Vernonia*

jordani Bouyer, 1997
Foodplants: *Tarchonanthus camphoratus*

obscura apora Jordan, 1922
Foodplants: *Zingiber*

obscura intermedia Jordan, 1922
Foodplants: *Zingiber*

obscura laeta Jordan, 1922
Foodplants: *Zingiber*

obscura obscura Aurivillius, 1893
sopponis Strand, 1914
Foodplants: *Zingiber*

orinoptena leonardo Stoneham, 1962
Foodplants: Unknown

orinoptena monroei Jordan, 1938
Foodplants: Unknown

orinoptena orinoptena Karsch, 1893
dentata Aurivillius, 1904
limbobrunnea Strand, 1911
Foodplants: *Acacia, Bothriocline longipes, Brachylaena discolor, Brachylaena rotundata, Coleus forskohlii, Cynara, Gynura, Hoslundia, Lactuca, Laggera alata, Melissa officinalis, Microglossa mespiloides, Mikania, Ocimum, Plectranthus barbatus, Plectranthus scutellarioides, Robinia pseudoacacia, Salvia runcinata, Senecio deltoideus, Syringa vulgaris, Tarchonanthus camphoratus, Tarchonanthus minor, Uapaca kirkiana, Vernonia natalensis, Vernonia oligocephala.*

orinoptena styx Darge, 1996
Foodplants: Unknown

pseudovetusta Rougeot, 1978
Foodplants: Unknown

pupillata Strand, 1911
Foodplants: Unknown

syngena Jordan, 1922
Foodplants: Unknown

tessmanni Strand, 1911
Foodplants: Unknown

vetusta Strand, 1911
Foodplants: Unknown

Micragone Walker, 1855

agathylla agathylla (Westwood, 1849)
camerunensis (Strand, 1909)
Foodplants: Unknown

agathylla flammostriata Rougeot, 1979
Foodplants: Unknown

allardi Darge, 1990
Foodplants: Unknown

ansorgei ansorgei (Rothschild, 1907)
bilineata (Rothschild, 1907)
trefurthi (Strand, 1909)
Foodplants: *Brachystegia spiciformis, Cercis, Julbernardia paniculata, Pseudoberlinia globiflora, Pseudoberlinia paniculata.*

ansorgei moxicoensis Darge, 2006
Foodplants: Unknown

avinoffi Bouvier, 1936
Foodplants: Unknown

cana cana (Aurivillius, 1893)
junodi (Oberthür, 1910)
Foodplants: *Pappea, Syzygium cordatum, Syzygium guineense, Syzygium owariense.*

cana politzari Basquin, 1989
Foodplants: Unknown

colettae Rougeot, 1959
Foodplants: *Pseudoberlinia paniculata, Brachystegia spiciformis, Julbernardia paniculata.*

ducorpsi (De Fleury, 1925)
Foodplants: *Annona senegalensis*

elisabethae Bouvier, 1930
phantasma Hering, 1930
Foodplants: Unknown

gaetani Bouyer, 2008
Foodplants: Unknown

herilla (Westwood, 1849)
herilloides Bouvier, 1936
Foodplants: *Annona*, *Malosma laurina*.

joiceyi Bouvier, 1930
Foodplants: Unknown

leonardi Bouyer, 2008
Foodplants: Unknown

lichenodes (Holland, 1893)
tholloni Bouvier, 1927
Foodplants: Unknown

martinae Rougeot, 1952
fluminis Darge, 1976
Foodplants: Unknown

mirei Darge, 1990
Foodplants: Unknown

morettoi Darge, 2001
Foodplants: Unknown

morini Rougeot, 1977
Foodplants: Unknown

nenia (Westwood, 1849)
nubifera Bouvier, 1936
Foodplants: *Annona senegalensis*, *Xylopia aethiopica*.

nenioide*s* Rougeot, 1979
Foodplants: *Annona senegalensis*

neonubifera Rougeot, 1979
Foodplants: Unknown

nyasae Rougeot, 1962
Foodplants: *Schinus molle*, *Schinus terebinthifolia*.

remota Darge, 2005
Foodplants: Unknown

rougeriei Bouyer, 2008
Foodplants: Unknown

Orthogonioptilum Karsch, 1893

Subgenus *Adafroptilum* Darge, 2004

acuminatum (Darge, 2003)
Foodplants: Unknown

acutum (Darge, Naumann & Brosch, 2003)
Foodplants: Unknown

austriorientale Darge, 2008
Foodplants: Unknown

bellum (Darge, Naumann & Brosch, 2003)
Foodplants: Unknown

coloratum (Darge, Naumann & Brosch, 2003)
Foodplants: Unknown

convictum Darge, 2007
Foodplants: Unknown

incana (Sonthonnax, 1899)
pancratia (Weymer, 1903)
Foodplants: *Brachystegia spiciformis*

hausmanni Darge, 2007
Foodplants: Unknown

indivisum (Darge, 2003)
Foodplants: Unknown

kalamboensis Darge, 2007
Foodplants: Unknown

kitongaensis Darge, 2006
Foodplants: Unknown

lejorai Darge, 2008
Foodplants: Unknown

mikessensis Darge, 2007
Foodplants: Unknown

occidaneum Darge, 2008
Foodplants: Unknown

permixtum (Darge, 2003)
Foodplants: Unknown

quinquevitreatum Darge, 2008
Foodplants: Unknown

rotundum (Darge, 2003)
Foodplants: Unknown

rougerii Darge, 2006
Foodplants: Unknown

scheveni (Darge, 2003)
Foodplants: Unknown

schmiti Darge, 2006
Foodplants: Unknown

septiguttata (Weymer, 1909)
Foodplants: Unknown

singularum (Darge, Naumann & Brosch, 2003)
Foodplants: Unknown

sommereri Darge, 2008
Foodplants: Unknown

tricoronatum (Darge, Naumann & Brosch, 2003)
Foodplants: Unknown

tuberculatum (Darge, Naumann & Brosch, 2003)
Foodplants: Unknown

Subgenus *Carnegia* (Holland, 1896)

mirabilis (Aurivillius, 1895)
impar (Aurivillius, 1899)
luminosa Bouvierd, 1930
Foodplants: *Annona senegalensis*

Subgenus *Orthogonioptilum* Karsch, 1893

adamaoensis Darge, 1988
Foodplants: Unknown

adiegetum adiegetum Karsch, 1893
neodollmani Darge, 1988
pseudadiegetum Darge, 1988
restrictum Darge, 1988
Foodplants: Unknown

adiegetum aequinoctalis Darge, 2008
Foodplants: Unknown

adiegetum baftuensis Darge, 1988
Foodplants: Unknown

adiegetum dollmani Jordan, 1922
Foodplants: Unknown

adiegetum mucronatum Darge, 1988
Foodplants: Unknown

adiegetum ugandensis Darge, 1988
Foodplants: Unknown

adustum Jordan, 1922
arnoldi (Hering, 1932)
Foodplants: Unknown

andreasum Rougeot, 1967
Foodplants: Unknown

apium Basquin, 1995
Foodplants: Unknown

bergeri Rougeot, 1962
Foodplants: Unknown

bernardii Bouyer, 1990
Foodplants: Unknown

bernaudi Darge, 1996
Foodplants: Unknown

bimaculatum Rougeot, 1972
Foodplants: Unknown

bouyeri Darge, 1990
Foodplants: Unknown

brunneum Jordan, 1922
Foodplants: Unknown

caecum Darge, 2003
Foodplants: Unknown

chalix Jordan, 1922
Foodplants: Unknown

chaminadei Darge, 1996
Foodplants: Unknown

concordia Bouyer, 1992
silvaticum Darge, 1992
Foodplants: Unknown

crystallinum Darge, 1993
Foodplants: Unknown

csomaense Bouyer, 1992
Foodplants: Unknown

dallastai Darge, 1988
Foodplants: Unknown

dargei Basquin, 1992
Foodplants: Unknown

deletum Jordan, 1922
Foodplants: Unknown

diabolicum Rougeot, 1971
Foodplants: Unknown

emmanuellae Basquin, 1995
Foodplants: Unknown

falcatissimum Rougeot, 1971
Foodplants: Unknown

fang Darge, 1990
Foodplants: Unknown

filippii Darge, 1992
Foodplants: Unknown

fontainei Rougeot, 1962
hodeberti Darge, 1988
Foodplants: Unknown

galleyi Basquin, 1992
Foodplants: Unknown

garmsi Bouyer, 1995
Foodplants: Unknown

herbuloti Darge, 1988
Foodplants: Unknown

ianthinum Rougeot, 1960
Foodplants: Unknown

infernarum Darge, 1990
Foodplants: Unknown

infinitum Darge, 1992
Foodplants: Unknown

kahli Holland, 1921
Foodplants: Unknown

kasaiensis Dufrane, 1953
Foodplants: Unknown

kivuensis Bouyer, 1990
Foodplants: Unknown

klinzigi Darge, 1988
Foodplants: Unknown

lemairei Darge, 1986
Foodplants: Unknown

loloense Basquin, 1995
Foodplants: Unknown

luminosum (Bouvier, 1930)
guilloteaui Darge, 1988
vitreatum Rougeot, 1958
Foodplants: Unknown

modestum Rougeot, 1965
Foodplants: Unknown

monochromum Karsch, 1893
Foodplants: Unknown

neglectum Darge, 1995
Foodplants: Unknown

neoprox Darge, 1992
Foodplants: Unknown

nigrescens Bouyer, 1994
magnificum Basquin, 1994
Foodplants: Unknown

nimbaense Rougeot, 1962
Foodplants: Unknown

obamba Darge, 1990
Foodplants: Unknown

occidentalis Bouyer, 1995
Foodplants: Unknown

ochraceum Rougeot, 1958
conspectum Rougeot, 1962
exspectatum Darge, 1975
Foodplants: Unknown

oremansi Darge, 2008
Foodplants: Unknown

pancratia (Weymer, 1903)
Foodplants: Unknown

paveci Darge, 1992
Foodplants: Unknown

perarcuatum Darge, 1993
Foodplants: Unknown

piersoni Bouyer, 1990
Foodplants: Unknown

prox Karsch, 1893
geniculipennis Strand, 1910
Foodplants: Unknown

rougeoti Darge, 1973
Foodplants: Unknown

sejunctum Darge, 1993
Foodplants: Unknown

silvaticum Darge, 1992
Foodplants: Unknown

solium Bouyer, 1990
Foodplants: Unknown

subueleense Rougeot, 1972
Foodplants: Unknown

tristis (Sonthonnax, 1899)
Foodplants: Unknown

ueleense Rougeot, 1967
Foodplants: Unknown

umbrulatum Basquin, 1995
Foodplants: Unknown

vestigiata (Holland, 1893)
servatia Weymer, 1909
olivescens Kiriakoff, 1963
Foodplants: Unknown

violascens garnieri Bouyer, 1987
Foodplants: Unknown

violascens violascens Rebel, 1914
Foodplants: Unknown

Subgenus *Pseudoludia* Strand, 1911

nyungwe Bouyer, 1987
Foodplants: *Vernonia*

suavis (Rothschild, 1907)
lilacina (Weymer, 1907)
Foodplants: Unknown

Subgenus *Vegetia* Karsch, 1893

dewitzi (Maassen & Weymer, 1885)
Foodplants: *Eriocephalus ericoides*

ducalis Jordan, 1922
Foodplants: *Artemisia californica, Eriocephalus africanus, Eriocephalus ericoides, Eriocephalus microcephalus, Eriocephalus umbellatus, Eriosema.*

grimmia (Geyer, 1831)
Foodplants: *Eriocephalus racemosus*

legraini Bouyer, 2004
Foodplants: *Eriocephalus ericoides*

Urotini

Antistathmoptera Tams, 1935

daltonae daltonae Tams, 1935
Foodplants: *Cordyla africana, Cordyla pinnata.*

daltonae granti Bouyer, 2006
Foodplants: *Cordyla pinnata*

daltonae rectangulata Pinhey, 1968
Foodplants: *Cordyla pinnata*

Eudaemonia Hübner, [1819]

argiphontes Kirby, 1877
barnsi Joicey & Talbot, 1921
batesi Bethune-Baker, 1927
Foodplants: *Albizia*

argus (Fabricius, 1781)
brachyura (Drury, 1782)
minor Holland, 1920
uroarge Hübner, [1819]
Foodplants: *Dialium guineense*

troglophylla hartfordi Roufeot, 1962
Foodplants: Unknown

troglophylla troglophylla Hampson, 1919
colini Bouvier, 1927
ghesquierii Talbot, 1925
pohli Dufrane, 1953
Foodplants: Unknown

Maltagorea Bouyer, 1993

andriai (Griveaud, 1962)
Foodplants: Unknown

ankaratra (Viette, 1954)
Foodplants: Unknown

auricolor auricolor (Mabille, 1879)
leporina (Oberthür, 1927)
Foodplants: Unknown

auricolor variabile (Viette, 1982)
Foodplants: Unknown

basquini Rougerie, 2003
Foodplants: Unknown

cincta (Mabille, 1879)
Foodplants: Unknown

dentata (Griveaud, 1962)
Foodplants: Unknown

dubiefi Bouyer, 2006
Foodplants: Unknown

dura (Keferstein, 1870)
lambertoni (Bouvier, 1927)
Foodplants: Unknown

fusicolor (Mabille, 1879)
cambouei (Oberthür, 1916)
fumosa (Sonthonnax, 1901)
madagascariensis (Sonthonnax, 1899)
subocellata (Butler, 1880)
Foodplants: *Aguiaria excelsa*

griveaudi Bouyer, 1996
Foodplants: Unknown

monsarrati (Griveaud, 1968)
Foodplants: Unknown

ornata (Griveaud, 1962)
Foodplants: Unknown

pseudovulpina Basquin & Rougerie, 2009
Foodplants: Unknown

rostaingi (Griveaud, 1962)
Foodplants: Unknown

rubriflava (Griveaud, 1962)
Foodplants: Unknown

sogai (Griveaud, 1962)
Foodplants: Unknown

vulpina (Butler, 1882)
subrufa (Griveaud, 1962)
Foodplants: Unknown

Parusta W. Rothschild, 1907

thelxinoe Fawcett, 1915
Foodplants: Unknown

xanthops Rothschild, 1907
Foodplants: Unknown

Pselaphelia Aurivillius, 1904

arenivaga Darge 2003
Foodplants: Unknown

arata Bouyer,1992
Foodplants: *Aframomum*

flavivitta (Walker, 1862)
conspersa (Aurivillius, 1893)
dentifera (Maassen & Weymer 1886)
Foodplants: *Aframomum, Nectandra lineata, Trichilia dregeana, Trichilia emetica, Trichilia martiana.*

gemmifera (Butler, 1878)
deficiens Dufrane, 1953
marginata Dufrane,1953
mariae Dufrane, 1953
pulverulenta Dufrane, 1953
Foodplants: *Aframomum*

kitchingi Darge 2007
Foodplants: Unknown

laclosi Darge 2002
Foodplants: Unknown

mariaetheresae Darge 2002
Foodplants: Unknown

neglecta Darge 2003
Foodplants: Unknown

noellae Bouyer, 2008
Foodplants: Unknown

oremansi Darge, 2008
Foodplants: Unknown

vandenberghei Bouyer, 1992
Foodplants: Unknown

vingerheodti Bouyer, 2008
Foodplants: Unknown

Pseudantheraea Weymer, 1892

discrepans (Butler, 1878)
arenosa Packard, 1914
arnobia (Westwood, 1881)
daphne Weymer, 1892
Foodplants: *Caloncoba welwitschii, Coelocaryon botryoides, Fagus, Mangifera, Ricinodendron heudelotii, Salix babylonica, Salix caprea, Terminalia, Uapaca guineensis.*

imperator Rougeot, 1962
Foodplants: *Fagus, Mangifera, Terminalia, Uapaca.*

Pseudaphelia Kirby, 1892

ansorgei Rothschild, 1892
Foodplants: Unknown

apollinaris (Boisduval, 1847)
alberici Dufrane, 1953
balanoal (Guérin-Méneville, 1847)
barotsina Rothschild, 1907
dialitha Tams, 1930
paleacea (Herrich-Schäffer, [1855])
Foodplants: *Combretum, Turraea floribunda, Turraea heterophylla, Turraea nilotica, Turraea obtusifolia, Turraea zambesica.*

flavomarginata Gaede, 1915
draudti Bryk, 1939
flava Bouvier, 1930
luteola Bouvier, 1930
Foodplants: Unknown

karemii Bouvier, 1927
Foodplants: Unknown

roseibrunnea Gaede, 1927
carnea Bouvier, 1930
mannowiana Bouvier, 1930
Foodplants: Unknown

simplex Rebel, 1906
basiflava Joicey & Talbot, 1924
marginilinea Joicey & Talbot, 1924
marginilutea Bouvier, 1927
ochracea Joicey & Talbot, 1924
rebeli Gaede, 1915
Foodplants: Unknown

Sinobirma Bryk, 1944

malaisei Bryk, 1944
Foodplants: Unknown

Tagoropsis C. Felder & R. Felder, 1874

flavinata flavinata (Walker, 1865)
natalensis C. Felder & R. Felder, 1874
Foodplants: *Allophylus africanus, Allophylus dregeanus, Allophylus melanocarpus, Allophylus natalensis.*

flavinata septentrionalis Bouyer, 2000
Foodplants: *Allophylus africanus, Allophylus melanocarpus.*

genoviefae Rougeot, 1950
centralis Dufrane, 1953
Foodplants: *Allophylus africanus, Allophylus melanocarpus.*

hanningtonii (Butler, 1883)
juncta Strand, 1911
nzoiana Stoneham, 1933
songeana Strand, 1911
Foodplants: *Allophylus africanus, Bauhinia.*

hecqui Bouyer, 1990
Foodplants: *Allophylus africanus, Allophylus melanocarpus.*

rougeoti Fletscher, 1952
Foodplants: *Allophylus africanus, Allophylus melanocarpus.*

sabulosa Rothschild, 1907
Foodplants: *Allophylus africanus*

Tagoropsiella Darge, 2008

expansa Darge, 2008
Foodplants: Unknown

ikondae (Rougeot, 1973)
Foodplants: Unknown

kagurensis Darge, 2008
Foodplants: Unknown

mbiziensis Darge, 2008
Foodplants: Unknown

mulanjensis Darge, 2008
Foodplants: Unknown

murphyi (Bouyer, 2000)
Foodplants: Unknown

nyikensis (Bouyer, 2000)
Foodplants: Unknown

rungwensis rungwensis Darge, 2008
Foodplants: Unknown

rungwensis tracenlis (Bouyer, 2002)
centralis (Bouyer, 2002)
Foodplants: Unknown

upembana Bouyer, 2008
Foodplants: Unknown

lupina (Rothschild, 1907)
Foodplants: Unknown

Urota Westwood, 1849

conjuncta Bouvier, 1930
Foodplants: Unknown

centralis Bouyer, 2008
Foodplants: *Erythrina abyssinica, Erythrina abyssinica* subsp. *abyssinica, Erythrina caffra, Erythrina lysistemon, Eucalyptus, Laburnum, Robinia pseudoacacia, Sclerocarya birrea* subsp. *caffra.*

herbuloti Darge, 1975
Foodplants: Unknown

melichari Bouyer, 2008
Foodplants: Unknown

sinope Westwood, 1849
Foodplants: *Erythrina abyssinica, Erythrina abyssinica* subsp. *abyssinica, Erythrina caffra, Erythrina lysistemon, Eucalyptus, Laburnum, Robinia pseudoacacia, Sclerocarya birrea* subsp. *caffra.*

Usta Wallengren, 1863

angulata Rothschild, 1895
Foodplants: *Commiphora woodii, Melia azedarach, Schinus molle, Sclerocarya birrea, Sclerocarya birrea* subsp. *caffra.*

terpsichore alba Terral & Lexqueux, 1991
Foodplants: Unknown

terpsichore abyssinica Aurivillius, 1898
Foodplants: Unknown

terpsichore grantae Terral & Lexqueux, 1991
Foodplants: Unknown

terpsichore subangulata Bouvier, 1930
Foodplants: *Commiphora woodii, Melia azedarach, Schinus molle, Sclerocarya birrea, Sclerocarya birrea* subsp. *caffra.*

terpsichore terpsichore Maassen & Weymer, 1885
Foodplants: *Commiphora woodii, Melia azedarach, Schinus molle, Sclerocarya birrea, Sclerocarya birrea* subsp. *caffra.*

terpsichore yaere Darge, 1994
Foodplants: *Commiphora woodii, Melia azedarach, Schinus molle, Sclerocarya birrea, Sclerocarya birrea* subsp. *caffra*

wallengrenii (C. Felder, 1859)
biplaga Rebel, 1912
terpsichorina (Westwood, 1881)
Foodplants: *Commiphora africana, Commiphora angolensis, Commiphora glandulosa, Commiphora glaucescens, Commiphora saxicola, Commiphora tenuipetiolata, Commiphora virgata, Schinus terebinthifolia, Sclerocarya birrea.*

Bunaeini

Athletes Karsch, 1896

albicans Rougeot, 1955
Foodplants: Unknown

ethra ethra (Westwood, 1849)
Foodplants: *Eucalyptus*

ethra ogoouensis Rougeot, 1962
Foodplants: Unknown

gigas Sonthonnax, 1903
steindachneri Rebel, 1904
Foodplants: *Acacia xanthophloea, Brachystegia randii, Brachystegia spiciformis, Brachystegia taxifolia, Cercis siliquastrum, Eucalyptus gunnii, Julbernardia paniculata, Pseudoberlinia globiflora, Quercus × hispanica.*

nyanzae bousqueti Darge, 1990
Foodplants: Unknown

nyanzae nyanzae Rebel, 1904
Foodplants: Unknown

semialba (Sonthonnax, 1904)
Foodplants: *Acacia polyacantha, Brachystegia randii, Brachystegia longifolia, Brachystegia spiciformis, Cassia, Julbernardia globiflora, Julbernardia paniculata, Prunus laurocerasus, Pyracantha coccinea, Quercus × hispanica, Quercus robur, Pseudoberlinia globiflora.*

Aurivillius Packard, 1902

arata (Westwood, 1849)
horsini Bouvier, 1927
Foodplants: *Acacia baileyana, Albizia adianthifolia, Albizia gummifera, Carpinus betulus, Gleditsia triacanthos, Malus »hillieri«, Parkia biglobosa, Peltophorum africanum, Newtonia buchanani, Quercus robur, Quercus × hispanica, Salix caprea.*

cadioui Bouyer, 2008
Foodplants: *Albizia, Quercus × hispanica.*

drumonti Bouyer, 2008
Foodplants: Unknown

fusca (Rothschild, 1895)
divaricatus Bouvier, 1927
Foodplants: *Albizia adianthifolia, Albizia gummifera, Calliandra haematocephala, Newtonia hildebrandtii, Parkia biglobosa, Peltophorum africanum, Quercus × hispanica.*

jolyanorum Bouvier, 1999
Foodplants: *Albizia, Quercus × hispanica.*

oberthueri Bouvier, 1927
Foodplants: *Quercus robur, Quercus × hispanica.*

orientalis Bouyer, 2007
Foodplants: *Albizia adianthifolia, Albizia gummifera, Newtonia buchanani, Parkia biglobosa, Peltophorum africanum, Salix caprea.*

seydeli Rougeot, 1962
Foodplants: *Albizia, Quercus × hispanica, Quercus robur.*

triramis Rothschild, 1907
Foodplants: *Acacia baileyana, Acacia dealbata, Albizia, Gleditsia triacanthos, Quercus × hispanica, Quercus robur.*

xerophilus Rougeot, 1977
cliftoni Rougeot, 1977
Foodplants: Unknown

Bunaea Hübner, [1819]

alcinoe alcinoe (Cramer, 1780)
alberici Dufrane, 1953
atlanta Dufrane, 1953
barbertonia (Oberthür, 1910)
caffra (Hübner, [1819])
caffraria (Stoll, 1790)
debeeri Oberthür, 1910
durbania (Oberthür, 1910)
katangensis Dufrane, 1953
nyctalops (Wallengren, 1860)
punctigera (Wallengren, 1860)
tricolor Rothschild, 1895
zanguebarica (Oberthür, 1910)
Foodplants: *Acacia auriculiformis, Anthocleista schweinfurthii, Balanites aegyptica, Bauhinia reticulata, Bauhinia thonningii, Celtis, Crataegus,*

Croton, Cussonia spicata, Dacryodes edulis, Ekebergia benguelensis, Ekebergia capensis, Gymnosporia senegalensis, Harpephyllum caffrum, Khaya senegalensis, Ligustrum ovalifolium, Ligustrum vulgare, Liquidambar styraciflua, Maesa lanceolata, Malus pumila, Mangifera indica, Persea americana, Prunus serotina, Quercus × hispanica, Quercus ilex, Quercus robur, Rhus glabra, Rosa, Sarcocephalus latifolius, Terminalia catappa.

alcinoe nubica Jordan, 1910
Foodplants: Unknown

aslauga (Kirby, 1877)
alcinoina Oberthür, 1910
cambouei Oberthür, 1916
densilineata Oberthür, 1916
diospyri (Mabille, 1879)
oberthueri Bouvier, 1926
plumicornis Butler, 1882
urania Oberthür, 1916
vulpes Oberthür, 1916
Foodplants: *Ligustrum ovalifolium, Ligustrum vulgare, Prunus spinosa.*

oremansi Bouyer, 2008
Foodplants: Unknown

Bunaeopsis Bouvier, 1927

angolanus (Le Cerf, 1918)
Foodplants: Unknown

annabellae Lemaire& Rougeot, 1975
Foodplants: *Dolichos*

ansorgei (Rothschild, 1898)
fenestricatula (Strand, 1911)
schoenheiti (Wichgraf, 1906)
Foodplants: Unknown

arabella (Aurivillius, 1893)
Foodplants: *Bridelia micrantha, Chloris, Sporobolus.*

aurantiaca aurantiaca (Rothschild, 1895)
Foodplants: *Bridelia micrantha, Hyperthelia, Imperata cylindrica*

aurantiaca thyene (Weymer, 1896)
Foodplants: Unknown

bomfordi Pinhey, 1962
Foodplants: *Dolichos*

clementi Lemaire& Rougeot, 1975
Foodplants: *Dolichos*

chromata Darge, 2003
Foodplants: Unknown

ferruginea Bouvier, 1927
Foodplants: Unknown

fervida Darge, 2003
Foodplants: Unknown

francottei balanitis Darge, 1993
Foodplants: Unknown

francottei francottei Darge 1992
Foodplants: Unknown

francottei neoparilis Darge, 1993
Foodplants: Unknown

gabonica Rougeot, 1955
Foodplants: *Bridelia micrantha*

hersilia (Westwood, 1849)
dido (Maassen & Weymer, 1881)
editha (Wichgraf, 1914)
elisa (Wichgraf, 1914)
hersilioides (De Fleury, 1924)
nigericola (Strand, 1916)
pagei Rougeot, 1962
rendalli (Rothschild, 1896)
rendalliana (Gaede, 1927)
Foodplants: *Annona senegalensis, Bambusa, Chrysophyllum, Englerophytum, Hyparrhenia bracteata, Irvingia, Pennisetum.*

jacksoni jacksoni (Jordan, 1908)
carabella (Strand, 1911)
oenopa (Grünberg, 1910)
rosea Bouvier, 1927)
Foodplants: *Scleria rehmanii*

jacksoni observabilis Darge, 1993
Foodplants: Unknown

licharbas (Maassen & Weymer, 1885)
Foodplants: *Fargesia murielae, Quercus, Salix babylonica, Salix caprea.*

lueboensis Bouvier, 1931
Foodplants: Unknown

macrophthalma (Kirby, 1881)
Foodplants: Unknown

murphyi Bouyer, 2007
Foodplants: Unknown

oubie (Guérin-Méneville, [1849])
aethiopica (Le Cerf, 1911)
afraorientalis Stoneham, 1959)
argillosa (Le Cerf, 1911)
callista (Le Cerf, 1927)
istsariensis (Stoneham, 1959)

neuvillei (Le Cerf, 1911)
rothschildi (Le Cerf, 1911)
tanganyikae Schüssler, 1927
Foodplants: *Euclea, Fagus, Pycreus muricatus, Salix caprea.*

phidias (Weymer, 1908)
maasseni (Strand, 1911)
Foodplants: Unknown

princeps aurata Rougeot, 1972
Foodplants: Unknown

princeps princeps (Le Cerf, 1918)
callichroma (Le Cerf, 1918)
Foodplants: *Dolichos*

saffronica Pinhey, 1972
Foodplants: Unknown

scheveniana Lemaire & Rougeot, 1974
Foodplants: Unknown

terrali Darge, 1993
Foodplants: Unknown

vau (Fawcett, 1915)
Foodplants: Unknown

zaddachii (Dewitz, 1879)
Foodplants: *Euclea, Heteropogon, Sporobolus.*

Cinabra Sonthonnax, 1901

bracteata (Distant, 1897)
bracteana (Sonthonnax, 1901)
Foodplants: *Acacia baileyana, Brachystegia randii, Brachystegia spiciformis, Brachystegia venosa, Ekebergia, Julbernardia, Protea, Prunus, Pseudoberlinia paniculata, Robinia pseudoacacia.*

hyperbius (Westwood, 1881)
gaedei Schüssler, 1927
Foodplants: *Acacia baileyana, Brachystegia randii, Brachystegia spiciformis, Brachystegia venosa, Ekebergia, Julbernardia paniculata, Protea, Prunus serotina, Pseudoberlinia paniculata, Quercus × hispanica, Robinia pseudoacacia*

kitalei clementi Darge, 1975
Foodplants: Unknown

kitalei kitalei Bouvier, 1930
Foodplants: Unknown

Cirina Walker, 1855

butyrospermi Vuillot, 1911
Foodplants: Unknown

forda amieti Darge, 1975
Foodplants: Unknown

forda forda (Westwood, 1849)
cana (C. Felder & R. Felder, 1874)
cervina (Westwood, 1881)
invenustus (Wallengren, 1860)
orientalis Bouvier, 1927
semicaeca (Walker, 1855)
similis (Distant, 1897)
venusta (Walker, 1865)
Foodplants: *Acacia nilotica, Albizia versicolor, Aloe, Betula, Burkea africana, Carissa macrocarpa, Carissa spinarum, Crossopteryx febrifuga, Cylicodiscus gabunensis, Cynometra inaequifolia, Ekebergia capensis, Erythrophleum africanum, Erythrophleum suaveolens, Euclea undulata, Faidherbia albida, Hippobromus pauciflorus, Malosma laurina, Malus »hillieri«, Malus pumila, Pappea capensis, Protorhus longifolia, Prunus, Quercus × hispanica, Schinus molle, Sideroxylon inerme, Warburgia.*

Eochroa C. Felder & R. Felder, 1874

trimenii (C. Felder & R. Felder 1874)
Foodplants: *Melianthus dregeanus* subsp. *insignis, Melianthus dregianus, Melianthus minor, Melianthus pectinatus.*

Gynanisa Walker, 1855

albescens Sonthonnax, 1904
carcassoni Rougeot, 1974
Foodplants: *Acacia, Brachystegia venosa, Julbernardia, Khaya nyasica.*

arba (Darge, 2008)
Foodplants: *Acacia, Colophospermum mopane.*

ata Strand, 1911
daula Tams, 1930
Foodplants: *Acacia, Brachystegia venosa, Quercus × hispanica, Quercus robur, Julbernardia, Khaya nyasica.*

basquini Bouyer, 2008
Foodplants: Unknown

commixta (Darge, 2008)
Foodplants: Unknown

germeauxi festa Rougeot, 1978
Foodplants: *Acacia, Colophospermum mopane.*

germeauxi germeauxi Rougeot, 1973
Foodplants: *Acacia, Colophospermum mopane.*

germeauxi hecqui Darge, 1992
Foodplants: *Acacia, Colophospermum mopane.*

jama Rebel, 1915
minettii Darge, 2002
Foodplants: *Acacia tortilis, Quercus × hispanica.*

kenya (Darge, 2008)
Foodplants: Unknown

maja maja (Klug, 1836)
campiona Girard, 1882
campionea (Signoret, 1845)
isis (Westwood, 1841)
isis Walker, 1855
macromaja Strand, 1920
simplex Bouvier, 1927
tata (Wallengren, 1860)
Foodplants: *Acacia baileyana, Acacia erioloba, Acacia karroo, Acacia mearnsii, Acacia pubescens, Brachystegia venosa, Cassia, Colophospermum mopane, Elephantorrhiza elephantina, Julbernardia, Laburnum, Malus »hillieri«, Peltophorum, Prunus persica, Pseudoberlinia paniculata, Pterocarpus, Quercus gambelii, Quercus × hispanica, Quercus robur, Robinia pseudoacacia, Schotia brachypetala.*

maja terrali Bouyer, 2001
Foodplants: *Acacia, Colophospermum mopane.*

meridei (Darge, 2008)
Foodplants: Unknown

murphyi Bouyer, 2001
Foodplants: Unknown

nigra Bouvier, 1927
Foodplants: *Acacia erioloba, Acacia karroo, Acacia pubescens, Brachystegia venosa, Cassia, Colophospermum mopane, Elephantorrhiza elephantina, Julbernardia, Laburnum, Peltophorum, Prunus persica, Pseudoberlinia paniculata, Pterocarpus, Quercus gambelii, Quercus × hispanica, Quercus robur, Robinia pseudoacacia, Schotia.*

thiryi Darge, 1992
Foodplants: *Acacia, Brachystegia venosa, Julbernardia, Khaya nyasica.*

uganda (Darge, 2008)
Foodplants: Unknown

westwoodi Rothschild, 1895
Foodplants: *Acacia, Acacia tortilis, Brachystegia venosa, Julbernardia, Khaya nyasica, Quercus × hispanica, Quercus robur.*

zimba (Darge, 2008)
Foodplants: Unknown

Gonimbrasia Butler, 1878

Subgenus *Gonimbrasia* Butler, 1878

abayana (Rougeot, 1977)
Foodplants: Unknown

alcestris (Weymer, 1906)
scheveni Lemaire & Rougeot, 1974
Foodplants: *Rubus fructicosus*

balachowskyi Rougeot, 1973
Foodplants: Unknown

belina (Westwood, 1849)
annulata (Bouvier, 1936)
cocaulti Darge & Terral, 1992
cytherea (Hübner, 1806)
helena (Kruck, 1939)
huebneri (Kirby, 1877)
junodi Oberthür, 1910
sardane Sonthonnax, 1911
ufipana Strand, 1911
ukerewensis (Rebel, 1905)
Foodplants: *Burkea africana, Carissa macrocarpa, Colophospermum mopane, Copaifera, Diospyros, Eucalyptus gunnii, Ficus, Ficus microcarpa, Malosma laurina, Malus »hillieri«, Mangifera indica, Medicago, Ozoroa crassinervia, Ozoroa longipes, Pinus radiata, Platanus, Protorhus longifolia, Quercus ilex, Rhus lancea, Salix caprea, Schotia brachypetala, Sclerocarya birrea, Sclerocarya birrea* subsp. *caffra, Terminalia schweinfurthi, Terminalia velutina, Trema bracteolata, Trema orientalis.*

birbiri (Bouvier, 1929)
Foodplants: Unknown

congolensis Bouvier, 1927
Foodplants: *Diplorhynchus condylocarpon, Erythrophleum africanum, Julbernardia paniculata.*

conradsi (Rebel, 1905)
bijuga (Bouvier, 1929)
jefferyi (Bouvier, 1930)
ochracea (Rebel, 1917)
parva (Bouvier, 1930)
ruandana Grünberg, 1927
Foodplants: *Mangifera indica, Corymbia ficifolia, Quercus robur, Rhus.*

ellisoni Lemaire, 1962
Foodplants: Unknown

felderi (Rothschild, 1895)
Foodplants: Unknown

fletcheri Rougeot, 1977
Foodplants: Unknown

fucata Rougeot, 1978
Foodplants: Unknown

godarti Lemaire, 1971
Foodplants: Unknown

hecate Rougeot, 1955
Foodplants: *Annona, Bauhinia, Berlinia, Combretum psidioides, Ligustrum ovalifolium, Liquidambar styraciflua, Malus »hillieri«, Rhus typhina, Terminalia.*

hoehnelii (Rogenhofer, 1891)
Foodplants: *Calluna vulgaris, Malus »hillieri«, Prunus laurocerasus.*

miranda Darge, 2005
Foodplants: Unknown

occidentalis Rothschild, 1907
vinosa (Riel, 1910)
Foodplants: *Malus »hillieri«, Quercus × hispanica.*

osiris (Druce, 1896)
alicia Stoneham, 1933
aniger Stoneham, 1962
chevalia Stoneham, 1933
deborah (Weymer, 1896)
Foodplants: *Anacardium occidentale, Carpinus betulus, Fagus sylvatica, Quercus ilex, Quercus robur, Rhus typhina, Terminalia schweinfurthi.*

pales (Weymer, 1908)
Foodplants: Unknown

rectilineata Sonthonnax, 1899
kasaiensis (Bouvier, 1927)
mpalensis Sonthonnax, 1899
richelmanni Weymer, 1908
Foodplants: *Diplorhynchus condylocarpon, Erythrophleum africanum, Julbernardia paniculata, Salix, Syzygium owariense, Uapaca kirkiana.*

tyrrhea (Cramer, 1776)
rubrescens Rebel, 1931
Foodplants: *Acacia karroo, Acacia pubescens, Crataegus, Diospyros lycioides, Euclea pseudebenus, Malosma laurina, Malus »hillieri«, Malus pumila, Ozoroa concolor, Padus virginiana, Prunus laurocerasus, Quercus × hispanica, Quercus robur, Salix babylonica, Salix caprea, Salix cinerea, Schinus molle.*

zambesina (Walker, 1865)
barcas (Maassen & Weymer, 1881)
rectilinea (Oberthür, 1910)
ringleri Wichgraf, 1906
said (Oberthür, 1878)
zambesia (C. Felder & R. Felder, 1874)
zanguebarica Oberthür, 1910
Foodplants: *Diospyros mespiliformis, Eucalyptus gunnii, Fagus, Liquidambar styraciflua, Malosma laurina, Malus »hillieri«, Mangifera, Nerium oleander, Poinciana, Quercus × hispanica, Quercus robur, Rhus typhina, Salix caprea, Salix cinerea, Strychnos potatorum.*

Subgenus *Nudaurelia* Rothschild, 1895

affinis Bouvier, 1927
Foodplants: Unknown

alcestris Weymer, 1906
Foodplants: Unknown

alopia (Westwood, 1849)
antigone Staudinger, 1904
intermiscens (Walker, 1869)
obliqua Bouvier, 1930
rhodophila (Walker, 1869)
sonthonnaxi Weymer, 1906
waterloti Bouvier, 1930
Foodplants: *Salix caprea, Trema orientalis.*

anna allardi Rougeot & Allard, 1966
Foodplants: Unknown

anna anna (Maassen & Weymer, 1885)
Foodplants: *Hyphaene coriacea, Hyphaene thebaica, Washingtonia filifera.*

anna mirei Darge 1973
Foodplants: *Hyphaene thebaica*

antelata Darge, 2003
Foodplants: Unknown

anthina (Karsch, 1892)
michaelae Darge, 1975
persephone Staudinger, 1901
preussi Staudinger, 1901
Foodplants: *Mangifera, Polyscias fulva.*

bamendana (Schultze, 1914)
Foodplants: *Maesa lanceolata*

belayneshae Rougeot, 1978
Foodplants: Unknown

bicolor Bouvier, 1930
Foodplants: Unknown

bouvieri (Le Moult, 1933)
Foodplants: Unknown

broschi Darge, 2002
Foodplants: Unknown

bubo Bouvier, 1930
clarki Geertsma, 1971
Foodplants: *Pinus radiata, Rhus tumulicola.*

capdevillei Rougeot, 1979
Foodplants: *Malosma laurina, Prunus laurocerasus, Quercus robur, Quercus × hispanica, Rhus tumulicola, Salix.*

carnegiei Janse, 1918
Foodplants: *Pseudoberlinia globiflora, Protea, Quercus × hispanica.*

cleoris (Jordan 1910)
Foodplants: *Crataegus, Ligustrum ovalifolium, Prunus, Rosa.*

cytherea (Fabricius, 1775)
capensis (Cramer, 1780)
pauper Bouvier, 1927
unicolor Bouvier, 1930
Foodplants: *Acacia cyclops, Acacia pubescens, Acacia saligna, Cupressus macrocarpa, Crataegus, Cydonia, Eucalyptus globulus, Euclea polyandra, Euclea racemosa, Euclea racemosa* subsp. *daphnoides, Leucospermum cuneiforme, Liriodendron tulipifera, Malus domestica, Malus »hillieri«, Malus pumila, Morella serrata, Myrica cordifolia, Pinus sylvestris, Pinus taeda, Protea mundii, Protea repens, Psidium, Pyrus communis, Quercus, Rapanea melanophloeos, Rhamnus prunifolia, Rhus lucida, Robinia pseudoacacia, Salix caprea, Searsia crenata, Searsia tomentosa, Watsonia meriana* var. *meriana.*

dargei Bouyer, 2008
Foodplants: Unknown

dione (Fabricius, 1793)
butleri Aurivillius, 1901
camerunensis Bouvier, 1930
dionysiae Rougeot, 1948
flavescens (Rothschild, 1895)
lutea Bouvier, 1930
petiveri (Guérin-Méneville, 1845)
simplicia (Maassen & Weymer, 1872)
Foodplants: *Anacardium occidentale, Bauhinia petersiana, Carpinus betulus, Chlorophora, Diplorhynchus condylocarpon, Erythroxylum havanense, Eugenia, Jatropha curcas, Khaya, Malus domestica, Mimosa, Psidium, Pyracantha coccinea, Quercus × hispanica, Quercus robur, Rhus, Ricinus communis, Salix babylonica, Salix caprea, Spondias mombin, Theobroma cacao, Vernicia fordii, Vernicia montana.*

eblis buchholzi (Plötz, 1880)
Foodplants: Unknown

eblis eblis (Strecker, 1876)
Foodplants: *Alchornea, Cassia, Citrus aurantium, Mangifera indica, Salix caprea.*

emini (Butler, 1888)
germaini Bouvier, 1927
gschwandneri Rebel, 1917
latifascia Sonthonnax, 1901
nyassanae Rothschild, 1907
Foodplants: *Bridelia, Liquidambar, Prosopis affinis, Psidium guajava, Psorospermum, Rhus.*

fasciata (Gaede, 1927)
zaodeae Rougeot, 1975
Foodplants: Unknown

flammeola Darge, 2002
Foodplants: Unknown

gueinzii (Staudinger, 1872)
Foodplants: *Bersama abyssinica, Cajanus cajan, Cerasus vulgaris, Combretum, Cotoneaster, Croton macrostachyus, Cupressus lusitanica, Eriosema, Eucalyptus camaldulensis, Eucalyptus cladocalyx, Eucalyptus diversicolor, Eucalyptus drepanophylla, Eucalyptus melliodora, Eucalyptus paniculata, Eucalyptus saligna, Euclea, Khaya grandifoliola, Maesa alnifolia, Maesa lanceolata, Malus pumila, Myrica, Pinus patula, Prunus serotina, Pseudarthria, Rosa, Rubus fructicosus, Schinus molle.*

hurumai Darge, 2003
Foodplants: Unknown

jamesoni aethiops (Rothschild, 1907)
batesi Bouvier, 1930
Foodplants: *Quercus × hispanica*

jamesoni ardesiaca Rougeot, 1975
Foodplants: Unknown

jamesoni jamesoni (Druce, 1890)
rubricostalis (Kirby, 1892)
Foodplants: Unknown

kiliensis Darge, 2009
Foodplants: Unknown

kilumilorum Darge, 2002
Foodplants: Unknown

kohlii Darge, 2009
Foodplants: Unknown

krucki Hering, 1930
brunneonigra Hering, 1930
Foodplants: *Akokanthera, Betula, Corymbia citriodora, Eucalyptus tereticornis, Fagus, Malosma laurina, Prunus armeniaca, Rhus trilobata, Schinus.*

lucida (Rothschild, 1907)
Foodplants: Unknown

macrops Rebel, 1917
benguelensis (Oberthür, 1921)
Foodplants: *Brachystegia venosa, Eucalyptus, Pinus.*

macrothyris (Rothschild, 1906)
heroum Oberthür, 1910
lubumbashii Seydel, 1933
regalis Rebel, 1906
Foodplants: *Albizia adianthifolia, Amphimas pterocarpoides, Brachystegia randii, Brachystegia spiciformis, Gossweilerodendron balsamiferum, Julbernardia, Malus pumila, Pentaclethra macrophylla, Piptadeniastrum africanum, Prunus, Quercus × hispanica.*

mariae Bouyer, 2007
Foodplants: Unknown

melanops (Bouvier, 1930)
oyemensis Rougeot, 1955
Foodplants: *Entandrophragma candollei, Fagus, Hymenocardia acida, Laburnum, Pteridium, Robinia pseudoacacia.*

mitfordi (Kirby, 1892)
Foodplants: Unknown

nictitans (Fabricius, 1775)
Foodplants: Unknown

nyassana (Rothschild, 1907)
murphyi Darge, 1992
reducta (Rebel, 1917)
Foodplants: *Crataegus, Ligustrum ovalifolium, Prunus, Quercus robur, Rosa.*

perscitus Darge, 1992
herbuloti Darge, 1992
Foodplants: Unknown

renvazorum Darge, 2002
Foodplants: Unknown

rhodina (Rothschild, 1907)
rubra Bouvier, 1927
Foodplants: *Corylus, Quercus × hispanica, Uapaca kirkiana, Uapaca nitida.*

staudingeri allardiana Bouyer, 1999
allardi Rougeot, 1971
Foodplants: Unknown

staudingeri gabunica Rougeot, 1971
Foodplants: Unknown

staudingeri kafubuensis Bouvier, 1931
Foodplants: Unknown

staudingeri staudingeri (Aurivillius, 1893)
Foodplants: Unknown

ungemachti Bouvier, 1927
orientalis Bouvier, 1936
Foodplants: Unknown

venus Rebel, 1905
myrtea Rebel, 1917
orientalis Rebel, 1917
Foodplants: *Trema orientalis*

wahlbergii callophthalma Rougeot, 1979
Foodplants: *Salix caprea*

wahlbergii wahlbergii (Boisduval, 1847)
ochracea Rebel, 1917
Foodplants: *Croton sylvatica, Ligustrum ovalifolium, Liquidambar styraciflua, Maerua, Malosma laurina, Mangifera indica, Prunus persica, Psidium guajava, Quercus × hispanica, Quercus robur, Ricinus communis, Salix caprea, Schinus molle, Schinus terebinthifolia, Trema bracteolata.*

wahlbergina Rougeot, 1972
Foodplants: *Trema orientalis*

xanthomma amathusia Weymer, 1908
Foodplants: Unknown

xanthomma xanthomma (Rothschild, 1907)
anthinoides Rougeot, 1978
Foodplants: Unknown

Heniocha Hübner, [1819] 1816

apollonia (Cramer, 1780)
apollina Maassen, 1927
divisa Bryk, 1939
flavida (Butler, 1877)
lindti Grünberg, 1910
Foodplants: *Acacia karroo, Acacia mearnsii, Acacia pubescens.*

digennaroi Bouyer, 2008
Foodplants: *Elephantorrhiza elephantina*

distincta Bryk, 1939
Foodplants: Unknown

dyops (Maassen & Weymer, [1872])
Foodplants: *Acacia burkei, Acacia erubescens, Acacia hereoensis, Acacia mellifera, Acacia pubescens, Acacia nigrescens.*

hassoni Bouyer, 2008
Foodplants: *Elephantorrhiza elephantina*

marnois Rogenhofer, 1891
bioculata Aurivillius, 1879
Foodplants: *Acacia baileyana, Acacia burkei, Acacia dealbata, Acacia erubescens, Acacia fleckii, Acacia mellifera, Acacia pubescens, Acacia nigrescens, Acacia polyacantha.*

vingerhoedti Bouyer, 1992
pudorosa Darge, 2005
Foodplants: *Elephantorrhiza elephantina*

Imbrasia Hübner, [1819] 1816

epimethea biokoensis Darge, 1988
Foodplants: Unknown

epimethea epimethea (Drury, 1773)
crameri Kirby, 1892
dorcas (Walker, 1855)
nictitans (Fabricius, 1787)
pallescens Strand, 1915
paradoxa Dufrane, 1953
Foodplants: *Acacia decurrens, Acacia lahai, Acacia mearnsii, Albizia, Annona senegalensis, Bauhinia reticulata, Berlinia, Brachystegia spiciformis, Canarium schweinfurtii, Combretum, Holarrhena floribunda, Maesopsis eminii, Malus »hillieri«, Newtonia, Quercus robur, Salix caprea, Terminalia schweinfurthi.*

ertli Rebel, 1904
vesperina Stoneham, 1962
Foodplants: *Acacia, Berlinia, Brachystegia randii, Brachystegia spiciformis, Caloncoba welwitschii, Crossopteryx febrifuga, Hyparrhenia diplandra, Julbernardia globiflora, Julbernardia paniculata, Laburnum, Millettia laurentii, Petersianthus macrocarpus, Pteridium centrali-africanum, Robinia pseudoacacia.*

longicaudata Holland, 1893
truncata Bouvier, 1927
Foodplants: *Malus »hillieri«, Salix caprea.*

obscura (Butler, 1878)
affinis Bouvier, 1927
convexa Bouvier, 1927
hebe Maassen & Weymer, 1886
mopsa Walker, (1855)
Foodplants: *Carpinus betulus, Euphorbia, Quercus robur, Rhamnus.*

orientalis Rougeot, 1962
Foodplants: Unknown

senegalensis (Olivier, 1792)
Foodplants: Unknown

truncata Aurivillius, 1909
pumila Bouvier, 1930
Foodplants: *Corylus, Gossweilerodendron balsamiferum, Petersianthus macrocarpus, Ricinodendron heudelotii, Uapaca kirkiana, Uapaca nitida.*

Leucopteryx Packard, 1903

ansorgei (Rothschild, 1897)
reticolens (Sonthonnax, 1927)
Foodplants: Unknown

mollis (Butler, 1889)
Foodplants: Unknown

Lobobunea Packard, 1901

acetes (Westwood, 1849)
alberici Dufrane, 1953
leopoldi (Bouvier, 1930)
rautenstrauchorum Lampe, 1994
Foodplants: *Crataegus, Ligustrum ovalifolium, Mangifera, Quercus × hispanica, Quercus robur, Salix babylonica.*

cadioui Bouyer, 2004
Foodplants: Unknown

goodii (Holland, 1893)
ansorgei (Rothschild, 1898)
erythrotes (Karsch, 1893)
karschi Schüssler, 1933
Foodplants: *Fagus sylvatica, Quercus × hispanica, Quercus robur.*

jeanneli dargei Lemaire, 1971
Foodplants: Unknown

jeanneli jeanneli Rougeot, 1959
dallastai Bouyer, 1984
Foodplants: Unknown

kuehnei Naumann, 2008
Foodplants: Unknown

melanoneura (Rothschild, 1907)
basquini Rougeot, 1972
Foodplants: Unknown

minutissima Darge, 1998
Foodplants: Unknown

niepelti Strand, 1912
reginae Bouvier, 1930
Foodplants: Unknown

phaedusa christyi (Sharpe, 1899)
molleti Darge, 1975
phaeax Jordan, 1910
rexnoctuae Stoneham, 1962
Foodplants: *Euphorbia*, *Quercus* × *hispanica*, *Quercus ilex*.

phaedusa desfontainei Darge, 1998
Foodplants: *Euphorbia*

phaedusa falcatissima Rougeot, 1962
Foodplants: *Eucalyptus gunnii*, *Fagus*, *Quercus* × *hispanica*.

phaedusa phaedusa (Drury, 1780)
elegans Bouvier, 1928
laestrygon (Mabille, 1877)
laurae Strand, 1912
thomsonii (Kirby, 1877)
Foodplants: *Euphorbia*, *Ligustrum ovalifolium*, *Prunus domestica*, *Quercus robur*, *Rosa*.

rosea rosea (Sonthonnax, 1899)
Foodplants: *Akokanthera*, *Anisophyllea laurina*, *Apodytes dimidiata*, *Eucalyptus*, *Eugenia cordata*, *Protorhus longifolia*, *Uapaca nitida*.

rosea thoirei Bouvier, 1926
Foodplants: *Akokanthera*, *Anisophyllea laurina*, *Apodytes dimidiata*, *Eucalyptus*, *Eugenia cordata*, *Protorhus longifolia*, *Uapaca nitida*.

rosea werneri Bouyer, 2001
Foodplants: *Protorhus longifolia*

sangha Darge, 2002
Foodplants: Unknown

saturnus (Fabricius, 1793)
ammon (Karsch, 1898)
angasana (Westwood, 1849)
hannibal (Karsch, 1898)
orientis Le Cerf, 1922
tanganicae (Sonthonnax, 1899)
weymeri Aurivillius, 1914
Foodplants: *Akokanthera*, *Anisophyllea laurina*, *Apodytes dimidiata*, *Eucalyptus gunnii*, *Eugenia cordata*, *Protorhus longifolia*, *Pyracantha coccinea*, *Quercus robur*, *Quercus* × *hispanica*, *Protorhus longifolia*, *Syzygium cordatum*, *Uapaca kirkiana*, *Uapaca nitida*.

turlini Lemaire, 1977
vandenberghei Dall'Asta, 1982
Foodplants: Unknown

vingerhoedti Bouyer, 2004
Foodplants: Unknown

Melanocera Sonthonnax, 1901

dargei Terral, 1991
emmanuellae Rougeot, 1991
Foodplants: *Ochna arborea* var. *arborea*, *Ochna natalita*.

menippe (Westwood, 1849)
fumosa (Rothschild, 1895)
habenichti Wichgraf, 1911
transiens Wichgraf, 1911
Foodplants: *Acacia mearnsii*, *Bauhinia purpurea*, *Dichrostachys cinerea*, *Ficus cordata*, *Ochna natalita*, *Ochna schweinfurthiana*, *Ochna serrulata*, *Quercus* × *hispanica*, *Salix caprea*, *Salix cinerea*, *Terminalia sericea*.

nereis (Rothschild, 1898)
microphtalma Darge, 1994
Foodplants: Unknown

parva Rothschild, 1907
Foodplants: *Ochna schweinfurthiana*

pinheyi Lemaire & Rougeot, 1974
Foodplants: *Ochna schweinfurthiana*

pujoli Lemaire & Rougeot, 1974
Foodplants: Unknown

sufferti (Weymer, 1896)
Foodplants: *Ochna schweinfurthiana*

widenti Terral & Darge, 1991
Foodplants: *Quercus robur*, *Quercus* × *hispanica*.

Protogynanisa Rougeot, 1971

athletoides Rougeot, 1971
Foodplants: *Acacia dealbata*

probsti Bouyer, 2001
Foodplants: Unknown

Pseudimbrasia Rougeot, 1962

deyrollei (Thomson, 1858)
colorata Rougeot, 1980
congolensis (Oberthür, 1910)
gerresheimi (Strand, 1911)
intermedia (Strand, 1911)
senegalensis (Kirby, 1892)
Foodplants: *Acacia mearnsii, Erythrophleum guineense, Euphorbia, Gleditsia triacanthos, Ligustrum ovalifolium, Morus, Quercus × hispanica, Quercus robur, Rosa, Rubus, Shirakiopsis elliptica, Salix caprea, Tamarindus indica.*

Pseudobunea Bouvier, 1927

alinda alinda (Drury, 1870)
cremeri (Oberthür, 1919)
schoenstedti (Rothschild, 1895)
sjoestedti (Aurivillius, 1893)
Foodplants: *Bauhinia reticulata, Fagus, Malus »hillieri«, Prunus cerasifera, Quercus robur.*

alinda deaconi Stoneham, 1962
Foodplants: Unknown

bjornstadi Bouyer, 2006
Foodplants: *Acacia mearnsii, Bauhinia thonningii, Brachystegia spiciformis, Erythrina lysistemon, Eucalyptus, Julbernardia globiflora, Pinus patula, Populus deltoides, Quercus × hispanica.*

callista (Jordan, 1910)
Foodplants: *Brachystegia, Julbernardia, Quercus robur.*

catochra catochra (Karsch, 1893)
cleopatra (Aurivillius, 1893)
catochra orientalis Rougeot, 1972
Foodplants: Unknown

citrinarius citrinarius (Gaede, 1927)
Foodplants: *Berlinia, Brachystegia randii, Brachystegia spiciformis, Populus, Pterocarpus, Uapaca.*

citrinarius dayensis Rougeot, 1991
Foodplants: Unknown

cyrene (Weymer, 1908)
Foodplants: Unknown

illustris (Weymer, 1901)
Foodplants: *Quercus robur, Rosa.*

irius (Fabricius, 1793)
bilineata Bouvier, 1928
bruneli Rougeot, 1975
epithyrena (Maassen & Weymer, 1886)
heyeri (Weymer, 1896)
immaculata Bouvier, 1930
inornata (Sonthonnax, 1899)
melinde (Maassen & Weymer, 1886)
morlandi (Rothschild, 1907)
natalensis (Aurivillius, 1893)
pallens (Sonthonnax, 1899)
patruelis (Distant, 1903)
redlichi (Weymer, 1901)
simpex Bouvier, 1928
Foodplants: *Acacia mearnsii, Annona senegalensis, Bauhinia reticulata, Bauhinia thonningii, Brachystegia randii, Brachystegia spiciformis, Erythrina lysistemon, Eucalyptus, Fagus, Julbernardia globiflora, Julbernardia paniculata, Pinus patula, Populus deltoides, Prunus laurocerasus, Pseudoberlinia paniculata, Pterocarpus tinctorius, Quercus emoryi, Quercus × hispanica, Quercus robur, Uapaca nitida.*

meloui (Riel, 1910)
Foodplants: Unknown

paratyrrhena Bouvier, 1927
barnsi Bouvier, 1930
Foodplants: Unknown

seydeli Le Cerf, 1923
Foodplants: Unknown

tyrrhena fumida Darge, 2003
Foodplants: Unknown

tyrrhena gauthieri Darge, 1988
Foodplants: Unknown

tyrrhena maculata Bouvier, 1930
Foodplants: *Malus »hillieri«*

tyrrhena neglecta Darge, 2003
Foodplants: Unknown

tyrrhena perchromatum Darge, 2003
Foodplants: *Prunus laurocerasus*

tyrrhena pratorum Darge, 2003
Foodplants: Unknown

tyrrhena tyrrhena (Westwood, 1849)
Foodplants: *Acacia karroo, Albizia adianthifolia, Albizia gummifera, Celtis africana, Celtis ehrenbergiana, Diospyros lycioides, Ekebergia capensis, Euclea pseudebenus, Fagus, Ozoroa concolor, Psychotria capensis, Quercus × hispanica, Salix babylonica, Schinus molle, Sclerocarya birrea, Trema bracteolata, Ulmus.*

vingerhoedti Bouyer, 2004
Foodplants: Unknown

Rohaniella Bouvier, 1927

pygmaea jebbae (Sharpe, 1899)
guineensis Bouvier, 1928
Foodplants: *Burkea africana*

pygmaea pygmaea (Maassen & Weymer, 1886)
pygela (Druce, 1886)
Foodplants: *Burkea africana, Quercus × hispanica.*

Ubaena Karsch, 1900

dolabella (Druce, 1886)
Foodplants: *Pinus patula, Pinus radiata.*

fuelleborniana Karsch, 1900
moesta Rebel, 1917
Foodplants: *Malus »hillieri«, Tilia cordata.*

lequeuxi Darge & Terral, 1988
Foodplants: *Pinus patula, Pinus radiata.*

periculosa Darge & Terral, 1988
Foodplants: Unknown

sabunii Darge & Kilumile, 2004
Foodplants: Unknown

Saturniini

Actias Leach in Leach & Nodder, 1815

angulocaudata Naumann & Bouyer, 1998
Foodplants: Unknown

arianeae Brechlin, 2007
Foodplants: *Larix decidua, Picea, Pinus.*

artemis aliena (Butler, 1879)
apollo Röber, 1923
flavicollis Jordan, 1911
xenia Jordan, 1911
Foodplants: *Acer, Alnus firma, Alnus japonica, Armeniaca mume* var. *mume, Betula platyphylla, Carya, Castanea crenata, Castanea mollissima, Cerasus pseudocerasus, Cerasus serrulata* var. *serrulata, Fagus crenata, Ilex verticillata, Juglans regia, Malus pumila, Prunus armeniaca, Prunus persica, Padus virginiana, Punica, Pyrus communis, Quercus aliena, Quercus serrata, Rosa, Salix babylonica, Salix caprea, Salix gracilistyla.*

artemis artemis (Bremer & Grey, 1853)
sjoqvisti Bryk, 1948
Foodplants: *Acer, Alnus firma, Alnus japonica, Armeniaca mume* var. *mume, Betula platyphylla, Carya, Castanea crenata, Castanea mollissima, Cerasus pseudocerasus, Cerasus serrulata* var. *serrulata, Crataegus monogyna, Eucalyptus gunnii, Fagus crenata, Ilex verticillata, Juglans regia, Liquidambar styraciflua, Malus pumila, Padus virginiana, Prunus armeniaca, Prunus persica, Punica, Pyrus communis, Quercus aliena, Quercus serrata, Rhododendron, Rosa, Salix babylonica, Salix caprea, Salix gracilistyla, Schinus terebinthifolia.*

artemis jordani Niepelt, 1936
Foodplants: Unknown

artemis yakushimaensis Kishida, 1994
Foodplants: *Juglans regia*

australovietnama Brechlin, 2000
Foodplants: Unknown

callandra Jordan, 1911
Foodplants: *Lannea coromandelica, Rhizophora.*

chapae bezverkhovi Sochivko & Ivshin, 2008
Foodplants: *Pinus*

chapae chapae (Mell, 1950)
Foodplants: *Picea abies, Pinus mugo, Pinus sylvestris.*

chrisbrechlinae Brechlin, 2007
Foodplants: *Larix decidua, Picea, Pinus.*

dubernardi (Oberthür, 1897)
Foodplants: *Abies alba, Abies nordmanniana, Cedrus deodora, Larix decidua, Liquidambar styraciflua, Metasequoia glyptostroboides, Picea abies, Picea glauca, Pinus brutia* var. *eldarica, Pinus contorta, Pinus massoniana, Pinus nigra, Pinus sylvestris, Rhododendron, Salix babylonica, Thuja.*

felicis (Oberthür, 1896)
Foodplants: *Larix decidua, Pinus sylvestris.*

gnoma gnoma (Butler, 1877)
dulcinea (Butler, 1881)
Foodplants: *Alnus cordata, Alnus glutinosa, Alnus japonica, Castanea crenata, Corylus avellana, Crataegus, Liquidambar styraciflua, Quercus acutissima, Quercus robur.*

gnoma mandschurica (Staudinger, 1892)
tomariactias Bryk, 1942
Foodplants: *Acer mono, Acer palmatum, Alnus japonica, Betula, Castanea crenata, Cinnamomum*

camphora, Quercus acutissima, Quercus × hispanica, Quercus robur, Salix babylonica.

gnoma miyatai Inoue, 1976
Foodplants: Unknown

groenendaeli acutapex Kishida 2000
Foodplants: Unknown

groenendaeli groenendaeli Roepke, 1954
Foodplants: *Liquidambar styraciflua, Rhus typhina.*

groenendaeli sumbawaensis Paukstadt, Paukstadt & Rougerie, 2010
Foodplants: Unknown

groenendaeli timorensis Paukstadt, Paukstadt & Rougerie, 2010
Foodplants: Unknown

ignescens Moore, 1877
Foodplants: *Crypteronia paniculata*

isis (Sonthonnax, [1897])
latona Rothschild & Jordan, 1901
Foodplants: *Eucalyptus gunnii, Ligustrum ovalifolium, Liquidambar styraciflua, Malus »hillieri«, Malus sylvestris, Quercus × hispanica, Rhus typhina, Salix caprea.*

kongjiaria kongjiaria Zhu & Wang, 1993
Foodplants: Unknown

kongjiaria shaanxiana Brechlin, 2007
Foodplants: Unknown

laotiana Testout, 1936
Foodplants: Unknown

luna azteca Packard, 1869
Foodplants: *Liquidambar styraciflua*

luna luna (Linnaeus, 1758)
bolli (Wagner, 1876)
dictynna (Walker, 1855)
maasseni (Kirby, 1892)
rossii (Ross, 1872)
Foodplants: *Acer rubrum, Acer saccharum, Alnus, Betula papyrifera, Betula pendula, Carpinus, Carya cordiformis, Carya glabra, Carya illinoinensis, Carya ovata, Castanea dentata, Castanea sativa, Corylus, Crataegus monogyna, Diospyros virginiana, Fagus grandifolia, Humulus lupulus, Juglans cinerea, Juglans nigra, Juglans regia, Liquidambar styraciflua, Liriodendron tulipifera, Nyssa sylvatica, Ostrya virginiana, Platanus mexicana, Populus tremuloides, Prunus domestica, Quercus alba, Quercus macrocarpa, Quercus palustris, Quercus rubra, Rhus glabra, Rhus typhina, Salix caprea, Tilia cordata, Ulmus.*

maenas diana Maassen & Weymer, [1872]
recta Bouvier, 1928
rieli Testout, 1942
rosenbergii Kaup, 1895
Foodplants: *Adinandra dumosa, Averrhoa bilimbi, Canarium, Eucalyptus gunnii, Ligustrum ovalifolium, Liquidambar styraciflua, Quercus robur, Rhus typhina, Schima wallichii.*

maenas maenas Doubleday, 1847
leto Doubleday, 1848
Foodplants: *Adinandra dumosa, Averrhoa bilimbi, Canarium, Castanea, Eucalyptus gunnii, Liquidambar styraciflua, Malus sylvestris, Prunus, Quercus × hispanica, Quercus robur, Rhus glabra, Schmaltzia copallinum, Rhus typhina, Schima wallichii, Toxicodendron radicans, Turpina pomifera, Turpina sphaerocarpa.*

maenas saja Van Eecke, 1913
Foodplants: Unknown

neidhoeferi Ong & Yu, 1968
Foodplants: *Picea morrisonicola*

ningpoana C. Felder, 1862
malaisei Bryk, 1944
omeishana Watson, 1912
Foodplants: *Alnus cremastogyne, Alnus formosana, Betula, Castanea mollissima, Cerasus pseudocerasus, Cinnamomum camphora, Crataegus monogyna, Eucalyptus gunnii, Hibiscus syriacus, Juglans regia, Lagerstroemia subcostata, Liquidambar formosana, Liquidambar styraciflua, Malus asiatica, Malus »hillieri«, Malus pumila, Prunus armeniaca, Pterocarya stenoptera, Pyrus communis, Quercus ilex, Rhododendron, Salix babylonica, Salix caprea, Triadica sebifera.*

parasinensis Brechlin, 2009
Foodplants: *Eucalyptus gunnii, Liquidambar styraciflua.*

philippinica bulbosa Nässig & Treadaway, 1997
Foodplants: Unknown

philippinica philippinica Nässig & Treadaway, 1997
Foodplants: *Fagus sylvatica*

rhodopneuma Röber, 1925
distincta Niepelt, 1932
Foodplants: *Cotinus coggygria, Eucalyptus gunnii, Liquidambar styraciflua, Malosma laurina, Rhus typhina.*

selene brevijuxta Nässig & Treadaway, 1997
Foodplants: *Liquidambar styraciflua*

selene eberti Rougeot, 1969
Foodplants: Unknown

selene seitzi Kalis, 1934
miae Toxopeus, 1944
Foodplants: Unknown

selene selene (Hübner, 1806)
astarte Maassen & Weymer, [1872]
Foodplants: *Acer campestre, Alnus glutinosa, Andromeda, Andromeda glaucophylla, Azadirachta indica, Carpinus betulus, Carya, Castanea, Cedrela tonduzii, Cerasus vulgaris, Cinnamomum camphora, Coriaria nepalensis, Corylus avellana, Corylus colurna, Crataegus laevigata, Crataegus monogyna, Frangula dodonei, Hibiscus, Juglans regia, Lannea coromandelica, Lawsonia inermis, Ligustrum ovalifolium, Ligustrum vulgare, Liquidambar styraciflua, Malosma laurina, Malus domestica, Malus »hillieri«, Malus pumila, Mangifera, Moringa oleifera, Padus avium* var. *avium, Prunus amygdalus, Prunus laurocerasus, Prunus serotina, Pyrus communis, Quercus × hispanica, Quercus robur, Rhododendron, Salix babylonica, Salix caprea, Salix fragilis, Schinus terebinthifolia, Syringa vulgaris, Terminalia crenulata, Triadica sebifera, Zanthoxylum acanthopodium, Zanthoxylum armatum* var. *armatum.*

selene taprobanis Paukstadt & Paukstadt, 1999
Foodplants: *Liquidambar styraciflua, Padus avium* var. *avium, Quercus × hispanica, Rhododendron.*

selene vandenbeghi Roepke, 1956
Foodplants: Unknown

sinensis sinensis Walker, 1855
heterogyna Mell, 1914
Foodplants: *Eucalyptus gunnii, Liquidambar formosana, Liqidambar styraciflua.*

sinensis subaurea Kishida, 1993
Foodplants: Unknown

truncatipennis (Sonthonnax, 1899) **rev. stat.**
Foodplants: *Juglans regia, Liquidambar formosana, Liquidambar styraciflua, Schinus terebinthifolia.*

uljanae Brechlin, 2007
Foodplants: *Abies nordmanniana, Larix decidua, Picea abies, Pinus sylvestris.*

winbrechlini Brechlin, 2007
Foodplants: *Picea abies, Pinus sylvestris*

witti Brechlin, 2007
Foodplants: Unknown

Antheraea Hübner, 1819 ["1816"]

Subgenus *Antheraea* Hübner, 1819 ["1816"]

alleni Holloway, 1987
Foodplants: Unknown

andamana Moore, 1877
cernyi Brechlin, 2002
Foodplants: *Ficus*

acehensis Paukstadt & Paukstadt, 2010
Foodplants: Unknown

angustomarginata Brechlin & Meister, 2009
Foodplants: Unknown

billitonensis Moore, 1878
Foodplants: Unknown

broschi Naumann, 2001
Foodplants: Unknown

brunei Allen & Holloway, 1986
Foodplants: Unknown

cadioui Naumann, 2010
Foodplants: Unknown

celebensis Watson, 1928
Foodplants: *Shorea*

cihangiri Naumann & Nässig, 1998
Foodplants: Unknown

cingalesa Moore, 1883
Foodplants: *Terminalia*

compta Rothschild, 1899
Foodplants: Unknown

cordifolia Weymer, 1906
fickei Weymer, 1909
Foodplants: *Betula pendula, Fagus grandifolia* subsp. *grandifolia, Liqidambar styraciflua, Malus »hillieri«, Salix caprea.*

diehli Lemaire, 1979
Foodplants: Unknown

exspectata Brechlin, 2000
Foodplants: *Betula utilis*

frithi frithi Moore, 1859
confusa Niepelt, 1932
falloui Schüssler, 1883
pedunculata Bouvier, 1936

Foodplants: *Cinnamomum camphora, Crataegus, Dipterocarpus alatus, Eucalyptus gunnii, Fagus, Liquidambar styraciflua, Quercus × hispanica, Quercus robur, Shorea robusta, Triadica sebifera.*

frithi tonkinensis Bouvier, 1936
Foodplants: Unknown

gephyra Niepelt, 1926
Foodplants: Unknown

gschwandneri Niepelt, 1918
subcaeca Bouvier, 1928
Foodplants: Unknown

gulata Nässig & Treadaway, 1998
Foodplants: Unknown

hagedorni Naumann, & Lourens, 2008
Foodplants: *Quercus × hispanica*

halconensis Paukstadt & Brosch, 1996
Foodplants: *Pyracantha coccinea, Quercus × hispanica, Quercus robur.*

harndti Naumann, 1999
Foodplants: Unknown

helferi borneensis Moore, 1892
Foodplants: *Malus »hillieri«, Quercus × hispanica, Quercus robur.*

helferi helferi Moore, 1859
Foodplants: *Quercus × hispanica, Quercus robur.*

hollowayi Nässig & Naumann, 1998
Foodplants: *Cinnamomum camphora, Triadica sebifera.*

imperator Watson, 1913
javanensis Bouvier, 1930
Foodplants: *Betula utilis, Canarium indicum, Carpinus, Cinnamomum camphora, Fagus, Prunus, Quercus alba, Quercus × hispanica, Quercus robur, Triadica sebifera.*

insularis Watson, 1914
Foodplants: *Carpinus, Fagus, Prunus, Quercus alba, Quercus × hispanica, Quercus robur.*

jakli Naumann, 2008
Foodplants: Unknown

jana (Stoll, 1782)
surakarta Moore, 1862
Foodplants: *Betula pendula, Betula utilis, Canarium indicum, Canarium vulgare, Cupania lessertiana, Malus »hillieri«, Quercus × hispanica, Quercus robur, Schleichera oleosa, Terminalia catappa.*

jawabaratensis Brechlin & Paukstadt, 2010
Foodplants: Unknown

kageri Paukstadt, Paukstadt & Suhardjono, 1997
Foodplants: *Betula pendula, Betula utilis, Malus »hillieri«.*

kalabahiensis Naumann, 2010
Foodplants: Unknown

kalangensis Brechlin & Meister, 2009
Foodplants: Unknown

kelimutuensis Paukstadt, Paukstadt & Suhardjono, 1997
Foodplants: *Betula utilis*

korintjiana Bouvier, 1928
Foodplants: *Betula alnoides, Cyperus esculentus, Daphniphyllum himalayense, Euodia fraxinifolia, Cerasus cerasoides, Lithocarpus dealbatus, Madhuca longifolia, Madhuca longifolia* var. *Latifolia, Malus, Malus pumila, Pyrus communis, Quercus glauca, Quercus floribunda, Quercus × hispanica, Quercus ilex, Quercus incana, Quercus robur, Quercus semecarpifolia, Quercus semiserrata, Quercus serrata, Quercus turneri.*

lampei Nässig & Holloway, 1989
Foodplants: *Crataegus monogyna, Quercus × hispanica, Quercus robur.*

larissa larissa (Westwood, 1847)
javanensis Bouvier, 1928
Foodplants: *Betula utilis, Liquidambar, Shorea glauca.*

larissa mindoroensis Brosch & Paukstadt 1996
Foodplants: *Fagus sylvatica*

larissa philippirissa Nässig & Treadaway, 1997
Foodplants: *Shorea glauca*

larissa ridlyi Moore, 1892
delegata Swinhoe, 1893
Foodplants: *Shorea robusta*

larissoides Bouvier, 1928
crypta Zhu & Wang, 1993
Foodplants: Unknown

loeffleri Naumann, 2010
Foodplants: Unknown

lorosae M. D. Lane, Naumann & D. A. Lane, 2004
Foodplants: Unknown

meisteri Brechlin, 2002
Foodplants: Unknown

mentawai Nässig, Lampe & Kager, 2002
Foodplants: Unknown

minahassae Niepelt, 1926
Foodplants: *Shorea glauca*

moultoni Watson, 1927
Foodplants: Unknown

myanmarensis Paukstadt, Paukstadt & Brosch, 1998 **rev. stat.**
Foodplants: Unknown

pahangensis Brechlin & Paukstadt, 2010
Foodplants: Unknown

paphia (Linnaeus, 1758)
fasciata Moore, 1892
fraterna Moore, 1888
kausalia Moore, 1877
kolisurra Sykes, 1834
mylitta (Drury, 1773)
nebulosa Hutton, 1869
ochripicta Moore, 1892
olivescens Moore, 1892
pulchra Moore, 1892
salthi Stack, 1884
sivalica Hutton, 1881
versicolor Moore, 1892
Foodplants: *Anogeissus latifolia, Barringtonia racemosa, Bauhinia variegata, Betula pendula, Betula pubescens, Betula nigra, Betula utilis, Bombax ceiba, Careya arborea, Carissa carandas, Carpinus betulus, Chamaecrista desvauxii* var. *mollissima, Celastrus paniculatus, Chloroxylon swietenia, Cipadessa baccifera, Corylus avellana, Dodonaea viscosa, Eucalyptus gunnii, Fagus grandifolia* subsp. *grandifolia, Ficus benjamina, Ficus religiosa, Ficus retusa, Lagerstroemia indica, Lagerstroemia parviflora, Liquidambar styraciflua, Madhuca longifolia, Madhuca longifolia* var. *Latifolia, Malus domestica, Malus »hillieri«, Mangium caseolaris, Melastoma, Pentaptera glabra, Pentaptera tomentosa, Prunus domestica, Prunus serotina, Pseudobombax septenatum, Psydrax dicoccos, Quercus alba, Quercus × hispanica, Quercus robur, Ricinus communis, Salix babylonica, Salix caprea, Salix fragilis, Shorea robusta, Shorea talura, Sonneratia caseolaris, Syzygium cumini, Tectona grandis, Terminalia alata, Terminalia arjuna, Terminalia bellirica, Terminalia catappa, Terminalia tomentosa, Tarenna asiatica, Ziziphus jujuba.*

pasteuri Bouvier, 1928
Foodplants: *Betula utilis, Quercus robur.*

paukpelengensis Brechlin & Meister, 2009
Foodplants: Unknown

paukstadtorum Naumann, Holloway & Nässig, 1996
Foodplants: *Betula pendula, Betula utilis, Crataegus monogyna, Fagus grandifolia* subsp. *grandifolia, Malus »hillieri«, Quercus robur, Salix caprea.*

pelengensis Brechlin, 2000
Foodplants: Unknown

pernyi pernyi (Guérin-Méneville, 1855)
confuci Moore, 1874
constans Staudinger, 1892
fantoni Rondot, 1887
francki Watson, 1928
hartii Moore, 1892
Foodplants: *Betula pendula, Carpinus, Castanea crenata, Cinnamomum camphora, Crataegus pinnatifida, Eucalyptus gunnii, Fagus grandifolia* subsp. *grandifolia, Lithocarpus hancei, Malus pumila, Prunus, Quercus acutissima, Quercus alba, Quercus aliena, Quercus dentata, Quercus falcata, Quercus × hispanica, Quercus lyrata, Quercus macrocarpa, Quercus myrsinifolia, Quercus palustris, Quercus phellos, Quercus robur, Quercus serrata, Quercus stellata, Quercus variabilis, Quercus wutaishanica, Salix alba, Salix babylonica, Salix caprea, Croton congestus.*

pernyi roylii Moore, 1859
knyvetti Hampson, 1892
Foodplants: *Betula alnoides, Betula pendula, Cerasus cerasoides, Cyperus esculentus, Daphniphyllum himalayense, Euodia fraxinifolia, Evodia, Fagus grandifolia* subsp. *grandifolia, Liquidambar styraciflua, Lithocarpus dealbatus, Madhuca longifolia, Madhuca longifolia* var. *Latifolia, Malus »hillieri«, Malus pumila, Pyrus communis, Quercus glauca, Quercus floribunda, Quercus × hispanica, Quercus ilex, Quercus incana, Quercus robur, Quercus semecarpifolia, Quercus semiserrata, Quercus serrata, Quercus wutaishanica, Salix viminalis.*

perrottetii (Guérin-Méneville, 1843)
Foodplants: Unknown

platessa Rothschild, 1903
fusca Rothschild, 1903
ornata Bouvier, 1929
Foodplants: *Quercus robur, Salix caprea, Salix viminalis.*

pratti Bouvier, 1928
Foodplants: *Quercus* × *hispanica*

raffrayi Bouvier, 1928
Foodplants: *Betula utilis, Salix caprea.*

ranakaensis Paukstadt, Paukstadt & Suhardjono, 1997
Foodplants: Unknown

ranongensis Paukstadt & Paukstadt, 2010
Foodplants: Unknown

rolfei Naumann & Löffler, 2010
Foodplants: Unknown

rosemariae Holloway, Nässig & Naumann,1995
Foodplants: *Betula utilis, Fagus grandifolia* subsp. *grandifolia, Liquidambar styraciflua, Malus »hillieri«, Quercus* × *hispanica, Quercus robur.*

rubicunda rubicunda Brechlin, 2009
Foodplants: *Betula alnoides, Cyperus esculentus, Daphniphyllum himalayense, Euodia fraxinifolia, Cerasus cerasoides, Lithocarpus dealbatus, Madhuca longifolia, Madhuca longifolia* var. *Latifolia, Malus pumila, Pyrus communis, Quercus glauca, Quercus floribunda, Quercus* × *hispanica, Quercus ilex, Quercus incana, Quercus robur, Quercus semecarpifolia, Quercus semiserrata, Quercus serrata.*

rubicunda rubiorientalis Brechlin, 2009
Foodplants: *Betula, Crataegus, Quercus.*

rumphii (C. Felder, 1860)
buruensis Bouvier, 1928
ceramensis Bouvier, 1930
Foodplants: *Quercus* × *hispanica, Quercus robur, Sonneratia alba.*

schroederi Paukstadt, Brosch & Paukstadt, 1999
Foodplants: Unknown

semperi noeli Nässig & Treadaway, 1998
Foodplants: *Quercus* × *hispanica.*

semperi semperi C. Felder & R. Felder, 1861
Foodplants: *Liquidambar styraciflua, Quercus* × *hispanica.*

steinkeorum Paukstadt, Paukstadt & Brosch, 1999
Foodplants: Unknown

sumatrana Niepelt, 1926
mylittoides Bouvier, 1928
prelarissa Bouvier, 1928
Foodplants: *Carpinus, Fagus, Prunus, Quercus alba, Quercus robur.*

sumbaensis Naumann, 2010
Foodplants: Unknown

sumbawaensis Brechlin, 2000
Foodplants: Unknown

superba Inoue, 1965
Foodplants: *Quercus acutissima, Quercus robur.*

taripaensis Naumann, Nässig & Holloway, 1996
Foodplants: Unknown

tenggarensis Brechlin, 2000
Foodplants: *Quercus acutissima*

ulrichbroschi Paukstadt & Paukstadt, 1999
Foodplants: Unknown

vietnamensis Brechlin & Paukstadt, 2010
Foodplants: Unknown

viridiscura Holloway, Nässig & Naumann, 1996
Foodplants: Unknown

yamamai titan Mell, 1958
Foodplants: *Quercus robur*

yamamai ussuriensis Schachbazov, 1953
Foodplants: *Castanea crenata, Malus »hillieri«, Malus pumila, Prunus, Quercus acutissima, Quercus serrata.*

yamamai yamamai (Guérin-Méneville, 1861)
bergmani Bryk, 1948
calida Butler, 1881
fentoni Butler, 1881
hazina Butler, 1881
morosa Butler, 1881
sergestus Westwood, 1881
Foodplants: *Betula pendula, Betula pubescens, Carya, Castanea crenata, Castanea sativa, Crataegus monogyna, Fagus sylvatica, Fagus grandifolia* subsp. *grandifolia, Malus »hillieri«, Malus pumila, Morus, Prunus, Quercus acutissima, Quercus coccinea, Quercus dentata, Quercus* × *hispanica, Quercus palustris, Quercus petraea, Quercus robur, Quercus rubra, Quercus serrata, Quercus wutaishanica.*

yoshimotoi Inoue, 1965
Foodplants: *Quercus*

zwicki Nässig & Treadaway, 1998
Foodplants: Unknown

Subgenus *Antheraeopsis* Wood-Mason, 1886

assamensis assamensis (Helfer, 1837)
assama (Westwood, 1848)
biedermanni (Niepelt, 1932)
mezankooria (Moore, 1862)
subvelata Bouvier, 1930
chengtuana (Watson, 1923)
yunnanensis Zhu & Wang, 1993
Foodplants: *Actinodaphne citrata, Cinnamomum bejolghota, Fagus, Laurus nobilis, Litsea glutinosa* var. *glutinosa, Litsea monopetala, Machilus gamblei, Magnolia champaca, Magnolia pterocarpa, Malus pumila, Persea americana, Persea odoratissima, Quercus × hispanica, Quercus robur, Sarcostemma acidum, Symplocos grandiflora, Tetranthera diglottica, Tetranthera glauca, Tetranthera macrophylla, Tetranthera monopetala.*

assamensis formosana Sonon, 1937
Foodplants: *Cinnamomum camphora*

castanea (Jordan, 1910)
mezops Bryk, 1944
Foodplants: *Fagus sylvatica, Laurus nobilis.*

paniki Nässig & Treadaway, 1998
Foodplants: Unknown

rudloffi Brechlin, 2002
Foodplants: Unknown

sahi Nässig & Treadaway, 1998
Foodplants: Unknown

youngei Watson, 1915
brunnea (Van Eecke, 1928)
rubiginea Toxopeus, 1940
Foodplants: *Betula utilis*

Subgenus *Loepantheraea* Toxopeus, 1940

rosieri (Toxopeus, 1940)
imeldae Nässig & Treadaway, 1998
Foodplants: Unknown

Subgenus *Telea* Hübner, 1819 ["1816"]

godmani (Druce, 1892)
columbiana (Draudt, 1930)
Foodplants: *Quercus nigra, Quercus palustris, Quercus petraea, Quercus robur.*

montezuma (Sallé, 1856)
Foodplants: *Quercus nigra, Quercus palustris, Quercus robur.*

oculea Neumoegen, 1883
aurelia Druce, 1892
Foodplants: *Betula pendula, Fagus grandifolia* subsp. *grandifolia, Juglans major, Quercus emoryi, Quercus oblongifolia, Quercus robur, Quercus turbinella.*

polyphemus mexicana Hoffmann, 1942
Foodplants: *Quercus nigra, Quercus robur.*

polyphemus polyphemus (Cramer, 1775)
fenestra Perry, 1811
olivacea Cockerell, 1914
Foodplants: *Acer campestre, Acer macrophyllum, Acer negundo, Acer pensylvanicum, Acer saccharinum, Acer rubrum, Alnus incana, Alnus serrulata, Arbutus menziesii, Baptisia tinctoria, Betula lenta, Betula papyrifera, Betula pendula, Betula populifolia, Betula pubescens, Capsicum annuum, Carpinus betulus, Carpinus caroliniana, Carya alba, Carya cordiformis, Carya glabra, Carya ovata, Castanea dentata, Castanea pumila, Castanea sativa, Cerasus vulgaris, Corylus americana, Corylus cornuta, Cornus racemosa, Crataegus chapmanii, Crataegus crus-galli, Crataegus douglasii, Cydonia oblonga, Fagus crenata, Fagus grandifolia, Fagus grandifolia* subsp. *grandifolia, Fraxinus, Gaylussacia dumosa, Gleditsia triacanthos, Hamamelis virginiana, Juglans cinerea, Juglans nigra, Liriodendron tulipifera, Malus domestica, Malus pumila, Mespilus coccinea, Myrica cerifera, Ostrya virginiana, Pinus, Platanus occidentalis, Platanus orientalis, Populus tremula* subsp. *grandidentata, Populus tremuloides, Prunus domestica, Prunus emarginata* subsp. *mollis, Prunus persica, Prunus serotina, Padus virginiana, Pyrus canadensis, Pyrus communis, Quercus alba, Quercus buckleyi, Quercus coccinea, Quercus falcata, Quercus laurifolia, Quercus lyrata, Quercus macrocarpa, Quercus nigra, Quercus palustris, Quercus petraea, Quercus phellos, Quercus robur, Quercus rubra, Quercus stellata, Quercus velutina, Quercus virginiana, Rhamnus, Grossularia cynosbati, Ribes sanguineum, Ribes sativum, Rosa, Salix alba, Salix babylonica, Salix fragilis, Salix lasiopes, Salix interior, Sambucus, Sassafras albidum, Sorbus, Tilia americana, Tilia cordata, Tilia europaea, Ulmus americana, Ulmus crassifolia, Vaccinium myrtillus, Vitis, Weigela.*

polyphemus tuxtlasensis Balcázar-Lara & Vázquez, 1994
Foodplants: Unknown

Antherina Sonthonnax 1901

suraka comorana Viette, 1965
Foodplants: Unknown

suraka suraka (Boisduval, 1833)
Foodplants: *Acer campestre, Carpinus betulus, Crataegus laevigata, Crataegus monogyna, Eugenia, Fagus, Ligustrum japonicum, Ligustrum ovalifolium, Ligustrum vulgare, Liquidambar, Maesa, Malus »hillieri«, Nerium oleander, Salix babylonica, Salix caprea, Schinus molle, Strophanthus, Vitis vinifera.*

Argema Wallengren 1958

besanti (Rebel, 1895)
Foodplants: *Commiphora mollis, Eucalyptus, Juglans regia, Liquidambar, Sclerocarya birrea* subsp. *caffra, Spirostachys africana.*

fournieri Darge, 1972
Foodplants: *Liquidambar styraciflua*

kuhnei katangensis Bouyer, 2008
Foodplants: Unknown

kuhnei kuhnei Pinhey, 1969
Foodplants: *Monotes katangensis*

mimosae (Boisduval, 1847)
bouvieri (Ghesquiére, 1934)
elucidata (Grünberg, 1910)
occidentalis (Gschwandner, 1923)
Foodplants: *Commiphora mollis, Cotinus coggygria, Eucalyptus, Euphorbia, Juglans regia, Liquidambar styraciflua, Malosma laurina, Rhus typhina, Schinus terebinthifolia, Sclerocarya birrea, Sclerocarya birrea* subsp. *caffra, Spirostachys africana.*

mittrei (Guérin-Méneville, 1847)
cometes (Guenée, 1865)
idae (C. Felder & R. Felder, 1874)
madagascariensis (Bartlett, 1873)
Foodplants: *Castanea sativa, Cotinus coggygria, Eucalyptus gunnii, Ligustrum ovalifolium, Liquidambar styraciflua, Malosma laurina, Mimosa, Quercus × hispanica, Schmaltzia copallinum, Rhus typhina, Schinus molle, Schinus terebinthifolia, Syzygium emirnense, Toxicodendron pubescens, Uapaca, Weinmannia eriorcampa.*

Austrocaligula Cockerell 1914

carnea (Sonthonnax, 1899)
Foodplants: *Alphitonia excelsa, Callicoma serratifolia, Flindersia australis, Flindersia maculosa, Flindersia schottiana.*

engaea Turner, 1922
Foodplants: *Amyema*

eucalypti (Scott, 1864)
Foodplants: *Betula pendula, Eucalyptus blakelyi, Eucalyptus botryoides, Eucalyptus bridgesiana, Eucalyptus cinerea, Eucalyptus eugenioides, Eucalyptus fastigata, Eucalyptus globulus, Eucalyptus globulus* subsp. *maidenii, Eucalyptus gunnii, Eucalyptus leucoxylon, Eucalyptus macarthurii, Eucalyptus macrorhyncha, Eucalyptus mannifera* subsp. *maculosa, Eucalyptus melliodora, Eucalyptus nitens, Eucalyptus obliqua, Eucalyptus paniculata, Eucalyptus pauciflora, Eucalyptus regnans, Eucalyptus saligna, Eucalyptus viminalis, Geijera salicifolia, Ligustrum ovalifolium, Liquidambar styraciflua, Lophostemon confertus, Olea paniculata, Pinus radiata, Podocarpus spinulosus, Populus, Schinus molle, Tristaniopsis laurina.*

helena (White, 1843)
intermedia (Lucas, 1890)
Foodplants: *Betula, Eucalyptus baxteri, Eucalyptus bicolor, Eucalyptus gunnii, Eucalyptus obliqua, Eucalyptus odorata, Eucalyptus ovata, Eucalyptus viminalis.*

loranthi Lucas, 1891
Foodplants: *Amyema, Eucalyptus, Loranthus, Viscum.*

Ceranchia Butler 1878

apollina Butler, 1878
cribrelli Butler, 1882
cribricollis Silbermann, 1897
mangiferae (Sganzin, 1833)
mucida Saalmüller, 1884
reticolens Butler, 1882
Foodplants: *Cynanchum acutum, Gomphocarpus fruticosus, Plumeria, Stephanotis floribunda, Uapaca bojeri, Vincetoxicum nigrum.*

Crysodesmia Röber 1925

flavipicta Röber, 1925
Foodplants: Unknown

Copaxa Walker, 1855

andensis Lemaire, 1971
Foodplants: *Persea americana, Salix babylonica.*

anikae Brechlin & Meister, 2010
Foodplants: Unknown

apollinairei Lemaire, 1978
Foodplants: Unknown

australoescalantei Brechlin & Meister, 2010
Foodplants: Unknown

bachuea Wolfe, 2005
Foodplants: *Coriaria ruscifolia* subsp. *microphylla, Persea americana.*

bella Wolfe, Naumann, Brosch, Wenczel & Nässig, 2005
Foodplants: *Persea americana*

canella Walker, 1855
vitellina Draudt, 1929
Foodplants: *Laurus nobilis, Nectandra reticulata, Nectandra venulosa, Persea americana, Ocotea porphyria.*

carolinae Brechlin, Käch & Meister, 2011
Foodplants: Unknown

chrisbrechlinae Brechlin & Meister, 2010
Foodplants: Unknown

cineracea W. Rothschild, 1895
Foodplants: *Persea americana*

conlani Brechlin & Meister, 2010
Foodplants: Unknown

copaxoides Dyar, 1912
Foodplants: *Persea americana*

curvilinea Schaus, 1912
Foodplants: *Licaria misantlae, Nectandra cuspidata, Nectandra hihua, Nectandra martinicensis, Nectandra purpurea, Nectandra umbrosa, Ocotea atirrensis, Ocotea insularis, Ocotea mollifolia, Persea americana, Persea schiedeana.*

cuscoandensis Brechlin & Meister, 2010
Foodplants: Unknown

cuscoexpandens Brechlin & Meister, 2010
Foodplants: Unknown

cydippe Druce, 1894
Foodplants: *Persea americana, Pinus nigra, Pinus radiata, Pinus thunbergii, Salix.*

decrescens Walker, 1855
Foodplants: *Carpinus betulus, Fagus sylvatica, Malus »hillieri«, Parrotia persica, Persea americana, Quercus ilex, Salix caprea.*

denda Druce, 1894
Foodplants: *Persea americana, Quercus robur.*

denhezi Lemaire, 1971
Foodplants: *Persea americana, Salix caprea.*

escalantei Lemaire, 1971
Foodplants: *Malus »hillieri«, Nectandra hihua, Nectandra martinicensis, Ocotea gomezii, Ocotea whitei, Persea americana.*

evelynae Wolfe & Lemaire, 1993
Foodplants: *Persea americana, Quercus.*

expandens brunnaeocaeca Lemaire, 1982
Foodplants: Unknown

expandens expandens Walker, 1855
anestios Weymer, 1909
arpi Gschwandner, 1925
trimacula Rothschild, 1895
Foodplants: *Malus »hillieri«, Quercus, Persea americana, Salix babylonica, Salix caprea.*

flavina Draudt, 1929
Foodplants: *Persea americana, Salix caprea, Schmidelia.*

flavobrunnea Bouvier, 1930
Foodplants: *Laurus nobilis, Nectandra reticulata, Nectandra venulosa, Persea americana, Ocotea porphyria.*

garciorum Brechlin & Meister, 2010
Foodplants: *Fagus sylvatica, Persea americana, Salix babylonica, Salix caprea, Salix viminalis.*

haxairei Herbin, 2009
Foodplants: *Pinus*

herbuloti Lemaire, 1971
Foodplants: *Persea americana*

ignescens Lemaire, 1978
Foodplants: Unknown

intermediata Wolfe, 2005
Foodplants: *Hieracium brachiatum, Persea americana, Tibouchina semidecandra.*

joinvillea Schaus, 1921
draudti Niepelt, 1933
Foodplants: *Persea americana*

kitchingi Brechlin & Meister, 2010
Foodplants: Unknown

koenigi Lemaire, 1974
Foodplants: Unknown

lavendera (Westwood, 1853)
chapata (Westwood, 1853)
conjuncta Bouvier, 1925
plenkeri C. Felder & R. Felder, 1860
pygmaea Bouvier, 1925
vetusta Hoffmann, 1942
Foodplants: *Cinnamomum camphora, Malosma laurina, Persea americana, Quercus ilex, Quercus robur, Rhus trilobata, Salix babylonica, Schinus molle.*

lavenderoguatemalensis Brechlin & Meister, 2010
Foodplants: Unknown

lavenderohidalgensis Brechlin & Meister, 2010
Foodplants: Unknown

lavenderojaliscensis Brechlin & Meister, 2010
Foodplants: Unknown

litensis Wolfe & Conlan, 2002
Foodplants: *Persea americana*

lunula Wolfe & Conlan, 2003
Foodplants: *Persea americana*

madrediosiana Brechlin & Meister, 2010
Foodplants: Unknown

mannana Dyar, 1914
heringi Draudt, 1929
Foodplants: *Quercus agrifolia, Quercus chrysolepis, Quercus gambelii, Quercus ilex, Quercus kelloggii, Quercus lobata, Quercus suber, Quercus robur.*

marona Schaus, 1906
Foodplants: Unknown

mazaorum Lemaire, 1982
Foodplants: *Fagus sylvatica, Liquidambar styraciflua, Persea americana, Prunus.*

medea (Maassen, 1890)
miles (Jordan, 1911)
Foodplants: *Persea americana, Pinus radiata*

mielkeorum Brechlin & Meister, 2010
Foodplants: Unknown

miranda Lemaire, 1971
Foodplants: *Persea americana*

moinieri Lemaire, 1974
Foodplants: *Cinnamomum triplinerve, Lauraceae, Nectandra belizensis, Nectandra salicifolia, Nectandra salicina, Ocotea, Ocotea cernua, Ocotea atirrensis, Ocotea veraguensis, Ocotea whitei, Persea americana, Salix caprea.*

muellerana Dyar, 1920
Foodplants: *Quercus × hispanica, Quercus robur.*

multifenestrata (Herrich-Schäffer, 1858)
flavescens Draudt, 1929
olivina Draudt, 1929
Foodplants: *Liquidambar styraciflua, Persea americana, Salix caprea, Salix cinerea, Salix viminalis.*

novocineracea Brechlin & Meister, 2009
Foodplants: *Persea americana*

ockendeni (Druce, 1906)
carina (Jordan, 1911)
charila (Jordan, 1911)
orios (Dyar, 1913)
Foodplants: Unknown

orientalis Lemaire, 1975
Foodplants: *Brachyotum ledifolium, Persea americana.*

pararufinans Brechlin & Meister, 2010
Foodplants: *Fagus sylvatica, Persea americana.*

parvohidalgensis Brechlin & Meister, 2010
Foodplants: Unknown

pascoandensis Brechlin & Meister, 2010
Foodplants: Unknown

rudloffi Brechlin & Meister, 2010
Foodplants: Unknown

rufa Draudt, 1929
Foodplants: *Liquidambar styraciflua, Persea americana, Salix babylonica, Salix caprea.*

rufijaliscensis Brechlin & Meister, 2010
Foodplants: Unknown

rufimichoacanensis Brechlin & Meister, 2010
Foodplants: Unknown

rufinans Schaus, 1906
Foodplants: *Beilschmiedia brenesii, Cinnamomum brenesii, Cinnamomum camphora, Cinnamomum triplinerve, Fagus sylvatica, Malosma laurina, Nectandra hihua, Nectandra cuspidata, Nectandra martinicensis, Nectandra salicifolia, Nectandra salicina, Nectandra umbrosa, Ocotea, Ocotea insularis, Ocotea leucoxylon, Ocotea mollifolia,*

Ocotea smithiana, Ocotea veraguensis, Ocotea whitei, Persea americana, Persea povedae, Persea schiedeana, Prunus, Quercus.

rufotincta Rothschild, 1895
Foodplants: Unknown

sapatoza (Westwood, [1854])
Foodplants: *Alnus acuminata, Alnus rhombifolia, Persea americana.*

satellita Walker, 1865
Foodplants: *Persea americana*

satipoexpandens Brechlin & Meister, 2010
Foodplants: Unknown

schmiti Brechlin & Meister, 2010
Foodplants: Unknown

semioculata (C. Felder & R. Felder 1874)
Foodplants: *Brachyotum ledifolium*

simoni Brechlin, Käch & Meister, 2011
Foodplants: Unknown

simson Maassen & Weymer, 1881
Foodplants: *Ceiba, Persea americana.*

sinjaevi Brechlin & Meister, 2010
Foodplants: Unknown

sophronia Schaus, 1921
Foodplants: *Persea americana*

syntheratoides W. Rothschild, 1895
Foodplants: *Crataegus*

troetschi Druce, 1886
Foodplants: Unknown

trottierorum Bénéluz, 1986
Foodplants: Unknown

vanschaycki Brechlin & Meister, 2010
Foodplants: Unknown

wernermeisteri Brechlin & Meister, 2010
Foodplants: Unknown

winbrechlini Brechlin, Käch & Meister, 2011
Foodplants: Unknown

wolfei Meister, Naumann, Brosch & Wenczel, 2005
Foodplants: Unknown

yungaskoenigoides Brechlin & Meister, 2010
Foodplants: Unknown

yungaspandens Brechlin & Meister, 2010
Foodplants: Unknown

yungescens Brechlin & Meister, 2010
Foodplants: Unknown

Cricula Walker, 1855

agria Jordan, 1909
nadari Bouvier, (1929)
Foodplants: *Malus »hillieri«*

andamanica Jordan, 1909
Foodplants: Unknown

andrei Jordan, 1909
vinosa Watson, 1912
Foodplants: *Crataegus, Ligustrum ovalifolium, Ligustrum vulgare, Liquidambar styraciflua, Malus »hillieri«, Malus pumila, Padus avium* var. *avium, Parthenocissus quinquefolia, Prunus domestica, Prunus serotina, Pyrus communis, Quercus × hispanica, Quercus robur, Rhododendron ponticum, Salix babylonica, Salix caprea, Vitis.*

aungsansuukyiae Naumann & Löffler, 2010
tonkintrifenestratoides Brechlin, 2010
Foodplants: Unknown

australosinica Brechlin, 2004
Foodplants: Unknown

bornea Watson, 1913
Foodplants: Unknown

buruensis Jordan, 1939
Foodplants: Unknown

cameronensis Paukstadt & Paukstadt, 1998
Foodplants: *Eucalyptus gunnii, Liquidambar styraciflua, Quercus ilex, Rhododendron.*

ceylonica Jordan, 1909
Foodplants: Unknown

elaezia elaezia Jordan, 1909
Foodplants: Unknown

elaezia pelengensis Paukstadt & Paukstadt, 1998
baliensis Naumann & Löffler, 2010
Foodplants: *Eucalyptus gunnii, Liquidambar styraciflua, Quercus ilex.*

falcata Löffler & Naumann, 2010
australovietnama Brechlin, 2010
Foodplants: Unknown

fansipanensis Löffler & Naumann, 2010
Foodplants: Unknown

flavoglena Zhu & Wang, 1993
Foodplants: Unknown

frederkingi Löffler & Naumann, 2010
Foodplants: Unknown

griseorubescens Löffler & Naumann, 2010
Foodplants: Unknown

halconensis Nässig & Treadaway, 2011
Foodplants: Unknown

hoabinhnguyeni Löffler & Naumann, 2010
Foodplants: Unknown

jordani Bryk, 1944
Foodplants: *Malus »hillieri«*, *Prunus domestica*, *Prunus serotina*, *Rhododendron*.

hainanensis Brechlin, 2004
Foodplants: Unknown

hayatiae Paukstadt & Suhardjono, 1992
Foodplants: *Persea americana*, *Psidium guajava*.

kalimantanensis Brechlin, 2010
Foodplants: Unknown

kareli Nässig & Treadaway, 1997
Foodplants: *Malus »hillieri«*, *Prunus domestica*.

leyteana bayani Nässig & Treadaway, 2011
Foodplants: Unknown

leyteana leyteana Nässig & Treadaway, 1997
Foodplants: Unknown

luzonica Jordan, 1909
Foodplants: *Malus »hillieri«*, *Persea americana*, *Prunus domestica*, *Salix babylonica*, *Salix caprea*.

magnifenestrata elaeziopahangensis Brechlin, 2010
Foodplants: Unknown

magnifenestrata magnifenestrata Naumann & Löffler, 2010
elaezioborneensis Brechlin, 2010
Foodplants: Unknown

maxalorensis Naumann & Löffler, 2010
Foodplants: Unknown

mindanaensis Nässig & Treadaway, 1997
Foodplants: Unknown

palawanica Brechlin, 2001
Foodplants: Unknown

pseudandrei Naumann & Löffler, 2010
hunana Brechlin, 2010
Foodplants: Unknown

quinquefenestrata Roepke, 1940
Foodplants: Unknown

schintlmeisteri Naumann & Löffler, 2010
Foodplants: Unknown

separata Naumann & Löffler, 2010
elaeziosumatrana Brechlin, 2010
Foodplants: Unknown

sichuana Brechlin, 2010
Foodplants: Unknown

sponai Löffler &Naumann, 2010
Foodplants: Unknown

sumatrensis Jordan, 1939
afenestra Watson, 1912
Foodplants: *Malus »hillieri«*

timorensis Naumann & Lane, 2010
Foodplants: Unknown

trifenestrata agrioides Nässig, 1989
Foodplants: *Persea americana*, *Psidium guajava*, *Semecarpus anacardium*.

trifenestrata banggaiensis Naumann & Paukstadt, 1997
Foodplants: *Malus »hillieri«*, *Persea americana*, *Psidium guajava*, *Semecarpus anacardium*.

trifenestrata burmana Swinhoe, 1890
Foodplants: Unknown

trifenestrata halmaheraensis Paukstadt & Paukstadt, 2010
Foodplants: Unknown

trifenestrata javana Watson, 1913
Foodplants: *Acrocarpus*, *Malus »hillieri«*, *Salix pentandra*.

trifenestrata kransi Jurriaanse & Lindemans, 1920
Foodplants: *Anacardium occidentale*, *Liquidambar styraciflua*, *Mangifera indica*, *Persea*.

trifenestrata serama Nässig, 1989
Foodplants: Unknown

trifenestrata tenggarensis Paukstadt, Paukstadt & Suhardjono, 1998
Foodplants: Unknown

trifenestrata treadawayi Nässig, 1989
Foodplants: Unknown

trifenestrata trifenestrata (Helfer, 1837)
haumpottonee Hugon, 1837
zuleika (Westwood, 1848)

Foodplants: *Acrocarpus, Anacardium occidentale, Artocarpus, Bischofia, Canarium album, Canarium indicum, Careya arborea, Cinchona pubescens, Cinnamomum, Crataegus, Eucalyptus gunnii, Malus »hillieri«, Malus pumila, Mangifera indica, Persea americana, Persea odoratissima, Prunus domestica, Prunus serotina, Psidium guajava, Pyrus communis, Quercus × hispanica, Quercus robur, Rhododendron, Salix babylonica, Salix caprea, Schleichera oleosa, Spondias, Syzygium aqueum, Syzygium jambos, Ziziphus.*

trifenestratoides Brechlin, 2010
Foodplants: Unknown

variabilis Naumann & Löffler, 2010
Foodplants: Unknown

vietnama Brechlin, Nässig & Naumann, 1999
Foodplants: *Liquidambar styraciflua*

zubsiana Nässig, 1985
Foodplants: unknown

Graellsia Grote, 1896

isabellae (Graëlls 1849)
ceballosi Gómez-Bustillo & Fernández-Rubio, 1974
diana (Fairmaire, 1849)
galliaegloria Oberthür, 1923
neuroflexaclara Abös-Castel, 1983
paradisea Marten, 1955
roncalensis Gomez-Bustillo, Gomez de Aizpürua & Fernandez-Rubio, 1974
Foodplants: *Abies alba, Abies nordmanniana, Cedrus atlantica, Cedrus deodora, Picea abies, Pinus halepensis, Pinus mugo, Pinus nigra, Pinus pinaster* subsp. *escarena, Pinus pinea, Pinus radiata, Pinus salzmanii, Pinus strobus, Pinus sylvestris.*

Lemaireia Nässig & Holloway, 1987

chrysopeplus (Toxopeus, 1940)
Foodplants: Unknown

hainana Nässig & Min, 2006
Foodplants: Unknown

inexspectata Nässig, 1996
Foodplants: Unknown

loepoides (Butler, 1880)
Foodplants: Unknown

luteopeplus aureopeplus Nässig & Holloway, 1988
Foodplants: *Liquidambar styraciflua, Rhus typhina.*

luteopeplus luteopeplus Nässig & Holloway, 1988
Foodplants: *Liquidambar styraciflua, Rhus typhina.*

naessigi Brechlin, 2001
Foodplants: Unknown

schintlmeisteri Lampe & Nässig, 1989
Foodplants: Unknown

Loepa Moore, 1859

anthera Jordan, 1911
Foodplants: *Parthenocissus tricuspidata, Vitis vinifera.*

cynopis Nässig & Suhardjono, 1989
Foodplants: Unknown

damartis Jordan, 1911
szechuana Zhu & Wang, 1993
Foodplants: *Parthenocissus, Vitis.*

diehli Brechlin, 2010
Foodplants: Unknown

diffundata Naumann, Nässig & Löffler, 2008
Foodplants: *Cissus, Parthenocissus quinquefolia.*

diffunoccidentalis Brechlin, 2010
Foodplants: Unknown

duffunorientalis Brechlin, 2010
Foodplants: Unknown

finnackermanni Brechlin, 2010
Foodplants: Unknown

formosensis Mell, 1939
formosibia Bryk, 1944
Foodplants: *Ampelopsis glandulosa* var. *brevipedunculata, Tetrastigma formosanum.*

javanica Mell, 1939
Foodplants: *Parthenocissus tricuspidata*

katinka Westwood, 1847
diversiocellata Bryk, 1944
sikkimensis Silbermann, 1897
sivalica Moore, 1881
Foodplants: *Ampelopsis, Cissus, Crataegus, Dillenia pentagyna, Leea, Parthenocissus tricuspidata, Vitis vinifera.*

kuangtungensis kuangtungensis Mell, 1939
Foodplants: *Parthenocissus tricuspidata, Vitis vinifera.*

kuangtungensis septentrionalis Mell, 1939
Foodplants: *Citrus reticulata, Vitis vinifera.*

martinii Brechlin & Paukstadt, 2010
Foodplants: Unknown

megacore Jordan, 1911
Foodplants: *Cissus antarctica, Crataegus monogyna, Parthenocissus, Vitis vinifera.*

meyi Naumann, 2003
Foodplants: Unknown

microocellata Naumann & Kishida, 2001
Foodplants: *Crataegus monogyna, Malus »hillieri«, Parthenocissus quinquefolia, Parthenocissus tricuspidata.*

minahassae Mell, 1939
Foodplants: *Parthenocissus, Rhus, Salix caprea, Salix viminalis, Tetracera scandens.*

mindanaensis Schüssler, 1933
Foodplants: *Parthenocissus tricuspidata*

miranda Moore, 1865
septentrionalis Mell, 1939
Foodplants: *Parthenocissus quinquefolia, Parthenocissus tricuspidata, Vitis vinifera.*

mirandula Yen, Nässig, Naumann & Brechlin, 2000
Foodplants: Unknown

nepalensis Brechlin, 2010
Foodplants: Unknown

nigropupillata Nässig & Treadaway, 1988
Foodplants: Unknown

oberthueri Leech, 1890
dogninia Sonthonnax, 1892
Foodplants: *Citrus reticulata, Parthenocissus tricuspidata, Vitis vinifera.*

obscuromarginata Naumann, 1998
Foodplants: *Malus »hillieri«, Parthenocissus quinquefolia, Parthenocissus tricuspidata, Vitis vinifera.*

orientomiranda mirella Brechlin & Kitching, 2010
Foodplants: Unknown

orientomiranda orientomiranda Brechlin & Kitching, 2010
Foodplants: Unknown

palawana Nässig & Treadaway, 1997
Foodplants: Unknown

paramiranda Brechlin & Kitching, 2010
Foodplants: Unknown

peggyae hainanensis Brechlin, 2010
Foodplants: Unknown

peggyae peggyae Brechlin, 2010
Foodplants: Unknown

roseomarginata Brechlin, 1997
Foodplants: Unknown

sakaei Inoue, 1965
Foodplants: Unknown

schintlmeisteri Brechlin, 2000
Foodplants: Unknown

siamensis malayensis Brechlin, 2010
Foodplants: Unknown

siamensis siamensis Brechlin, 2010
Foodplants: Unknown

sikkima Moore, 1865
Foodplants: *Cissus, Dillenia, Leea, Parthenocissus tricuspidata, Saurauia.*

sinjaevi Brechlin, 2004
Foodplants: Unknown

sumatrana Nässig, Lampe & Kager, 1989
Foodplants: Unknown

taipeishanis Mell, 1939
Foodplants: Unknown

tibeta Naumann, 2003
Foodplants: Unknown

visayana Brechlin, 2000
Foodplants: Unknown

wlingana Yang, 1978
Foodplants: Unknown

yunnana Mell, 1939
Foodplants: Unknown

Neodiphthera Fletcher, 1982

albicera (Rothschild & Jordan, 1907)
Foodplants: Unknown

arfakiana Brechlin, 2011
Foodplants: Unknown

aruensis Paukstadt, Paukstadt & Suhardjono, 2003
Foodplants: Unknown

buruensis Brechlin, 2005
Foodplants: Unknown

ceramensis (Bouvier, 1928)
Foodplants: Unknown

decellei (Nässig & Lemaire, 2002)
Foodplants: Unknown

elleri (Eckerlein, 1935)
Foodplants: Unknown

excavus (Lane, 1995)
Foodplants: *Erythrophleum chlorostachys*

foucheri (Bouvier, 1928)
grisea (Bouvier, 1928)
Foodplants: Unknown

gazella (Niepelt, 1934)
Foodplants: Unknown

goodgeri (D'Abrera, 1998)
Foodplants: Unknown

habemana Brechlin, 2005
Foodplants: Unknown

intermedia (Bouvier, 1928)
Foodplants: Unknown

joiceyi (Bouvier, 1928)
Foodplants: Unknown

magodensis Brechlin, 2011
Foodplants: Unknown

mapiana Brechlin, 2011
Foodplants: Unknown

monacha (Niepelt, 1934)
Foodplants: Unknown

nabiriana Brechlin, 2011
Foodplants: Unknown

papuana (Rothschild, 1904)
Foodplants: Unknown

paukstadtorum Brechlin, 2011
Foodplants: Unknown

pelepagana Brechlin, 2011
Foodplants: Unknown

rhythmica (Turner, 1936)
Foodplants: *Litsea reticulata, Doryphora.*

roicki Brechlin, 2005
Foodplants: Unknown

saccopoea (Turner, 1928)
Foodplants: *Planchonia careya*

sahulensis Paukstadt, Paukstadt & Suhardjono, 2003
Foodplants: Unknown

schaarschmidti Brechlin, 2005
Foodplants: Unknown

sciron (Westwood, 1881)
alephostra (Swinhoe, 1892)
inversa (Rothschild, 1896)
strandi (Niepelt, 1934)
Foodplants: Unknown

strigata (Bethune-Baker, 1908)
Foodplants: Unknown

sulphurea Lane & Naumann, 2003
Foodplants: Unknown

talboti (Bouvier, 1928)
Foodplants: Unknown

tennenti (Nässig & Lemaire, 2002)
Foodplants: Unknown

venusta (Rothschild & Jordan, 1907)
Foodplants: Unknown

Opodiphthera Wallengren, 1858

astrophela (Walker, 1855)
simplex (Walker, 1855)
varicolor Wallengren, 1859
Foodplants: *Alphitonia excelsa, Callicoma serratifolia, Elaeocarpus obovatus, Flindersia maculosa, Flindersia australis, Flindersia schottiana, Podocarpus elatus.*

fervida Jordan, 1910
Foodplants: *Maesa muelleri, Rapanea porosa.*

juriaansei Van Eecke, 1933
tenimberensis Niepelt, 1934
Foodplants: Unknown

Pararhodia Cockerell, 1914

daviesorum Lemaire, 1979
Foodplants: Unknown

gyra (Rothschild & Jordan, 1905)
Foodplants: Unknown

meeki (Jordan, 1908)
Foodplants: Unknown

rotalis Paukstadt, Paukstadt & Suhardjono, 1992
Foodplants: Unknown

setekwa D'Abrera, 1998
Foodplants: Unknown

Rhodinia Staudinger, 1892

benedeki Brechlin, 2011
Foodplants: Unknown

broschi Brechlin, 2001
Foodplants: *Celtis sinensis, Quercus × hispanica.*

davidi (Oberthür, 1886)
Foodplants: *Celtis sinensis*

fugax diana (Oberthür, 1886)
Foodplants: *Acer, Castanea crenata, Cerasus serrulata* var. *serrulata, Padus avium* var. *avium, Populus maximowiczii, Prunus avium, Quercus acutissima, Salix caprea, Ulmus, Zelkova serrata.*

fugax flavescens Brechlin, 2007
Foodplants: *Padus avium* var. *avium, Salix caprea.*

fugax fugax (Butler, 1877)
Foodplants: *Acer, Castanea, Celtis ehrenbergiana, Crataegus monogyna, Fagus, Hevea brasiliensis, Ilex verticillata, Malus »hillieri«, Padus avium* var. *avium, Phellodendron amurense, Platanus, Prunus domestica, Populus nigra, Quercus acutissima, Quercus dentata, Quercus glauca, Quercus × hispanica, Quercus robur, Quercus serrata, Salix babylonica, Salix caprea, Zelkova serrata.*

fugax guangdongensis Brechlin, 2007
Foodplants: *Tilia cordata*

fugax jiangxiana Brechlin, 2007
Foodplants: Unknown

fugax shaanxiana Brechlin, 2007
Foodplants: Unknown

fugax szechuanensis Mell, 1939
Foodplants: *Padus avium* var. *avium, Salix caprea.*

grigauti Le Moult, 1933
Foodplants: Unknown

jankowskii hattoriae Inoue, 1965
Foodplants: *Malus »hillieri«*

jankowskii hokkaidoensis Inoue, 1965
Foodplants: Unknown

jankowskii jankowskii (Oberthür, 1880)
Foodplants: *Crataegus monogyna, Fagus, Phellodendron amurense, Platanus occidentalis, Quercus dentata, Salix caprea.*

newara (Moore, 1872)
nissori Watson, 1935
Foodplants: *Acer saccharum, Crataegus monogyna, Juglans regia, Malus »hillieri«, Padus avium* var. *avium, Salix babylonica, Salix caprea, Salix elegans.*

rudloffi rudloffi Brechlin, 2001
Foodplants: *Padus avium* var. *avium, Quercus × hispanica, Salix caprea*

rudloffi sinensis Brechlin, 2007
Foodplants: *Padus avium* var. *avium, Quercus × hispanica, Salix caprea, Tilia cordata*

silkae Brechlin & Van Schayck, 2010
Foodplants: *Prunus, Tilia*

tenzingyatsoi Naumann, 2001
Foodplants: *Crataegus, Liquidambar styraciflua, Padus avium* var. *avium, Salix caprea*

verecunda Inoue, 1984
Foodplants: *Celtis sinensis, Prunus avium, Padus avium* var. *avium, Quercus acutissima, Quercus aliena, Quercus glauca, Salix caprea*

viksinjaevi Brechlin, 2004
Foodplants: Unknown

Saturnia Schrank, 1802

Subgenus *Agapema* Neumoegen & Dyar, 1894

anona anona Ottolengui, 1903
Foodplants: *Celtis ehrenbergiana, Condalia ericoides, Condalia hookeri, Condalia spathulata, Condalia viridis, Lycium, Rhus microphylla, Sarcobatus vermiculatus.*

anona dyari Cockerell, 1914
Foodplants: *Condalia ericoides*

anona platensis Peigler & Kendall, 1993
Foodplants: *Condalia ericoides*

dentifasciata Lemaire, 1973
Foodplants: *Condalia ericoides*

galbina (Clemens, 1860)
solita Ferguson, 1972
Foodplants: *Celtis, Condalia ericoides, Condalia hookeri, Condalia spathulata, Ebenopsis ebano, Larrea, Ligustrum, Prunus serotina, Ribes, Sarcobatus vermiculatus.*

homogena Dyar, 1908
Foodplants: *Frangula californica, Frangula dodonei, Malus »hillieri«, Ribes alpinum, Ribes cereum, Rhus, Salix exigua.*

pelora (Rindge, 1966)
Foodplants: *Condalia hookeri*

Subgenus *Calosaturnia* J. B. Smith, 1886

albofasciata J. W. Johnson, 1938
Foodplants: *Ceanothus cordulatus, Ceanothus cuneatus, Ceanothus leucodermis, Ceanothus spinosus, Cercocarpus betuloides, Cercocarpus montanus, Fremontodendron californicum, Malus »hillieri«.*

mendocino (Behrens, 1876)
Foodplants: *Arbutus menziesii, Arctostaphylos glauca, Arctostaphylos tomentosa, Arctostaphylos uva-ursi, Ceanothus cuneatus, Comarostaphylos polifolia, Malosma laurina, Rhus trilobata, Rhus typhina, Schmaltzia integrifolia.*

walterorum Hogue & J. W. Johnson, 1958
meridionalis Johnson, 1940
Foodplants: *Arbutus unedo, Arctostaphylos uva-ursi, Crataegus laevigata, Liquidambar styraciflua, Malosma laurina, Rhus trilobata, Rhus typhina, Schinus terebinthifolia, Schmaltzia integrifolia.*

Subgenus *Eudia* Jordan, 1911

cephalariae (Christoph, 1885)
Foodplants: *Cephalaria procera*

pavonia (Linnaeus, 1758)
alpina (Favre, 1897)
borealis (Rangnow, 1935)
carpini ([Denis & Schiffermüller], 1776)
iberligurica Gómez-Bustillo, Gómez de Aizpúrua & Mendez-Garnica,1982
irkutskana Sheljuzhko, 1943
josephinae Schawerda, [1924]
lappmarchica Bryk, 1948
lapponica Lingoniblad, 1936
media (Geoffroy, 1785)
minor (Linnaeus, 1758)
pavonus (Haworth, 1802)
tiefi Sheljuzhko, 1943
valcareli Agenjo, 1972
Foodplants: *Alnus glutinosa, Arbutus unedo, Betula pendula, Betula pubescens, Berberis aetnensis, Berberis cretica, Berberis vulgaris, Berberis vulgaris* subsp. *australis, Calluna vulgaris, Cornus mas, Crataegus monogyna, Eleagnus rhamnoides, Erica, Fagus sylvatica, Filipendula ulmaria, Filipendula vulgaris, Fragaria, Fraxinus, Humulus, Juglans regia, Lythrum salicaria, Malus domestica, Malus pumila, Myrica gale, Padus avium* var. *avium, Padus virginiana, Pistacia terebinthus, Persicaria acuminata, Persicaria bistorta, Persicaria vivipara, Polygonum aviculare, Potentilla tormentilla, Prunus cerasifera, Prunus domestica, Prunus serotina, Prunus spinosa, Pyrus communis, Quercus robur, Rhamnus, Rosa, Rubus idaeus, Rumex, Salvia officinalis, Salvia pratensis, Salix alba, Salix babylonica, Salix caprea, Salix fragilis, Sambucus, Schinus terebinthifolia, Sorbus aucuparia, Spiraea alba* var. *latifolia, Ulmus, Vaccinium myrtillus, Vaccinium vitis-idaea.*

pavoniella (Scopoli, 1763)
donauensis Seyer, 1991
ligurica (Weismann, 1876)
melichi Seyer, 1991
meridionalis (Calberla, 1887)
Foodplants: *Crataegus monogyna, Eleagnus rhamnoides, Malus »hillieri«, Padus avium* var. *avium, Prunus spinosa, Rubus fructicosus.*

spini ([Denis & Schiffermüller], 1776)
media (Esper, 1782)
Foodplants: *Alnus glutinosa, Cephalaria procera, Crataegus curvisepala, Crataegus monogyna, Malus domestica, Malus »hillieri«, Malus pumila, Planera aquatica, Populus, Prunus spinosa, Rosa, Rubus, Salix babylonica, Salix caprea.*

Subgenus *Perisomena* Walker, 1855

caecigena Kupido, (1825)
parviocellata Friedel, 1968
stroehlei Nässig, 2002
transcaucasica Bang-Haas, 1927
Foodplants: *Carpinus betulus, Fagus sylvatica, Fraxinus, Malus »hillieri«, Padus avium* var. *avium, Populus alba, Populus nigra, Prunus serotina, Pyrus communis, Quercus cerris, Quercus ilex, Quercus nigra, Quercus robur, Quercus petraea, Quercus pubescens, Quercus robur, Quercus suber, Rosa Ulmus pumila.*

codyi Peigler, 1996
Foodplants: Unknown

huttoni (Moore, 1862)
Foodplants: *Crataegus, Ligustrum ovalifolium, Malus »hillieri«, Padus avium* var. *avium, Prunus cerasifera, Prunus domestica, Pyrus communis, Salix babylonica, Spiraea.*

naessigi De Freina, 1992
Foodplants: *Crataegus, Cydonia, Fraxinus, Ligustrum ovalifolium, Ligustrum vulgare, Malus »hillieri«, Padus avium* var. *avium, Pyrus, Rhus, Spiraea, Salix caprea.*

shadulla galerope Püngeler, (1900)
Foodplants: *Malus »hillieri«, Padus avium* var. *avium, Rosa.*

shadulla shadulla (Moore, 1872)
alatauica (Bang-Haas, 1936)
oliva (Bang-Haas, 1910)
haraldi (Schawerda, 1922)
schencki (Staudinger, 1881)
stoliczkana C. Felder & R. Felder, 1874
svenihedini (Hering, 1936)
Foodplants: *Malus »hillieri«, Malus pumila, Padus avium* var. *avium, Pistacia chinensis, Prunus amygdalus, Prunus cerasifera, Prunus persica, Pyrus communis, Salix babylonica, Salix purpurea, Spiraea, Ulmus.*

Subgenus *Rinaca* Walker, 1855

anna Moore, 1865
tsinglingshanis Mell, 1939
bieti Oberthür, 1886
dejeani Bouvier & Real, 1931
diversa (Bryk, 1944)
yunnana Mell, 1939
Foodplants: *Acer saccharinum, Betula pendula, Corylus avellana, Juglans regia, Liquidambar styraciflua, Malus, Padus avium* var. *avium, Populus deltoides, Prunus avium, Pyrus calleryana, Rhus, Salix caprea, Tilia cordata.*

boisduvalii Eversmann, 1847
chinganensis Bang-Haas, 1936
fallax Jordan, 1911
intermediula Bryk, 1948
privata Krulikowsky, 1906
Foodplants: *Acacia, Acer, Armeniaca mume* var. *mume, Betula, Cornus controversa, Carpinus betulus, Castanea, Cerasus serrulata* var. *serrulata, Cornus controversa, Corylus heterophylla, Fagus crenata, Juglans neotropica, Juglans regia, Lonicera japonica, Malus »hillieri«, Malus pumila, Morus alba, Padus avium* var. *avium, Pyrus communis, Quercus × hispanica, Rosa, Salix caprea, Salix cinerea, Sorbus, Tilia cordata, Viburnum odoratissimum* var. *awabuki, Zelkova serrata.*

bonita (Jordan, 1911)
Foodplants: Unknown

cachara (Moore, 1872)
microcaligula Nässig, 1994
Foodplants: *Crataegus laevigata, Crataegus monogyna, Juglans regia, Ligustrum ovalifolium, Liquidambar styraciflua, Prunus laurocerasus, Quercus acuta, Quercus ilex, Rhus typhina, Salix caprea, Tilia platyphyllos.*

chinensis chinensis Rebel, [1926]
kansuensis Mell, 1939
meridionalis Bouvier, 1936
meridionalis Mell, 1939
Foodplants: *Carpinus betulus, Juglans regia.*

chinensis luteoshaanxiana Brechlin, 2011
Foodplants: Unknown

chinghaina Zhu & Wang, 1993
Foodplants: Unknown

florianii Brechlin, 2011
Foodplants: Unknown

fujiana Brechlin, 2007
Foodplants: Unknown

fukudai Sonan, 1937
Foodplants: *Betula pendula, Populus deltoides.*

grotei Moore, 1858
Foodplants: Unknown

heinrichi (Lemaire, 1976)
Foodplants: Unknown

japonica arisana (Shiraki, 1913)
Foodplants: *Castanea crenata, Cinnamomum camphora, Ginkgo biloba, Juglans mandshurica, Quercus acutissima.*

japonica japonica (Moore, 1862)
castanea Swinhoe, 1892
kurimushi Voelschow, 1902
manonis Matsumara, 1927
regina Staudinger, 1901
Foodplants: *Acer, Aesculus hippocastanum, Armeniaca mume* var. *mume, Carpinus betulus, Castanea crenata, Castanea mollissima, Catalpa bungei, Cerasus serrulata* var. *serrulata, Cercis siliquastrum, Cinnamomum camphora, Citrus junos, Corylus heterophylla, Crataegus monogyna, Crataegus laevigata, Cupania lessertiana, Diospyros kaki, Fagus, Ginkgo biloba, Juglans neotropica, Juglans regia, Lagerstroemia indica, Larix decidua, Lespedeza bicolor, Malus »hillieri«, Malus pumila, Malus sylvestris, Morus alba, Padus virginiana, Platanus occidentalis, Populus*

deltoides, Prunus armeniaca, Prunus cerasifera, Prunus persica, Prunus salicina, Punica granatum, Pyrus communis, Quercus acutissima, Quercus dentata, Quercus serrata, Quercus variabilis, Rosa, Toxicodendron vernicifluum, Salix caprea, Schleichera oleosa, Toxicodendron vernicifluum, Viburnum opulus, Zelkova serrata.

japonica ryukyuensis (Inoue, 1984)
Foodplants: *Crataegus*

jonasii Butler, 1877
Foodplants: *Carpinus betulus, Castanea crenata, Crataegus, Juglans neotropica, Malus »hillieri«, Padus avium* var. *avium, Populus deltoides, Quercus acutissima, Quercus mongolica, Quercus serrata, Quercus variabilis, Rosa.*

kitchingi Brechlin, 2001
Foodplants: Unknown

lindia (Moore, 1865)
hockingii (Moore, 1888)
sillemi Bouvier, 1936
Foodplants: *Acer saccharinum, Betula pendula, Corylus avellana, Juglans regia, Malus, Padus avium* var. *avium, Populus deltoides, Prunus avium, Prunus laurocerasus, Pyrus calleryana, Quercus × hispanica, Quercus robur, Salix caprea.*

minshanensis Brechlin, 2011
Foodplants: Unknown

nanlingensis Brechlin, 2004
Foodplants: Unknown

naumanni Brechlin, 2001
Foodplants: *Carpinus betulus, Juglans regia.*

paragrotei Naumann & Nässig, 2010
Foodplants: Unknown

pelelaensis Brechlin, 2009
Foodplants: Unknown

roselata Naumann & Nässig, 2010
Foodplants: Unknown

ruda Brechlin, 2011
Foodplants: Unknown

saldaitisi Brechlin, 2011
Foodplants: Unknown

shaanxiana Brechlin, 2009
Foodplants: Unknown

simla Westwood, 1847
francki Watson, 1919
jaintaiensis Watson, 1927
oberthueri Watson, 1919
Foodplants: *Betula pendula, Carpinus betulus, Carya, Castanea sativa, Crataegus monogyna, Crataegus curvisepala, Juglans neotropica, Juglans regia, Ligustrum ovalifolium, Liquidambar styryciflua, Malus »hillieri«, Malus pumila, Padus avium* var. *avium, Prunus cerasifera, Prunus domestica, Pyrus communis, Quercus robur, Salix babylonica, Salix caprea, Salix fragilis.*

sinanna Naumann & Nässig, 2010
Foodplants: Unknown

sinjaevi Brechlin, 2004
Foodplants: *Carpinus betulus, Juglans regia, Malus »hillieri«, Padus avium* var. *avium, Prunus serotina.*

thibeta Westwood, 1853
arctica Yang, 1978
bonhourei Le Moult, 1933
extensa Butler, 1881
okurai Okano, 1960
pahangensis Paukstadt & Paukstadt, 2005
Foodplants: *Andromeda, Crataegus monogyna, Fagus, Glochidion zeylanicum* var. *Zeylanicum, Ligustrum ovalifolium, Liquidambar styraciflua, Lyonia ovalifolia, Malus »hillieri«, Padus avium* var. *avium, Prunus cerasifera, Prunus serotina, Quercus variabilis, Salix caprea*

tibetanna Naumann & Nässig, 2010
Foodplants: Unknown

winbrechlini Brechlin, 2000
Foodplants: Unknown

witti Brechlin, 1997
Foodplants: Unknown

zuleika lesoudieri Le Moult, 1933
malaisei Bryk, 1944
Foodplants: *Acer caudatum, Actinodaphne sikkimensis, Cornus macrophylla, Quercus acutissima, Salix caprea.*

zuleika zuleika (Hope, 1843)
Foodplants: *Acer caudatum, Actinodaphne sikkimensis, Betula, Cornus macrophylla, Quercus acutissima, Salix caprea.*

Subgenus *Saturnia* Schrank, 1802

atlantica Lucas, 1848
maroccana Austaut, 1894
Foodplants: *Alnus, Crataegus monogyna, Fraxinus berlandieriana, Fraxinus excelsior, Ligustrum ovalifolium, Malus domestica, Malus »hillieri«, Malus pumila, Malus sylvestris, Olea europaea, Padus avium* var. *avium, Prunus armeniaca, Prunus domestica, Prunus persica, Prunus spinosa, Rhamnus, Rosa, Syringa vulgaris.*

cameronensis Lemaire, 1979
Foodplants: Unknown

centralis Naumann & Löffler, 2005
Foodplants: Unknown

cidosa Moore, 1865
Foodplants: Unknown

cognata (Jordan, 1911)
Foodplants: *Cinnamomum camphora, Crataegus monogyna, Eriobotrya japonica, Liquidambar formosana, Liquidambar styraciflua, Psidium guajava, Pterocarya stenoptera, Pyrus pyrifolia, Rosa multiflora, Ulmus pumila.*

koreanis Brechlin, 2009
Foodplants: Unknown

luctifera (Jordan, 1911)
Foodplants: Unknown

pinratanai Lampe, 1989
Foodplants: *Ligustrum ovalifolium*

pyretorum Westwood, 1847
fusca (Watson, 1915)
pearsoni Watson, 1911
Foodplants: *Cinnamomum camphora, Crataegus, Eriobotrya japonica, Juglans regia, Liquidambar formosana, Liquidambar styraciflua, Prunus cerasifera, Psidium guajava, Pterocarya stenoptera, Pyrus pyrifolia, Rosa multiflora, Ulmus pumila.*

roseata (De Joannis, 1929)
tonkinensis (Bouvier, 1936)
Foodplants: Unknown

pyri ([Denis & Schiffermüller], 1776)
alticola Denso, 1912
ilvana Tauber, 1969
junonia Schaw, 1806
major Linnaeus, 1758
pavoniella (Gmelin, 1788)
pinkerae Kobes, 1980
teherana Lének, [1971]
Foodplants: *Acer pseudoplatanus, Aesculus hippocastanum, Alnus glutinosa, Amygdalus, Amygdalus triloba, Betula pendula, Castanea, Corylus avellana, Crataegus monogyna, Cydonia, Fraxinus excelsior, Fraxinus ornus, Humulus, Juglans regia, Ligustrum ovalifolium, Ligustrum vulgare, Malus domestica, Malus »hillieri«, Malus pumila, Olea europaea, Padus avium* var. *avium, Planera aquatica, Platanus, Populus, Prunus amygdalus, Prunus armeniaca, Prunus avium, Prunus cerasifera, Prunus domestica, Prunus emarginata* subsp. *mollis, Prunus persica, Prunus serotina, Prunus spinosa, Pyrus communis, Salix alba, Salix babylonica, Salix caprea, Salix fragilis, Sorbus aucuparia, Syringa persica, Syringa vulgaris, Tilia cordata, Tilia platyphyllos.*

taibaishanis Brechlin, 2009
Foodplants: Unknown

Solus Watson, 1913

drepanoides (Moore, [1866])
Foodplants: Unknown

parvifenestratus gabaidanus Nässig, 1989
Foodplants: Unknown

parvifenestratus parvifenestratus Bryk, 1944
Foodplants: Unknown

parvifenestratus sinjaevi Nässig, 1994
Foodplants: Unknown

sichuanus Brechlin, 2007
Foodplants: Unknown

Syntherata Maassen, 1873

anettae Naumann, Lane & Löffler, 2009
Foodplants: Unknown

apicalis Bouvier, 1928
Foodplants: Unknown

arfakiana Brechlin, 2010
Foodplants: Unknown

bettinae Naumann, Lane & Löffler, 2009
Foodplants: Unknown

bretschneideri Naumann, Lane & Löffler, 2009
Foodplants: Unknown

brunnea Eckerlein, 1935
Foodplants: Unknown

cernyi Brechlin, 2010
Foodplants: Unknown

devosi Naumann, Lane & Löffler, 2009
Foodplants: Unknown

doboensis Paukstadt & Paukstadt, 2004
Foodplants: Unknown

engaiana Brechlin, 2010
Foodplants: Unknown

erici Naumann, Lane & Löffler, 2009
Foodplants: Unknown

escarlata Lane, Edwards & Naumann, 2010
Foodplants: Unknown

godeffroyi godeffroyi Butler, 1882
dahli Weymer, 1898
Foodplants: Unknown

godeffroyi michaschaarschmidti Brechlin, 2010
Foodplants: Unknown

gracilis Naumann, Lane & Löffler, 2009
Foodplants: Unknown

groenendaeli Naumann, Lane & Löffler, 2009
Foodplants: Unknown

haiaensis Naumann, Lane & Löffler, 2009
Foodplants: Unknown

hoffmanni Naumann, Lane & Löffler, 2009
Foodplants: Unknown

innescens Naumann & Brechlin, 2001
Foodplants: Unknown

janetta (White, 1843)
aliena Schüssler, 1934
disjuncta (Walker, 1865)
insignis (Walker, 1869)
melvilla (Westwood, 1854)
olivacea Niepelt, 1934
olivescens Schüssler, 1934
purpurascens (Walker, 1865)
weymeri Maassen, [1873]
Foodplants: *Aegiceras, Ceriops, Citrus, Corymbia torelliana, Eucalyptus deglupta, Geijera salicifolia, Glochidion ferdinandii, Ligustrum ovalifolium, Maesopsis eminii, Malus »hillieri«, Melicope elleryana, Olea europaea, Paullinia, Petalostigma quadriloculare, Podocarpus spinulosus, Prunus laurocerasus, Psidium, Quercus robur, Schinus molle, Timonius wallichianus, Zanthoxylum clava-herculis, Zanthoxylum gilletii.*

labriquei Naumann, Lane & Löffler, 2009
Foodplants: Unknown

lagariana Brechlin, 2010
Foodplants: Unknown

leonae Lane, 2003
Foodplants: Unknown

lilianae Naumann, Lane & Löffler, 2009
Foodplants: Unknown

malukuensis Paukstadt & Paukstadt, 2005
Foodplants: Unknown

marlenae Naumann, Lane & Löffler, 2009
Foodplants: Unknown

minoris Brechlin, 2010
Foodplants: Unknown

mirata Lane, Edwards & Naumann, 2010
Foodplants: Unknown

naessigi Peigler, 1992
Foodplants: Unknown

okapiana Brechlin, 2010
Foodplants: Unknown

papuensis Brechlin, 2010
Foodplants: Unknown

parvoantennata Brechlin, 2010
Foodplants: Unknown

pierrei Naumann, Lane & Löffler, 2009
Foodplants: Unknown

pristina (Walker, 1865)
Foodplants: *Alphitonia excelsa, Callicoma serratifolia, Flindersia australis, Flindersia maculosa, Flindersia schottiana.*

pseudescarlata Lane, Edwards & Naumann, 2010
Foodplants: Unknown

rudloffi Brechlin, 2010
Foodplants: Unknown

sinjaevi Naumann & Brechlin, 2001
Foodplants: Unknown

vincenti Naumann, Lane & Löffler, 2009
Foodplants: Unknown

Attacini

Archaeoattacus Watson, 1914

edwardsii (White, 1859)
malayanus Kurosawa & Kishida, [1985]
Foodplants: *Ailanthus altissima, Berberis vulgaris,*

Cinnamomum, Ilex chinensis, Laurus nobilis, Ligustrum ovalifolium, Machilus, Populus, Prunus domestica, Prunus laurocerasus, Salix, Syringa vulgaris, Tilia cordata.

staudingeri (Rothschild, 1895)
Foodplants: *Ailanthus altissima, Ligustrum ovalifolium, Prunus laurocerasus, Syringa vulgaris*

Archaeosamia Brechlin, 2007

watsoni (Oberthür, 1914)
formosana (Matsumura, 1931)
Foodplants: *Liquidambar formosana, Manihot esculenta.*

Attacus Linnaeus, 1767

atlas (Linnaeus, 1758)
atlantis Watson, 1910
baliensis Jurriaanse & Lindemans, 1921
burmaensis Jurriaanse & Lindemans, 1920
chinensis Bouvier, 1936
formosanus Villiard, 1969
gladiator Fruhstorfer, 1904
incerta Bouvier, 1936
javanensis Jurriaanse & Lindemans, 1920
major Dusuzeau & Sonthonnax, 1897
mannus Fruhstorfer, 1904
minor Dusuzeau & Sonthonnax, 1897
opaca Bouvier, 1936
pallida Bouvier, 1936
roseus Fruhstorfer, 1904
rotundus Jurriaanse & Lindemans, 1920
ryukyuensis Inoue, 1993
silhetica Helfer, 1837
simalurana Watson, 1915
simplex Bouvier, 1936
strandi Schüssler, 1933
sumatranus Fruhstorfer, 1904
talas Hübner, [1819]
tonkinensis Roepke, 1953
triumphator Fruhstorfer, 1904
varia Bouvier, 1936
vitrea Perry, 1811
Foodplants: *Ailanthus altissima, Alstonia scholaris, Anacardium, Annona muricata, Ardisia sieboldii, Artabotrys, Artemisia vulgaris, Averrhoa carambola, Berberis asiatica, Berberis thunbergii, Berberis vulgaris, Betula platyphylla, Bischofia javanica, Camellia sinensis, Cananga odorata, Canarium indicum, Carpinus betulus, Caryodaphnopsis tonkinensis, Ceiba pentandra, Chamaedaphne latifolia, Cinchona pubescens, Cinnamomum camphora, Cinnamomum iners, Cinnamomum verum, Citrus, Rotheca serrata, Clerodendrum infortunatum, Clidemia hirta, Coffea arabica, Combretum indicum, Cotoneaster horizontalis, Curcuma viridiflora, Dillenia indica, Dillenia pentagyna, Embelia, Erythrina subumbrans, Etlingera elatior, Euphorbia longispina, Falconeria insignis, Ficus variegata, Fraxinus, Glochidion rubrum, Glochidion velutinum, Hibiscus, Ilex chinensis, Ipomoea batatas, Lagerstroemia indica, Lannea coromandelica, Lantana camara, Leucosceptrum canum, Ligustrum ovalifolium, Ligustrum vulgare, Litchi chinensis, Lonicera japonica, Liquidambar styraciflua, Malus, Malus pumila, Mangifera indica, Melastoma malabathricum, Meyna grisea, Michelia, Morus, Mussaenda, Mitragyna rotundifolia, Nephelium lappaceum, Nerium oleander, Parkia intermedia, Parkia speciosa, Persea americana, Phyllanthus emblica, Piper, Populus, Prunus laurocerasus, Psidium guajava, Pyrus communis, Quercus glauca, Rhododendron, Ricinus communis, Rosa, Salix caprea, Sandoricum koetjape, Sarcostemma viminale* subsp. *Brunonianum, Sassafras albidum, Schefflera heptaphylla, Schleichera oleosa, Sesbania grandiflora, Setaria viridis, Spathodea campanulata, Spondias, Stachytarpheta cayennensis, Swietenia mahogani, Symplocos paniculata, Syringa vulgaris, Syzygium aqueum, Syzygium malaccense, Teucrium, Theobroma cacao, Triadica sebifera, Uncaria gambir, Vangueria, Vernicia montana.*

aurantiacus Rothschild, 1895
atlantis Strand, 1913
aurota Skell, 1942
Foodplants: Unknown

caesar Maassen & Weymer, 1872
imperator Kirby, 1892
Foodplants: *Annona muricata, Annona squamosa, Artocarpus integer, Cananga odorata, Caryodaphnopsis tonkinensis, Citrus, Erythrina subumbrans, Gmelina arborea, Ligustrum ovalifolium, Ligustrum vulgare, Liquidambar styraciflua, Myrica cerifera, Persea americana, Prunus laurocerasus, Salix caprea, Sandoricum koetjape, Swietenia mahogani, Syringa vulgaris.*

crameri C. Felder, 1861
amboinensis Hutton, 1863
Foodplants: Unknown

dohertyi W. Rothschild, 1895
Foodplants: *Ailanthus altissima, Annona muricata, Ligustrum ovalifolium, Persea americana, Syringa vulgaris.*

erebus Fruhstorfer, 1904
Foodplants: *Ficus, Ligustrum ovalifolium, Prunus laurocerasus, Sandoricum koetjape, Syringa vulgaris.*

inopinatus Jurriaanse & Lindemans, 1920
Foodplants: *Ligustrum ovalifolium, Syringa vulgaris.*

intermedius Jurriaanse & Lindemans, 1920
ruegeri Gschwandner, 1920
Foodplants: Unknown

lemairei Peigler, 1985
Foodplants: *Annona, Hibiscus, Ligustrum ovalifolium, Nauclea orientalis, Sandoricum koetjape.*

lorquinii C. Felder & R. Felder, 1861
banghaasi Gschwandner, 1920
philippina Bouvier, 1930
Foodplants: *Annona muricata, Annona squamosa, Artocarpus integer, Cananga odorata, Chrysophyllum, Citrus, Fraxinus, Heliconia, Ligustrum ovalifolium, Ligustrum vulgare, Magnolia virginiana, Mangifera indica, Myrica cerifera, Persea americana, Persea borbonia, Prunus laurocerasus, Psidium guajava, Quercus glauca, Saccharum officinarum, Salix, Sandoricum koetjape, Schefflera arboricola, Swietenia mahogani, Syringa vulgaris, Theobroma cacao.*

mcmulleni J. H. Watson, 1914
Foodplants: Unknown

paraliae Peigler, 1985
Foodplants: *Ligustrum ovalifolium, Syringa vulgaris.*

paukstadtorum Brechlin, 2010
Foodplants: Unknown

soembanus Van Eecke, 1933
dammermani Roepke, 1953
Foodplants: Unknown

suparmani U. Paukstadt & L. H. Paukstadt, 2002
Foodplants: *Ligustrum ovalifolium, Syringa vulgaris.*

tabrobanis Moore, 1882
mysorensis Bouvier, 1936
similis Bouvier, 1936
Foodplants: *Aglaia elaeagnoidea, Berberis asiatica, Berberis thunbergii, Berberis vulgaris, Cinnamomum camphora, Cinnamomum verum, Ligustrum ovalifolium, Aglaia elaeagnoidea, Ocimum.*

wardi Rothschild, 1910
Foodplants: Unknown

Callosamia Packard, 1864

angulifera (Walker, 1855)
atys (Dusuzeau & Sonthonnax, 1897)
Foodplants: *Liriodendron tulipifera, Magnolia tripetala, Malus »hillieri«, Prunus serotina, Sassafras albidum.*

promethea (Drury, 1773)
caeca Cockerell, 1914
Foodplants: *Acer campestre, Azalea, Berberis vulgaris, Betula papyrifera, Betula pendula, Cephalanthus occidentalis, Cinnamomum camphora, Corylus avellana, Cotinus coggygria, Diospyros, Fagus sylvatica, Fraxinus americana, Fraxinus excelsior, Fraxinus pennsylvanica, Gordonia lasianthus, Halesia carolina, Kalmia, Ligustrum ovalifolium, Ligustrum vulgare, Lindera benzoin, Liquidambar styraciflua, Liriodendron tulipifera, Magnolia acuminata, Magnolia virginiana, Malus domestica, Malus pumila, Myrica pensylvanica, Padus virginiana, Pinus, Populus, Prunus domestica, Prunus emarginata* subsp. *mollis, Prunus persica, Prunus serotina, Pyrus communis, Quercus robur, Rhododendron, Salix babylonica, Salix caprea, Sassafras albidum, Styrax americanus japonicus, Symplocos, Symplocos tinctoria, Syringa vulgaris, Thuja occidentalis, Tilia americana, Viburnum.*

securifera (Maassen & Weymer, 1873)
aurantiaca Igel, 1928
carolina Jones, 1908
Foodplants: *Cinnamomum camphora, Gordonia lasianthus, Liquidambar styraciflua, Liriodendron tulipifera, Magnolia grandiflora, Magnolia virginiana.*

Coscinocera Butler, 1879

anteus Bouvier, 1928
amputata Niepelt, 1936
brachyura Bouvier, 1936
Foodplants: *Ligustrum ovalifolium*

aruensis Naumann & Löffler, 2010
aruensis Brechlin, 2010
Foodplants: Unknown

eurystheus Rothschild, 1898
Foodplants: *Ligustrum ovalifolium*

hercules (Miskin, 1876)
butleri Rothschild, 1895
heraclides Joicey & Talbot, 1916
joiceyi Bouvier, 1927
heros Rothschild, 1899
Foodplants: *Dysoxylum mollissimum* subsp. *molle, Glochidion ferdinandii, Homalanthus populifolius, Homalanthus nutans, Ligustrum ovalifolium, Omalanthus novoguinensis, Polyscias elegans, Polyscias fulva, Prunus serotina, Syringa vulgaris, Timonius wallichianus, Timonius singularis.*

jakli Naumann, 2009
Foodplants: Unknown

niepelti Brechlin, 2004
Foodplants: Unknown

omphale Butler, 1879
titanus Niepelt, 1916
Foodplants: Unknown

rothschildi Le Moult, 1933
Foodplants: Unknown

Epiphora Wallengren, 1860

albida (Druce, 1886)
noigamagetana (Stoneham, 1933)
paradoxa Dufrane, 1953
rufa Dufrane, 1953
semialba Rebel, 1927
Foodplants: *Citrus, Maesopsis eminii, Paullinia, Zanthoxylum clava-herculis, Zanthoxylum gilletii.*

antinorii (Oberthür, 1880)
cotei Le Moult, 1940
ducalis Grünberg, 1927
rufa (Bouvier, 1929)
Foodplants: *Citrus, Fraxinus berlandieriana, Maesopsis eminii, Paullinia, Zanthoxylum clava-herculis, Zanthoxylum gilletii.*

atbarina Butler, 1877
sudanica Le Cerf, 1917
Foodplants: *Ziziphus*

bauhiniae (Guérin-Méneville, 1829)
damarensis Schultze, 1913
faidherbia (De la Rocha, 1874)
schultzei Aurivillius, 1905
vera Janse, 1918
Foodplants: *Ceanothus thyrsiflorus, Frangula dodonei, Ziziphus abyssinica, Ziziphus jujuba, Ziziphus mucronata, Ziziphus spina-christi.*

bedoci (Bouvier, 1930)
Foodplants: Unknown

boolana (Strand, 1909)
cordieri (Bouvier, 1928)
liberiensis (Bouvier, 1928)
Foodplants: *Discoglypremna caloneura*

cadioui Bouyer, 2008
Foodplants: *Discoglypremna caloneura*

feae (Aurivillius, 1908)
conjuncta (Bouvier, 1930)
niepelti (Gschwandner, 1925)
Foodplants: Unknown

fournierae Le Moult, 1946
elainae Rougeot, 1973
hassoni Bouyer, 2008
Foodplants: *Citrus, Discoglypremna caloneura, Maesopsis eminii, Paullinia, Zanthoxylum clava-herculis, Zanthoxylum gilletii.*

imperator (Stoneham, 1933)
imperator Rougeot, 1991
Foodplants: Unknown

kipengerensis Darge, 2007
Foodplants: Unknown

lugardi Kirby, 1897
marwitzi Weymer, 1894
Foodplants: *Ceanothus, Citrus, Discoglypremna caloneura, Maesopsis eminii, Paullinia, Zanthoxylum clava-herculis, Zanthoxylum gilletii, Ziziphus.*

macedoi Darge, Mendes & Bivar de Souza, 2006
Foodplants: Unknown

macrops (Bouvier, 1929)
Foodplants: Unknown

magdalena (Grünberg, 1909)
Foodplants: Unknown

manowiensis (Gschwandner, 1923)
brunnea (Bouvier, 1930)
Foodplants: *Helinus, Ziziphus mucronata.*

marginimacula (Joicey & Talbot, 1927)
ugandensis (Testout, 1948)
Foodplants: Unknown

mineti Darge, 1994
styrax Darge, 1994
Foodplants: Unknown

murphyi Bouyer, 2008
Foodplants: Unknown

mythimnia (Westwood, 1849)
scribonia Wallengren, 1860
Foodplants: *Ceanothus thyrsiflorus, Croton gratissimus, Croton sylvaticus, Fernandoa magnifica, Frangula dodonei, Helinus integrifolius, Ziziphus abyssinica, Ziziphus mucronata.*

newporti Bouyer, 2007
Foodplants: Unknown

pelosoma (Rothschild, 1907)
boursini (Testout, 1936)
bouvieri (Testout, 1936)
cotei (Testout, 1935)
nubilosa (Testout, 1938)
Foodplants: *Ceanothus thyrsiflorus*

oberprieleri Bouyer, 2008
Foodplants: Unknown

perspicua (Butler, 1878)
berliozi (Rougeot, 1947)
vacunoides (Testout, 1948)
Foodplants: *Xanthoxylum americanum, Zanthoxylum gilletii.*

ploetzi (Weymer, 1880)
aequitorialis (Testout, 1936)
congolana (Bouvier, 1929)
gabonensis (Testout, 1936)
intermedia (Rougeot, 1955)
obscura Dufrane, 1953
pygmaea (Bouvier, 1929)
testenoirei (Bouvier, 1929)
testouti Rougeot, 1947
Foodplants: *Citrus, Maesopsis eminii, Paullinia, Xanthoxylum americanum, Zanthoxylum clava-herculis, Zanthoxylum gilletii.*

rectifascia rectifascia (Rothschild, 1907)
ileshana Rougeot, 1955
Foodplants: *Croton, Maesopsis eminii, Ziziphus abyssinica.*

rectifascia watulegei Rougeot, 1974
Foodplants: Unknown

rotunda Naumann, 2006
Foodplants: Unknown

torquata (Bouvier, 1929)
Foodplants: Unknown

vacuna (Westwood, 1849)
getula (Maassen & Weymer, 1886)
lecerfi (Testout, 1936)
lineata (Bouvier, 1930)
modesta (Bouvier, 1936)
victoria (Maassen & Weymer, 1886)
weymeri (Druce, 1886)
Foodplants: *Croton oligandrus, Discoglypremna caloneura, Maesopsis eminii, Neoboutonia melleri, Zanthoxylum heitzii.*

werneri Darge, 2007
Foodplants: Unknown

Eupackardia Cockerell, 1912

calleta (Westwood, [1854])
caeca Draudt, 1929
digueti Bouvier, 1936
polyommata (Tepper, 1883)
Foodplants: *Betula, Forestiera angustifolia, Fouquieria columnaris, Fouquieria splendens, Fraxinus greggii, Leucophyllum frutescens, Ligustrum ovalifolium, Ligustrum sinense, Ligustrum vulgare, Malosma laurina, Malus pumila, Olea europaea, Platycladus orientalis, Prunus domestica, Prunus emarginata* subsp. *mollis, Prunus serotina, Salix babylonica, Schinus molle, Sebastiania bilocularis, Syringa vulgaris, Tecoma stans.*

Hyalophora Ducan, 1841

cecropia (Linnaeus, 1758)
albofasciata Sageder, 1933
obscura Sageder, 1933
Foodplants: *Acer negundo, Acer plantanoides, Acer rubrum, Acer spicatum, Acer saccharinum, Adenorachis arbutifolia, Ailanthus altissima, Alnus glutinosa, Alnus incana, Alnus incana* subsp. *Rugosa, Alnus serrulata, Aucuparia americana, Berberis vulgaris, Betula alleghaniensis, Betula lenta, Betula papyrifera, Betula papyrifera* var. *papyrifera, Betula pendula, Betula populifolia, Betula pubescens, Berberis, Carpinus caroliniana* subsp. *caroliniana, Carya illinoinensis, Ceanothus, Cephalanthus occidentalis, Cerasus vulgaris, Cornus sericea* subsp. *sericea, Corylus americana, Corylus cornuta, Crataegus calpodendron, Crataegus chapmanii, Crataegus crus-galli, Crataegus curvisepala, Crataegus pedicellata, Diospyros texana, Fagus crenata, Fagus grandifolia, Fraxinus americana, Fraxinus excelsior, Gaylussacia frondosa, Gleditsia triacanthos, Humulus lupulus, Juglans nigra, Juglans regia, Ju-*

niperus virginiana, Larix, Laurocerasus ilicifolia, Liquidambar styraciflua, Liriodendron tulipifera, Maclura pomifera, Malosma laurina, Malus domestica, Malus pumila, Mespilus coccinea, Morus alba, Myrica caroliniensis, Myrica cerifera, Negundo aceroides, Nyssa sylvatica, Opulaster opulifolius, Ostrya virginiana, Oxydendrum arboreum, Padus virginiana, Paeonia officinalis, Parthenocissus quinquefolia, Platanus occidentalis, Populus balsamifera, Populus tremuloides, Prunus avium, Prunus domestica, Prunus emarginata subsp. *mollis, Prunus lanata, Prunus laurocerasus, Prunus maritima, Prunus serotina, Prunus spinosa, Pyrus canadensis, Pyrus communis, Quercus alba, Ribes alpinum, Ribes americanum, Ribes nigrum, Ribes rubrum, Ribes sativum, Ribes uva-crispa, Rhus glabra, Rhus typhina, Rosa blanda, Rubus allegheniensis, Rubus idaeus, Rubus occidentalis, Salix alba, Salix eriocephala* var. *eriocephala, Salix interior, Salix lasiopes, Salix lucida, Salix viminalis, Sambucus aurea, Sambucus canadensis, Sambucus nigra, Sambucus pubens, Sambucus racemosa, Sassafras albidum, Sorbus coronaria, Sorbus coronaria* var. *angustifolia, Spiraea betulifolia, Spiraea alba* var. *latifolia, Spiraea salicifolia, Spiraea mongolica* var. *tomentulosa, Symphoricarpos albus, Syringa inodora, Syringa vulgaris, Tilia americana, Tilia europaea, Ulmus americana, Ulmus rubra, Ulmus rubra, Ulmus thomasii, Vaccinium, Viburnum nudum* var. *cassinoides, Wisteria sinensis.*

columbia columbia (S. I. Smith, 1865)
Foodplants: *Abies balsamea, Cerasus vulgaris, Ilex mucronata, Kalmia augustifolia, Larix decidua, Larix laricina, Myrica pensylvanica, Picea glauca, Picea mariana, Prunus domestica, Prunus emarginata* subsp. *mollis, Prunus serotina, Padus virginiana, Pseudotsuga menziesii, Rhododendron canadense, Salix alba, Salix babylonica, Salix caprea, Spiraea mongolica* var. *tomentulosa, Tsuga canadensis, Ulmus.*

columbia gloveri (Strecker, 1872)
nokomis (Brodie, 1884)
reducta (Neumoegen, 1891)
wenonah (Brodie, 1884)
Foodplants: *Acer negundo, Alnus, Arctostaphylos patula, Arctostaphylos pungens, Betula pendula, Ceanothus cordulatus, Ceanothus fendleri, Ceanothus vemutinus, Cercocarpus betuloides, Cercocarpus montanus, Elaeagnus augustifolia, Larix decidua, Larix laricina, Lepargyrea argentea, Ligustrum ovalifolium, Limnobotrya lacustris, Malus »hillieri«, Prunus domestica, Prunus emarginata, Prunus serotina, Padus virginiana, Padus virginiana* subsp. *demissa, Pseudotsuga, Purshia tridentata, Ribes alpinum, Ribes sativum, Rosa, Rhus trilobata, Salix alba, Salix babylonica, Salix caprea, Salix eriocephala* var. *eriocephala, Salix exigua, Shepherdia canadensis, Syringa vulgaris, Ulmus.*

euryalus cedrosensis (Cockerell, 1914)
Foodplants: *Acer glabrum, Alnus rubra, Amelanchier alnifolia* subsp. *florida, Amelanchier sanguinea* var. *alnifolia, Arctostaphylos glauca, Arctostaphylos patula, Betula occidentalis, Ceanothus cordulatus, Ceanothus integerrimus, Ceanothus sanguineus, Ceanothus thyrsiflorus, Ceanothus velutinus, Cephalanthus, Coffea, Frangula californica, Frangula purshiana, Heteromeles salicifolia, Larix laricina, Lepargyrea argentea, Liquidambar styraciflua, Malosma laurina, Malus pumila, Pachycormus discolor, Padus virginiana, Padus virginiana* subsp. *demissa, Picea, Prunus emarginata, Prunus serotina, Pseudotsuga menziesii, Purshia tridentata, Quercus robur, Rhamnus crocea, Ribes alpinum, Ribes sanguineum, Rosa, Salix babylonica, Salix eriocephala* var. *eriocephala, Salix exigua, Schinus molle, Schmaltzia integrifolia, Shepherdia canadensis, Syringa vulgaris*

euryalus euryalus (Boisduval, 1855)
california (Genzke, 1859)
californica (Grote, 1865)
ceanothi (Boiduval, 1869)
parvimacula (Grote, 1902)
rubra (Neumoegen & Dyar, 1894)
Foodplants: *Acer glabrum, Alnus rubra, Amelanchier alnifolia* subsp. *florida, Amelanchier sanguinea* var. *alnifolia, Arbutus, Arctostaphylos glauca, Arctostaphylos patula, Betula occidentalis, Ceanothus cordulatus, Ceanothus integerrimus, Ceanothus sanguineus, Ceanothus thyrsiflorus, Ceanothus velutinus, Cephalanthus, Cercocarpus betuloides, Coffea, Frangula californica, Frangula purshiana, Larix laricina, Lepargyrea argentea, Liquidambar styraciflua, Malosma laurina, Malus pumila, Heteromeles salicifolia, Padus virginiana* subsp. *demissa, Picea, Prunus domestica, Prunus emarginata, Prunus emarginata* subsp. *mollis, Prunus serotina, Padus virginiana, Pseudotsuga menziesii, Purshia tridentata, Quercus robur, Rhamnus crocea, Ribes alpinum, Ribes sanguineum, Rosa, Salix babylonica, Salix caprea, Salix eriocephala* var. *eriocephala, Salix exigua, Schinus molle, Shepherdia canadensis, Syringa vulgaris.*

kasloensis (Cockerell, 1914)
Foodplants: *Ceanothus cordulatus, Ceanothus integerrimus, Ceanothus sanguineus, Ceanothus velutinus, Elaeagnus augustifolia, Frangula rubra, Lepargyrea argentea, Malosma laurina, Padus virginiana* subsp. *demissa, Prunus emarginata, Prunus serotina, Pseudotsuga menziesii, Rhamnus persica, Salix exigua, Shepherdia canadensis, Spiraea douglasii.*

Rothschildia Grote, 1897

amoena Jordan, 1911
Foodplants: *Ailanthus altissima, Jatropha cinerea, Ligustrum ovalifolium, Ligustrum vulgare, Liquidambar styraciflua, Prunus domestica, Pyracantha, Quercus robur, Syringa vulgaris.*

anikae Brechlin & Meister, 2010
Foodplants: Unknown

arethusa arethusa (Walker, 1855)
speculum (Maassen, [1873])
Foodplants: *Ailanthus altissima, Ligustrum ovalifolium, Ligustrum vulgare, Quercus × hispanica, Syringa vulgaris.*

arethusa rhodina Jordan, 1911
Foodplants: *Ailanthus altissima, Ligustrum ovalifolium, Quercus × hispanica, Syringa vulgaris.*

aricia aricia (Walker, 1855)
Foodplants: *Ligustrum ovalifolium, Prunus serotina, Salix babylonica.*

aricia napoecuadoriana Brechlin & Meister, 2010
Foodplants: *Ligustrum ovalifolium, Syringa vulgaris.*

aurota aurota (Cramer, 1775)
andensis Rothschild, 1907
Foodplants: *Ailanthus altissima, Ilex, Ligustrum ovalifolium, Ligustrum vulgare, Liquidambar styraciflua, Prunus laurocerasus, Ricinus communis, Syringa vulgaris.*

aurota speculifera (Walker, 1855)
Foodplants: *Cephalanthus, Croton urucurana, Fraxinus, Ligustrum japonicum, Ligustrum ovalifolium, Liquidambar styraciflua, Prunus laurocerasus, Ricinus communis, Schinus molle, Syringa vulgaris.*

belus (Maassen, [1873])
mussehli Schaus, 1941
Foodplants: *Ailanthus altissima, Ligustrum ovalifolium, Prunus serotina.*

chiris W. Rothschild, 1907
Foodplants: *Ailanthus altissima, Ligustrum japonicum, Prunus serotina.*

cincta cincta (Tepper, 1883)
Foodplants: *Cephalanthus, Cnidoscolus tubulosus* var. *Tubulosus, Coursetia glandulosa, Croton urucurana, Dodonaea viscosa, Jatropha curcas, Ligustrum ovalifolium, Ligustrum sinense, Liquidambar styraciflua, Mimosa dysocarpa, Prunus serotina, Quercus × hispanica, Salix babylonica, Salix caprea, Salix humboldtiana, Schinus molle, Schinus terebinthifolia, Sebastiania bilocularis.*

cincta guerreronis Draudt, 1929
pseudoguerreronis C. C. Hoffmann, 1942
Foodplants: *Ailanthus altissima, Jatropha cinerea, Ligustrum japonicum, Ligustrum ovalifolium, Ligustrum sinense, Prunus serotina, Quercus × hispanica, Rhus typhina, Syringa vulgaris.*

condor (Staudinger, 1894)
schreiteri Draudt, 1930
stuarti Rothschild & Jordan, 1901
Foodplants: *Quercus robur*

erycina erycina (Shaw, [1796])
amazonia (Packard, 1871)
martha Rothschild, 1907
satyrus (C. Felder & R. Felder, 1874)
splendida (Palisot de Beauvois, [1811])
vinacea Rothschild, 1907
Foodplants: *Ailanthus altissima, Antonia ovata, Cenostigma macrophyllum, Coutarea hexandra, Exostema mexicanum, Ligustrum ovalifolium, Malus »hillieri«, Prunus serotina, Syringa vulgaris.*

erycina luciana W. Rothschild, 1907
balatana Bouvier, 1936
Foodplants: *Ailanthus altissima, Ligustrum, Prunus.*

erycina mexicana Draudt, 1929
Foodplants: *Ailanthus altissima, Coutarea hexandra, Ligustrum ovalifolium, Quercus, Syringa vulgaris.*

erycina nigrescens W. Rothschild, 1907
Foodplants: *Ailanthus altissima, Chiococca alba, Chiococca belizensis, Coutarea hexandra, Exostema mexicanum, Ligustrum ovalifolium, Lonicera, Machaonia erythrocarpa, Prunus serotina, Syringa vulgaris.*

hesperus betis (Walker, 1855)
angulatus Bouvier, 1936
augias (Girard, 1874)
hoffmanni Vogeler, 1933
paulista Vogeler, 1933
Foodplants: *Ailanthus altissima, Ligustrum ovalifolium, Quercus robur, Salix babylonica, Syringa vulgaris.*

hesperus hesperus (Linnaeus, 1758)
imitator Draudt, 1929
melini Bryk, 1953
olivenca Vogeler, 1933
procyon Jordan, 1929
Foodplants: *Ailanthus altissima, Betula pendula, Ligustrum ovalifolium, Liquidambar styraciflua, Malus domestica, Prunus armeniaca, Prunus serotina, Pyrus communis, Ribes aureum, Rubus, Salix babylonica, Syringa vulgaris.*

hesperus lutea Jordan, 1911
wagneri Bouvier, 1936
Foodplants: *Ailanthus altissima, Ligustrum ovalifolium, Quercus, Syringa vulgaris.*

hopfferi (C. Felder & R. Felder, 1859)
rhombifer (Burmeister, 1878)
Foodplants: *Ailanthus altissima, Jacaranda copaia, Ligustrum ovalifolium, Ligustrum vulgare, Prunus serotina, Quercus × hispanica, Quercus robur, Syringa vulgaris.*

interaricia Brechlin & Meister, 2010
Foodplants: Unknown

jacobaeae (Walker, 1855)
affinis (C. Felder & R. Felder, 1874)
encelades (Girard, 1874)
Foodplants: *Cephalanthus glabratus, Ligustrum ovalifolium, Ligustrum vulgare, Padus virginiana, Prunus laurocerasus, Prunus serotina, Salix babylonica, Salix caprea, Schinus terebinthifolia, Senecio, Syringa vulgaris.*

jorulla jorulla (Westwood, [1854])
Foodplants: *Ailanthus altissima, Fraxinus excelsior, Ligustrum ovalifolium, Prunus laurocerasus, Prunus serotina, Quercus robur, Syringa vulgaris.*

jorulla lichtenba Dyar, 1912
Foodplants: *Ailanthus altissima, Fraxinus excelsior, Ligustrum ovalifolium, Prunus laurocerasus, Syringa vulgaris.*

jurolloides (Dognin, 1895)
bucaya Schaus, 1941
cruentata Bouvier, 1930
micrinus Draudt, 1929
Foodplants: *Ailanthus altissima, Ligustrum ovalifolium, Prunus serotina, Salix caprea, Syringa vulgaris.*

lebeau amacurensis Lemaire, 1972
Foodplants: *Ailanthus altissima, Ligustrum ovalifolium, Quercus, Syringa vulgaris.*

lebeau aroma Schaus, 1905
benjamini C. C. Hoffmann, 1942
morana Schaus, 1921
Foodplants: *Ailanthus altissima, Allamanda cathartica, Annona purpurea, Casearia corymbosa, Casearia sylvestris, Casearia tremula, Cestrum racemosum, Citrus sinensis, Conocarpus erectus, Coutarea hexandra, Croton schiedeanus, Erythroxylum havanense, Erythroxylum rotundifolium, Exostema mexicanum, Hippomane mancinella, Hyeronima oblonga, Jatropha curcas, Ligustrum ovalifolium, Machaonia erythrocarpa, Oreopanax capitatus, Pachira quinata, Prunus annularis, Prunus laurocerasus, Psidium guajava, Quercus, Rhizophora mangle, Sapium macrocarpum, Spondias mombin, Spondias purpurea, Swietenia macrophylla, Syringa vulgaris, Trichospermum galeottii, Uncaria tomentosa, Zanthoxylum caribaeum, Zanthoxylum melanostictum, Zanthoxylum setulosum, Zuelania guidonia.*

lebeau forbesi Benjamin, 1934
draudti Benjamin, 1934
semiimmacula Gardiner, 1962
Foodplants: *Ailanthus altissima, Casearia corymbosa, Citrus, Enicospilus lebophagus, Exostema mexicanum, Fraxinus velutina, Ligustrum ovalifolium, Ligustrum sinense, Prunus laurocerasus, Prunus serotina, Malosma laurina, Salix caprea, Salix humboldtiana, Schinus terebinthifolia, Spondias mombin, Spondias purpurea, Syringa vulgaris, Zanthoxylum fagara, Zanthoxylum setulosum, Zuelania guidonia.*

lebeau inca W. Rothschild, 1907
rectilineata Bouvier, 1930
Foodplants: *Annona purpurea, Annona spraguei, Coutarea hexandra, Ligustrum ovalifolium, Ligustrum vulgare, Padus avium* var. *avium, Prunus laurocerasus, Rhizophora, Syringa vulgaris.*

lebeau lebeau (Guérin-Méneville, 1868)
bolivar (Maassen, 1873)
Foodplants: *Ailanthus altissima, Casearia corymbosa, Citrus, Exostema mexicanum, Fraxinus, Ligu-*

strum ovalifolium, Ligustrum vulgare, Malosma laurina, Prunus laurocerasus, Prunus serotina, Salix babylonica, Salix caprea, Salix exigua, Schinus terebinthifolia, Spondias mombin, Spondias purpurea, Syringa vulgaris, Zanthoxylum fagara, Zanthoxylum setulosum, Zuelania guidonia.

maurus (Burmeister, 1879)
maurusius Schreiter, 1925
Foodplants: *Bignoniaceae, Croton, Euphorbia.*

orizaba cauca W. Rothschild, 1907
Foodplants: *Ligustrum ovalifolium, Prunus serotina.*

orizaba equatorialis W. Rothschild, 1907
bogotana Rothschild, 1907
draudti Bouvier, 1936
poecilator Draudt, 1929
Foodplants: *Fraxinus excelsior, Ligustrum ovalifolium, Prunus serotina.*

orizaba meridana W. Rothschild, 1907
venezuelensis Bouvier, 1936
Foodplants: *Ligustrum ovalifolium, Syringa vulgaris.*

orizaba orizaba (Westwood, 1853)
ochracea Draudt, 1929
prionidia Draudt, 1929
paradoxa C. C. Hoffmann, 1942
Foodplants: *Cephalanthus, Fraxinus americana, Fraxinus excelsior, Hedera helix, Ligustrum japonicum, Ligustrum ovalifolium, Ligustrum vulgare, Liquidambar styraciflua, Malosma laurina, Padus virginiana, Prunus armeniaca, Prunus emarginata* subsp. *mollis, Prunus laurocerasus, Prunus serotina, Pyrus, Quercus robur, Salix babylonica, Salix caprea, Sassafras albidum, Schinus molle, Syringa vulgaris.*

orizaba peruviana W. Rothschild, 1907
coxeyi Schaus, 1932
Foodplants: *Ailanthus altissima, Ligustrum ovalifolium, Ligustrum vulgare, Quercus, Syringa vulgaris.*

orizaba uruapana C. C. Hoffmann, 1942
Foodplants: Unknown

paucidentata Lemaire, 1971
Foodplants: Unknown

prionia W. Rothschild, 1907
gounellei Bouvier, 1936
Foodplants: Unknown

renatae Lampe, 1985
Foodplants: Unknown

roxana Schaus, 1905
Foodplants: *Ligustrum ovalifolium, Prunus laurocerasus, Prunus serotina, Syringa vulgaris.*

schreiteriana Breyer & Orfila, 1945
maurosinus Kohler, 1924
maurusius Breyer, 1927
maurusius Draudt, 1929
Foodplants: *Ailanthus altissima, Ligustrum ovalifolium, Quercus × hispanica, Syringa vulgaris.*

triloba W. Rothschild, 1907
Foodplants: *Casearia arborea, Casearia corymbosa, Casearia tremula, Chimarrhis parviflora, Chiococca alba, Chiococca belizensis, Chiococca pachyphylla, Chione venosa* var. *Venosa, Cosmibuena macrocarpa, Coutarea hexandra, Croton schiedeanus, Dendropanax arboreus, Guarea rhopalocarpa, Guettarda macrosperma, Ilex skutchii, Jatropha curcas, Lacistema aggregatum, Ligustrum ovalifolium, Lozania pittieri, Machaonia erythrocarpa, Mosquitoxylum jamaicense, Nectandra hihua, Oreopanax capitatus, Prunus annularis, Psychotria berteroana, Sapium glandulosum, Sapium laurifolium, Sapium macrocarpum, Schefflera sphaerocoma, Simarouba amara, Spondias mombin, Swietenia macrophylla, Syringa vulgaris, Tapirira mexicana, Tetragastris panamensis, Uncaria tomentosa, Vernonanthura patens, Viburnum costaricanum, Zanthoxylum acuminatum, Zanthoxylum caribaeum, Zanthoxylum juniperinum, Zanthoxylum melanostictum, Zanthoxylum setulosum.*

tucumani (Dognin, 1901)
steinbachi Rothschild, 1904
vibidia (Druce, 1904)
Foodplants: *Baccharis, Eupatorium buniifolium.*

xanthina paraxanthina Brechlin & Meister, 2010
Foodplants: Unknown

xanthina xanthina W. Rothschild, 1907
Foodplants: *Ligustrum ovalifolium, Prunus serotina, Quercus × hispanica, Syringa vulgaris.*

yucatana Lemaire, 1971
Foodplants: *Ailanthus altissima, Ligustrum ovalifolium, Ligustrum vulgare, Quercus × hispanica, Syringa vulgaris.*

zacateca (Westwood, [1854])
Foodplants: *Ligustrum ovalifolium, Prunus serotina, Symplocos.*

Samia Hübner, 1819 (»1816«)

abrerai Naumann & Peigler, 2001
Foodplants: *Ligustrum ovalifolium, Persea americana.*

burica Brechlin, 2007
Foodplants: *Ligustrum ovalifolium, Persea americana.*

canningi (Hutton, 1860)
manillensis (Bouvier, 1927)
Foodplants: *Ailanthus altissima, Ailanthus excelsa, Berberis asiatica, Coriaria nepalensis, Cotinus coggygria, Evodia, Lagerstroemia indica, Ligustrum ovalifolium, Persea americana, Tilia cordata, Zanthoxylum acanthopodium, Zanthoxylum armatum* var. *armatum.*

ceramensis (Bouvier, 1927)
Foodplants: Unknown

cynthia (Drury, 1773)
advena (Cockerell, 1914)
ailanthi (Weil, 1863)
ailanti (Fauvel, 1895)
aylanti (Motschulsky, 1858)
eulouvaina (Watson, 1914)
parisiensis (Clément, 1901)
vera (Girard, 1863)
walkeri (C. Felder & R. Felder, 1862)
Foodplants: *Ailanthus altissima, Alcea rosa, Apium graveolens, Azadirachta indica, Berberis, Carica papaya, Carpinus betulus, Cassia fistula, Cinnamomum camphora, Coriaria nepalensis, Cornus, Cotinus coggygria, Eriophorum latifolium, Euodia fraxinifolia, Euscaphis japonica, Firmiana simplex, Forsythia, Fraxinus velutina, Glochidion zeylanicum* var. *Zeylanicum, Gmelina arborea, Heteropanax fragrans, Ilex chinensis, Ilex integra, Jatropha curcas, Juglans cinerea, Juglans regia, Lagerstroemia indica, Lawsonia inermis, Ligustrum lucidum, Ligustrum ovalifolium, Ligustrum sinense, Ligustrum vulgare, Lindera benzoin, Lindera megaphylla, Liquidambar styraciflua, Liriodendron tulipifera, Magnolia figo, Malus pumila, Manihot esculenta, Phellodendron amurense, Phellodendron chinense, Picrasma quassioides, Platanus occidentalis, Prunus serotina, Quercus acutissima, Rhus chinensis, Rhus typhina, Ricinus communis, Salix, Sambucus, Sassafras albidum, Styrax japonicus, Syringa vulgaris, Tilia, Triadica cochinchinensis, Triadica sebifera, Viburnum lentago, Zanthoxylum acanthopodium, Zanthoxylum armatum* var. *armatum, Zanthoxylum avicennae, Zanthoxylum hostile, Zanthoxylum javanicum, Ziziphus jujuba, Zanthoxylum simulans.*

fulva (Jordan, 1911)
Foodplants: *Heteropanax fragrans, Picrasma javanica, Zanthoxylum rhetsa.*

insularis (Snellen van Vollenhoven, 1862)
vaneeckei Watson, (1913)
vollenhoveni Bouvier, 1927
Foodplants: *Ailanthus altissima, Camellia sinensis, Camellia sinensis* var. *sinensis, Cinchona pubescens, Erythrina subrumbrans, Erythrina variegata, Ligustrum ovalifolium, Syringa vulgaris, Syzygium aqueum.*

kohlli Naumann & Peigler, 2001
Foodplants: *Ailanthus altissima, Cinnamomum camphora, Ligustrum ovalifolium, Syringa vulgaris.*

luzonica (Watson, 1913)
mindanaensis (Rebel, 1923)
Foodplants: *Ailanthus altissima, Cinnamomum camphora, Ligustrum ovalifolium, Persea americana, Syringa vulgaris, Thevetia.*

naessigi Naumann & Peigler, 2001
Foodplants: Unknown

naumanni Paukstadt, Peigler & Paukstadt, 1998
Foodplants: *Ligustrum ovalifolium*

peigleri Naumann & Nässig, 1995
Foodplants: *Ailanthus altissima, Quercus × hispanica, Syringa vulgaris.*

pryeri (Butler, 1878)
Foodplants: *Ailanthus altissima, Celtis sinensis, Cinnamomum camphora, Clerodendrum trichotomum, Firmiana simplex, Ilex rotunda, Ligustrum japonicum, Ligustrum ovalifolium, Osmanthus fragrans, Paulownia tomentosa, Phellodendron amurense, Picrasma, Styrax japonicus, Tilia cordata, Triadica, Zanthoxylum ailanthoides.*

ricini (Jones, 1791)
arrindia (Milne-Edwards, 1854)
depuiseti (Boiduval, 1854)
fuscofasciatus (Bouvier, 1927)
guerini (Moore, 1858)
iole (Westwood, 1881)
lunula (Walker, 1855)
lunuloides (Rebel, 1923)
obscura (Butler, 1879)
penelope (Buchanan, 1833)

Foodplants: *Ailanthus altissima, Carica papaya, Cichorium endivia, Cinchona, Cinnamomum glaucescens, Eriophorum latifolium, Evodia, Ficus religiosa, Fraxinus velutina, Gmelina arborea, Heteropanax fragrans, Hodgsonia heteroclita, Jatropha curcas, Jatropha multifida, Ligustrum ovalifolium, Ligustrum vulgare, Manihot esculenta, Padus virginiana, Prunus laurocerasus, Ricinus communis, Salix, Sapium eugeniifolium, Syringa vulgaris, Taraxacum officinale, Terminalia catappa, Zanthoxylum armatum* var. *armatum, Ziziphus jujuba.*

tetrica (Rebel, 1923)
borneensis (Rebel, 1926)
Foodplants: *Ailanthus altissima, Ligustrum ovalifolium.*

treadawayi Naumann, 1998
Foodplants: Unknown

vandenberghi (Watson, 1915)
Foodplants: *Ligustrum japonicum, Ligustrum ovalifolium, Malus »hillieri«, Syringa vulgaris.*

wangi Naumann & Peigler, 2001
Foodplants: *Ailanthus altissima, Cinnamomum camphora, Citrus, Euscaphis japonica, Firmiana simplex, Glochidion zeylanicum* var. *Zeylanicum, Ilex asprella, Ilex chinensis, Ilex rotunda, Lantana, Ligustrum ovalifolium, Lindera megaphylla, Magnolia figo, Michelia alba, Persea americana, Rhamnus formosana, Sapindus saponaria, Syringa vulgaris, Triadica cochinchinensis, Triadica sebifera, Zanthoxylum piperitum.*

yayukae Naumann & Nässig, 1995
kikibudiamini Paukstadt, Paukstadt & Suhardjono, 2002
Foodplants: *Ailanthus altissima, Ligustrum ovalifolium, Psidium guajava, Quercus, Toddalia asiatica, Syringa vulgaris, Xanthoxylum armatum.*

Literature / Literatur

Aue, A. U. E. (1933): Handbuch für den praktischen Entomologen. 1. Abteilung: Lepidoptera IV. Band. – 176 S.; Frankfurt am Main (Verlag des Internationalen Entomologischen Vereins e. V.).

Baxter, R. N. (1992): Rearing wild silkmoths. – 72 pp.; Essex (Cravitz Printing Company Limited).

Beeke, M., U. Brosch, R. E. J. Lampe & W. A Nässig. (2000): Beobachtungen zur Biologie von *Aglia tau* (Linnaeus, 1758) im Freiland (Lepidoptera: Saturniidae, Agliinae). – Nachrichten des Entomologischen Vereins Apollo, Frankfurt am Main, N.F. 21 (1): 11–18.

Balcazàr, L. M. A. (1991): Sistemática y biogeografia del Género *Antheraea* Hübner en Norteamérica (Lepidoptera: Saturniidae). – V + 71 pp.; Master's thesis, Univ. Nacional Autonoma de México.

Boisduval, Jean Baptiste Alphonse Dechauffour De (1875): Aperçu monographique du genre Io, l'un des démenbrements de la grande famille des Saturnides. – Séance du 5 septembre1874. – Annales de la Société entomologique de Belgique (Bruxelles); XVIII: 205–248.

Bouyer, T., R. E. J. Lampe & W. A. Nässig (2004): The life history of *Pseudantheraea discrepans* (Butler, 1878), with an ecological comparison with *P. imperator* Rougeot, 1962 (Lepidoptera: Saturniidae, Saturniinae, Urotini). – Nachrichten des Entomologischen Vereins Apollo, Frankfurt am Main, N.F. 25 (1/2): 27–37.

Brechlin, R. (1997): *Saturnia (Rinaca) witti* n. sp., eine neue Saturniide aus Nepal (Lepidoptera: Saturniidae). – Nachrichten des Entomologischen Vereins Apollo, Frankfurt am Main, N.F. 17 (4): 423–434.

– (1997): Zwei neue Saturniiden aus dem Gebirgsmassiv des Fan Si Pan (nördliches Vietnam): *Salassa fansipana* n. sp. und *Loepa roseomarginata* n. sp. (Lepidoptera: Saturniidae). – Nachrichten des Entomologischen Vereins Apollo, Frankfurt am Main, N.F. 18 (1): 75–87.

– (2000): Zwei weitere neue *Antheraea*-Arten von Sulawesi und den östlich anschließenden Inseln (Indonesien): *Antheraea (Antheraea) exspectata* n. sp. und *Antheraea (Antheraea) pelengensis* n. sp. (Lepidoptera: Saturniidae). – Nachrichten des Entomologischen Vereins Apollo, Frankfurt am Main, N.F. 20 (3/4): 291–310.

– (2000): *Saturnia (Rinaca) winbrechlini* n. sp., eine neue Saturniide aus China (Lepidoptera: Saturniidae). – Nachrichten des Entomologischen Vereins Apollo, Frankfurt am Main, N.F. 21 (1): 5–10.

– (2000): Eine neue Art der Gattung *Actias* Leach, 1815 aus Südvietnam: *Actias australovietnama* n. sp. (Lepidoptera: Saturniidae). – Nachrichten des Entomologischen Vereins Apollo, Frankfurt am Main, N.F. 21 (1): 33–37.

– (2000): Zwei neue Arten der Gattung *Antheraea* Hübner, 1819 [»1816«] von den westlichen Kleinen Sundainseln, Indonesien: *Antheraea (Antheraea) tenggarensis* n. sp. und *A. (A.) sumbawaensis* n. sp. (Lepidoptera: Saturniidae). – Nachrichten des Entomologischen Vereins Apollo, Frankfurt am Main, N.F. 21 (1): 38–44.

– (2000): Zwei neue Arten der Gattung *Loepa* Moore, 1859 (Lepidoptera: Saturniidae). – Nachrichten des Entomologischen Vereins Apollo, Frankfurt am Main, N.F. 21 (3): 165–170.

– (2001): *Lemaireia naessigi* n. sp., eine neue Saturniidae aus Südostasien (Lepidoptera: Saturniidae). – Nachrichten des Entomologischen Vereins Apollo, Frankfurt am Main, N.F. 21 (4): 207–212.

– (2001): Beschreibung des bisher unbekannten Weibchens von *Antheraea (Antheraeopsis) sahi* Nässig & Treadaway 1998 stat. nov. (Lepidoptera: Saturniidae). – Nachrichten des Entomologischen Vereins Apollo, Frankfurt am Main, N.F. 22 (1): 37–40.

– (2001): Eine neue Art der Gattung *Cricula* Walker, 1855 von Palawan, Philippinen (Lepidoptera: Saturniidae). – Nachrichten des Entomologischen Vereins Apollo, Frankfurt am Main, N.F. 22 (1): 41–44.

– (2001): *Rhodinia broschi* n. sp., eine neue Saturniide aus China (Lepidoptera: Saturniidae). – Nachrichten des Entomologischen Vereins Apollo, Frankfurt am Main, N.F. 22 (1): 45–48.

– (2001): Einige Anmerkungen zur Gattung *Saturnia* Schrank, 1802 mit Beschreibung von zwei neuen Arten (Lepidoptera: Saturniidae). – Nachrichten des Entomologischen Vereins Apollo, Frankfurt am Main, N.F. 22 (2): 89–100.

– (2001): *Rhodinia rudloffi* n. sp., eine neue Saturniide aus dem nördlichen Vietnam (Lepidoptera: Saturniidae). – Nachrichten des Entomologischen Vereins Apollo, Frankfurt am Main, N.F. 22 (2): 124–128.

– (2002): Neue Arten der Gattung *Antheraea* Hübner 1819 [»1816«] von den Andamanen (Lepidoptera: Saturniidae). – Nachrichten des Entomologischen Vereins Apollo, Frankfurt am Main, N.F. 22 (4): 219–224.

– (2004): Zwei neue Arten der Gattung *Saturnia* Schrank, 1802 aus China (Lepidoptera: Saturniidae). – Nachrichten des Entomologischen Vereins Apollo, Frankfurt am Main, N.F. 25 (1/2): 39–45.

– (2004): Drei neue Saturniiden (Lepidoptera) aus China: *Cricula australosinica* n. sp., *Cricula hainanensis* n. sp. und *Loepa sinjaevi* n. sp. – Arthropoda 12 (3): 17–25.

– (2004): Einige Anmerkungen zur Gattung *Coscinocera* Butler, 1879 mit Beschreibung einer neuen Art (Lepidoptera: Saturniidae). – Nachrichten des Entomologischen Vereins Apollo, Frankfurt am Main, N.F. 25 (4): 189–194

– (2005): Einige Anmerkungen zur Gattung *Neodiphthera* Fletcher in Fletcher & Nye (in Nye [edit.]), 1982 stat. rev. mit Beschreibung von vier neuen Arten (Lepidoptera: Saturniidae). – Nachrichten des Entomologischen Vereins Apollo, Frankfurt am Main, N.F. 26 (1/2): 17–29.
– (2007): Eine neue Art der Gattung *Saturnia* Schrank, 1802 aus dem Südosten Chinas (Lepidoptera: Saturniidae). – Nachrichten des Entomologischen Vereins Apollo, Frankfurt am Main, N.F. 28 (1/2): 93–96.
– (2007): Fünf neue Taxa der Gattung *Actias* Leach, 1815 aus China (Lepidoptera: Saturniidae). – Entomofauna, Monographie 1: 12–27.
– (2007): Einige Anmerkungen zur Gattung *Rhodinia* Staudinger, 1892 mit Beschreibung neuer Taxa aus China (Lepidoptera: Saturniidae). – Entomofauna, Monographie 1: 28–43.
– (2007): Anmerkungen zur Gattung *Solus* Watson, 1913 mit Beschreibung eines neuen Taxons aus China (Lepidoptera: Saturniidae). – Entomofauna, Monographie 1: 44–49.
– (2007): *Actias witti*, eine neue Saturniide von den Kleinen Andamanen (Lepidoptera: Saturniidae). – Entomofauna, Monographie 1: 50–55.
– (2007): Einige Anmerkungen zur Gattung *Samia* Hübner 1819 (»1816«) mit Beschreibung einer neuen Art (Lepidoptera: Saturniidae). – Entomofauna, Monographie 1: 56–63.
– (2009): An annotated checklist of the Saturniidae and Brahmaeidae (Lepidoptera) of the Kingdom of Bhutan. – Entomo-Satsphingia 2 (1): 47–55.
– (2009): *Antheraea rubicunda rubiorientalis* n. ssp., ein neues Taxon der Gattung *Antheraea* Hübner 1819 [»1816«] (Lepidoptera: Saturniidae). – Entomo-Satsphingia 2 (2): 25–29.
– (2009): Zwei neue Arten der Gattung *Saturnia* Schrank, 1802 (Subgenus *Rinaca* Walker 1855) (Lepidoptera: Saturniidae). – Entomo-Satsphingia 2 (2): 30–36.
– (2009): Zwei neue Arten des Genus (Subgenus) *Saturnia* Schrank, 1802 (Lepidoptera: Saturniidae). – Entomo-Satsphingia 2 (2): 37–42.
– (2009): Drei neue Arten der Gattung *Salassa* Moore, 1859 (Lepidoptera: Saturniidae, Salassinae). – Entomo-Satsphingia 2 (2): 43–49.
– (2010): Eine neue Art der Gattung *Coscinocera* Butler, 1879 (Lepidoptera: Saturniidae). – Entomo-Satsphingia 3 (1 [a (Erstversion; gedruckt 8.1.2010)]): 5–8.
– (2010): Eine neue Art der Gattung *Coscinocera* Butler, 1879 (Lepidoptera: Saturniidae). – Entomo-Satsphingia 3 (1 [b (Endversion; gedruckt 25.1.2010)]): 5–8.
– (2010): Neue Taxa der Gattung *Loepa* Moore, 1859 (Lepidoptera: Saturniidae). – Entomo-Satsphingia 3 (1 [a (Erstversion; gedruckt 8.1.2010)]): 22–33.
– (2010): Neue Taxa der Gattung *Loepa* Moore, 1859 (Lepidoptera: Saturniidae). – Entomo-Satsphingia 3 (1 [b (Endversion; gedruckt 25.1.2010)]): 23–35.
– (2010): Neue Taxa der Gattung *Cricula* Walker, 1855 (Lepidoptera: Saturniidae). – Entomo-Satsphingia 3 (1 [a (Erstversion; gedruckt 8.1.2010)]): 34–41.
– (2010): Neue Taxa der Gattung *Cricula* Walker, 1855 (Lepidoptera: Saturniidae). – Entomo-Satsphingia 3 (1 [b (Endversion; gedruckt 25.1.2010)]): 36–44.
– (2010): Neue Taxa der Gattung *Syntherata* Maassen, 1873 (Lepidoptera: Saturniidae). – Entomo-Satsphingia 3 (2): 13–22.
– (2010): Zwei neue Arten der Gattung *Salassa* Moore, 1859 (Lepidoptera: Saturniidae, Salassinae). – Entomo-Satsphingia 3 (3): 60–65.
– (2010): *Gamelioides meisteri* n. sp., eine neue Hemileucine aus Nordperu (Lepidoptera: Saturniidae). – Entomo-Satsphingia 3 (4): 36–38.
– (2010): Einige Anmerkungen zur Gattung *Attacus* Linnaeus, 1767 von den Kleinen Sundainseln (Indonesien) mit Beschreibung einer neuen Art (Lepidoptera: Saturniidae). – Entomo-Satsphingia 3 (5): 62–67.
– (2011): Fünf neue Taxa der Gattung *Saturnia* Schrank, 1802 (Subgenus *Rinaca* Walker, 1855) aus China (Lepidoptera: Saturniidae). – Entomo-Satsphingia 4 (2): 78–85.
– (2011): Sechs neue Taxa der Gattung *Neodiphthera* Fletcher, 1982 (Lepidoptera: Saturniidae). – Entomo-Satsphingia 4 (2): 86–93.
– (2011): *Rhodinia benedeki* n. sp., eine neue Saturniide (Lepidoptera). – Entomo-Satsphingia 4 (2): 105–107.
Brechlin, R. & I. J. Kitching (2001): Neue Informationen zu einigen Arten der Gattung *Antheraea* Hübner 1819 [»1816«] von den Andamanen, Indien (Lepidoptera: Saturniidae). – Nachrichten des Entomologischen Vereins Apollo, Frankfurt am Main, N.F. 22 (2): 119–123.
– (2010): Eine neue Art der Gattung *Salassa* Moore, 1859 (Lepidoptera: Saturniidae, Salassinae). – Entomo-Satsphingia 3 (1 [a (Erstversion; gedruckt 8.1.2010)]): 9–11.
– (2010): Eine neue Art der Gattung *Salassa* Moore, 1859 (Lepidoptera: Saturniidae, Salassinae). – Entomo-Satsphingia 3 (1 [b (Endversion; gedruckt 25.1.2010)]): 9–11.
– (2010): Zwei neue Taxa der *miranda*-Gruppe der Gattung *Loepa* Moore, 1859 (Lepidoptera: Saturniidae). – Entomo-Satsphingia 3 (1 [a (Erstversion; gedruckt 8.1.2010)]): 12–17.
– (2010): Drei neue Taxa der *miranda*-Gruppe der Gattung *Loepa* Moore, 1859 (Lepidoptera: Saturniidae). – Entomo-Satsphingia 3 (1 [b (Endversion; gedruckt 25.1.2010)]): 12–18.
Brechlin, R. & F. Meister (2002): Zwei neue Saturniiden aus Südamerika: *Leucanella anikae* Meister & Brechlin n. sp. und *Kentroleuca boliviensis* Brechlin & Meister n. sp. (Lepidoptera: Saturniidae, Hemileucinae). – Arthropoda 10 (2): 28–32.
– (2008): Vier neue Arten der Gattung *Pseudodirphia* Bouvier, 1928 aus Argentinien und Peru (Lepidoptera: Saturniidae, Hemileucinae). – Entomo-Satsphingia 1 (1): 5–10.

- (2008): Drei neue Hemileucinae aus Südamerika (Lepidoptera: Saturniidae). – Entomo-Satsphingia 1 (1): 11–16.
- (2008): Zwei neue Arten der Gattung *Molippa* Walker, 1855 aus Peru (Lepidoptera: Saturniidae, Hemileucinae). – Entomo-Satsphingia 1 (1): 17–20.
- (2008): Neue Arten der Gattung *Ptiloscola* Michener, 1949 (Lepidoptera: Saturniidae). – Entomo-Satsphingia 1 (1): 21–26.
- (2008): Neue Arten der Gattung *Meroleuca* Packard, 1904 (Lepidoptera: Saturniidae, Cercophaninae, Janiodini). – Entomo-Satsphingia 1 (1): 27–29.
- (2008): Zwei neue Arten der Gattung *Janiodes* Jordan; 1924 aus Peru (Lepidoptera: Saturniidae, Cercophaninae, Janiodini). – Entomo-Satsphingia 1 (1): 30–33.
- (2008): Einige Anmerkungen zur Zucht von *Caio harrietae* Forbes, 1944 mit Beschreibung der Präimaginalstadien (Lepidoptera: Saturniidae, Arsenurinae). – Entomo-Satsphingia 1 (1): 34–37.
- (2009): Eine neue, bisher verkannte Art der Gattung *Copaxa* Walker, 1855 (Lepidoptera: Saturniidae). – Entomo-Satsphingia 2 (1): 23–26.
- (2009): Einige Anmerkungen zu *Salassa lemaii* Le Moult, 1933 mit Beschreibung einer neuen Unterart (Lepidoptera: Saturniidae, Salassinae). – Entomo-Satsphingia 2 (1): 33–35.
- (2009): Drei neue Arten der Gattung *Antheraea* Hübner, 1819 [»1816«] aus Südostasien (Lepidoptera: Saturniidae). – Entomo-Satsphingia 2 (1): 42–46.
- (2009): Zwei neue Arten der Gattung *Dysdaemonia* Hübner, 1819 [»1816«] (Lepidoptera: Saturniidae). – Entomo-Satsphingia 2 (1): 56–61.
- (2009): Eine neue Art der Gattung *Bathyphlebia* C. Felder & R. Felder, 1874 (Lepidoptera: Saturniidae). – Entomo-Satsphingia 2 (2): 5–7.
- (2009): Zwei neue Arten der Gattung *Eacles* Hübner 1819 [»1816«] (Lepidoptera: Saturniidae). – Entomo-Satsphingia 2 (2): 8–10.
- (2010): Eine neue Art der Gattung *Ormiscodes* Blanchard, 1852 (Lepidoptera: Saturniidae). – Entomo-Satsphingia 3 (1 [a (Erstversion; gedruckt 8.1. 2010)]): 42–43.
- (2010): Eine neue Art der Gattung *Ormiscodes* Blanchard, 1852 (Lepidoptera: Saturniidae). – Entomo-Satsphingia 3 (1 [b (Endversion; gedruckt 25.1. 2010)]): 45–46.
- (2010): Zwei neue Taxa der Gattung *Arsenura* Duncan, 1841 aus Bolivien (Lepidoptera: Saturniidae). – Entomo-Satsphingia 3 (3): 5–9.
- (2010): Drei neue Arten der Gattung *Arsenura* Duncan, 1841 (Lepidoptera: Saturniidae). – Entomo-Satsphingia 3 (3): 10–14.
- (2010): *Arsenura paraorbignyana* n. sp., eine neue Art der Gattung *Arsenura* Duncan, 1841 (Lepidoptera: Saturniidae). – Entomo-Satsphingia 3 (3): 15–19.
- (2010): Einige Anmerkungen zur Gattung *Copaxa* Walker, 1855 mit Beschreibung von achtzehn neuen Taxa (Lepidoptera: Saturniidae). – Entomo-Satsphingia 3 (3): 27–55.
- (2010): Vier neue Taxa der Gattung *Rothschildia* Grote, 1896 (Lepidoptera: Saturniidae). – Entomo-Satsphingia 3 (3): 75–82.
- (2010): Einige Anmerkungen zur Gattung *Periphoba* Hübner, [1820] mit Beschreibung von sechs neuen Arten (Lepidoptera: Saturniidae). – Entomo-Satsphingia 3 (4): 5–12.
- (2010): Einige Anmerkungen zur Gattung *Paradirphia* Michener, 1949 mit Beschreibung von elf neuen Arten (Lepidoptera: Saturniidae). – Entomo-Satsphingia 3 (4): 13–25.
- (2010): Sieben neue Arten der Gattung *Copaxa* Walker, 1855 aus Mittelamerika (Lepidoptera: Saturniidae). – Entomo-Satsphingia 3 (4): 26–35.
- (2010): Zwei neue Arten der Gattung *Arsenura* Duncan, 1841 (Lepidoptera: Saturniidae). – Entomo-Satsphingia 3 (4): 39–43.
- (2010): Drei neue Arten der Gattung *Caio* Travassos & Noronha, 1968 aus Mexiko (Lepidoptera: Saturniidae). – Entomo-Satsphingia 3 (4): 44–49.
- (2010): Vier neue Arten der Gattung *Pseudautomeris* Lemaire, 1967 (Lepidoptera: Saturniidae). – Entomo-Satsphingia 3 (5): 5–12.
- (2010): Einige Anmerkungen zur Gattung *Hirpida* Draudt, 1930 mit Beschreibung von elf neuen Arten (Lepidoptera: Saturniidae). – Entomo-Satsphingia 3 (5): 13–25.
- (2010): Einige Anmerkungen zur Gattung *Hyperchiria* Hübner, [1819] mit Beschreibung von dreizehn neuen Arten (Lepidoptera: Saturniidae). – Entomo-Satsphingia 3 (5): 26–39.
- (2010): Vier neue Arten der Gattung *Coloradia* Blake, 1863 aus Mexiko (Lepidoptera: Saturniidae). – Entomo-Satsphingia 3 (5): 40–45.
- (2010): Zwei neue Arten der *sapatoza*-Gruppe der Gattung *Copaxa* Walker, 1855 (Lepidoptera: Saturniidae). – Entomo-Satsphingia 3 (5): 46–49.
- (2011): Neue Taxa der Gattung *Automeris* Hübner, [1819] (Lepidoptera: Saturniidae). – Entomo-Satsphingia 4 (1): 5–89.
- (2011): Vier neue Arten der Gattung *Eubergia* Bouvier, 1929 (Lepidoptera: Saturniidae). – Entomo-Satsphingia 4 (1): 90–94.
- (2011): Zwei neue Arten der Gattung *Hirpida* Draudt, 1930 aus Peru (Lepidoptera: Saturniidae). – Entomo-Satsphingia 4 (1): 95–98.
- (2011): Zwei neue Arten der Gattung *Leucanella* Lemaire, 1969 (Lepidoptera: Saturniidae). – Entomo-Satsphingia 4 (2): 30–41.
- (2011): Zehn neue Arten der Gattung *Janiodes* Jordan, 1924 (Lepidoptera: Saturniidae, Cercophaninae, Janiodini). – Entomo-Satsphingia 4 (2): 42–55.
- (2011): Vier neue Taxa der Gattung *Eacles* Hübner, 1819 [»1816«] (Lepidoptera: Saturniidae). – Entomo-Satsphingia 4 (2): 94–99.
- (2011): Zwei neue Ceratocampinae (Lepidoptera: Saturniidae). – Entomo-Satsphingia 4 (2): 100–104.

Brechlin, R. & U. Paukstadt (2010): Eine neue Art der Gattung *Loepa* Moore, 1859 von Borneo (Lepidopte-

ra: Saturniidae). – Entomo-Satsphingia 3 (1 [a (Erstversion; gedruckt 8.1.2010)]): 18–21.
– (2010): Eine neue Art der Gattung *Loepa* Moore, 1859 von Borneo (Lepidoptera: Saturniidae). – Entomo-Satsphingia 3 (1 [b (Endversion; gedruckt 25.1.2010)]): 19–22.
– (2010): Einige Anmerkungen zur *pernyi*-Gruppe der Gattung *Antheraea* Hübner, 1819 [»1816«] mit Beschreibung neuer Arten (Lepidoptera: Saturniidae). – Entomo-Satsphingia 3 (2): 5–12.
Brechlin, R. & E. Van Schayck (2010): Eine neue Art der Gattung *Rhodinia* Staudinger, 1892 aus China (Lepidoptera: Saturniidae). – Entomo-Satsphingia 3 (5): 56–61.
Brechlin, R. & F. Meister & H. Käch (2009): Zwei neue Arten der Gattung *Janiodes* Jordan, 1924 aus Ecuador (Lepidoptera: Saturniidae, Cercophaninae, Janiodini). – Entomo-Satsphingia 2 (1): 27–32.
Brechlin, R., E. Van Schayck & P. Ackermann (2009): Einige Anmerkungen zur Zucht von *Actias rhodopneuma* Röber, 1925 mit Beschreibung der Präimaginalstadien (Lepidoptera: Saturniidae). – Entomo-Satsphingia 2 (1): 36–39.
Brechlin, R., H. Käch & F. Meister (2010): *Gamelioides kattyae* n. sp., eine neue Hemileucine aus Ecuador (Lepidoptera: Saturniidae). – Entomo-Satsphingia 3 (3): 56–59.
Brechlin, R. & F. Meister & E. Van Schayck (2010): Sechs neue Arten der Gattung *Catacantha* Bouvier, 1930 (Lepidoptera: Saturniidae). – Entomo-Satsphingia 3 (5): 50–55.
Brechlin, R., H. Käch & F. Meister (2011): Zwei neue Arten der Gattung *Automeris* Hübner, [1819] aus Ecuador (Lepidoptera: Saturniidae). – Entomo-Satsphingia 4 (2): 5–8.
Brechlin, R., H. Käch & F. Meister (2011): Drei neue Arten der Gattung *Copaxa* Walker, 1855 aus Ecuador (Lepidoptera: Saturniidae). – Entomo-Satsphingia 4 (2): 9–15.
Brechlin, R., H. Käch & F. Meister (2011): Drei neue Arten der Gattung *Gamelioides* Lemaire, 1988 aus Ecuador (Lepidoptera: Saturniidae). – Entomo-Satsphingia 4 (2): 56–60.
Brechlin, R., F. Meister, C. G. C. Mielke & E. Van Schayck (2011): Fünfzehn neue Arten der Gattung *Lonomia* Walker, 1855 (Lepidoptera: Saturniidae). – Entomo-Satsphingia 4 (2): 61–77.
Brechlin, R. & F. Meister & E. Van Schayck (2011): Neun neue Arten der Gattung *Automeris* Hübner, [1819] (Lepidoptera: Saturniidae). – Entomo-Satsphingia 4 (2): 16–29.
Burmeister, C. H. C. (1878): Description physique de la République Argentine, d'après des observations personelles et étrangères. Tome cinquième. Lépidoptères. Première partie. Contenant les Diurnes, Crépusculaires et Bombycoïdes. – Pp. [i]–vi, 1–524, 2 pp.; Buenos Aires (Imprimerie de Paul-Émile Coni), Paris (en commission – F. Savy), ed. Anton (Halle).
– (1879–[1880]): Atlas de la Description physique de la République Argentine. Contenant des vues pittoresques et des figures d'histoire naturelle. Cinquième section, seconde partie. Lépidoptères. – 2 [+1] pp. without pagination, pp. 1–64 pp. 1 « Planche Supplémentaire », 24 pls.; Buenos Aires (l'imprimerie de Paul-Emile Coni), Paris (en commission – F. Savy), ed. Anton (Halle).
D'Abrera, B. (1995): Saturniidae mundi. Saturniid moths of the world. Part I. – 177 pp.; Keltern (Goecke & Evers).
– (1998): Saturniidae mundi. Saturniid moths of the world. Part III. – 171 pp.; Keltern (Goecke & Evers).
D'Almeida, R. F. (1975): Biologia de duas espécies de Saturnioideae (Lepidoptera). 2 – Evolução de *Copiopteryx montei* Gagarin, 1933, Adelocephalidae. Atas Soc. Biol. Rio de Janeiro 17 (3): 105–109.
Dias, M. M. (1978) : Morfologia e biologia de *Dysdaemonia boreas* (Cramer, 1775) (Lepidoptera, Adelocephalidae). Revta bras. Ent., São Paulo, 22 (2): 83–90.
Draudt, M. (1929–1930): 12. Familie: Saturnidae [sic]. – In Seitz, A. (ed.): Die Gross-Schmetterlinge der Erde. Eine systematische Bearbeitung der bis jetzt bekannten Gross Schmetterlinge. Die Amerikanischen Spinner und Schwärmer; 6. Band. – S. 713–827; 45 Taf.; Stuttgart (Alfred-Kernen-Verlag).
Drechsel, U. & R. E. J. Lampe (1996): Die Präimaginalstadien von *Neorcarnegia basirei* (Schaus, 1892) mit Anmerkungen zur Biogeographie (Lepidoptera, Saturniidae, Ceratocampinae). – Nachrichten des Entomologischen Vereins Apollo (Frankfurt am Main), N.F. 17 (2): 143–152.
Friedrich, E. (1975): Handbuch der Schmetterlingszucht – Europäische Arten. – 186 S., 32 Taf.; Stuttgart (Franckh'sche Verlagshandlung, W. Keller & Co.).
Furtado, E. (2004): *Almeidaia aidae* Mielke & Casagrande: seus estágios imaturos e notas taxonômicas (Lepidoptera, Saturniidae, Arsenurinae, Almeidaiini). – Rev. Bras. Zool. [online]; 21 (3): 663–669.
Furtado, E. & C. Lemaire (1999): The biology and immature stages of *Automeris granulosa* (Lepidoptera: Saturniidae: Hemileucinae). – Tropical Lepidoptera (Gainesville, FL); 10 (1): 27–29.
Gardiner, B. O. C. (1982): A silkmoth rearer's handbook (Third Edition). The Amateur Entomologist, 12 [revised and extended edit. of Crotch (1956)]. – xiii + 255 pp.; Pindar Print Ltd. (Scarborough).
Hoffmann, C. C. (1942): El género *Rothschildia* en la República Mexicana. – Anales del Instituto de Biología 13 (1): pp 165–194.
Holloway, J. D. (1976): Moths of Borneo with special reference to Mount Kinabalu. – [12 +] 263 pp.; Kuala Lumpur (Sun U Book Co.).
– (1987): The moths of Borneo, part 3, Lasiocampidae, Eupterotidae, Bombycidae, Brahmaeidae, Saturniidae, Sphingidae. – 199 pp.; Kuala Lumpur (Southdene Sdn. Bhd.).

ICZN [International Commission on Zoolocigal Nomenclature] (2000): Internationale Regeln für die Zoologische Nomenklatur. Vierte Auflage. Angenommen von International Union of Biological Sciences. Offizieller Deutscher Text. Ausgearbeitet von Otto Kraus. Mitglied der Internationalen Kommission für Zoologische Nomenklatur. – Abhandlungen des Naturwissenschaftlichen Vereins in Hamburg (Hamburg); N.F. 34: 232 S.; Keltern-Weiler (Goecke & Evers).

Janzen, D. H. 1993. Caterpillar seasonality in a Costa Rican dry forest. – Pp. 448–477 in Stamp, N. E. & T. M. Casey, (eds.): Caterpillars. Ecological and evolutionary constraints on foraging. – New York (Chapman and Hall).

Kegel, B. (1999): Die Ameise als Tramp – Von biologischen Invasionen. – 420 S.; Zürich (Ammann-Verlag & Co.).

Lampe, R. E. J. (1983): Eine Winterzucht von *Antheraea roylei korintjana* Bouvier 1928 (Lepidoptera, Saturniidae). – Entomologische Zeitschrift (Essen), 93 (12): 161–166.

– (1983): Eine Winterzucht von *Antheraea pratti* Bouvier 1928 (Lepidoptera, Saturniidae). – Entomologische Zeitschrift (Essen), 93 (15): 219–223.

– (1983): Eine Doppelzucht von *Actias maenas* Doubleday (1847) (Lepidoptera, Saturniidae). – Entomologische Zeitschrift (Essen), 93 (21): 305–310.

– (1984): Eine Zucht von *Antheraea roylei roylei* Moore 1858, mit taxonomischen Bemerkungen (Lepidoptera, Saturniidae). – Entomologische Zeitschrift (Essen), 94 (22): 330–334.

– (1984): Die Saturniiden der Cameron – und Genting-Highlands in West Malaysia (Lepidoptera: Saturniidae). – Neue Entomologische Nachrichten (Keltern); 15 [+17] S.

– (2010): Saturniidae of the World – Pfauenspinner der Welt. – 368 pp.; München (Pfeil).

Lampe, R. E. J. & W. A. Nässig (1989): Neue Erkenntnisse über die Gattung *Lemaireia*: 1. Beschreibung der Präimaginalstadien von *L. luteopeplus aureopeplus* Nässig & Holloway (Lepidoptera, Saturniidae). – Nachrichten des Entomologischen Vereins Apollo, (Frankfurt am Main), N.F. 10 (3): 225–238.

Lampe, R. E. J. & W. A. Nässig (1994): Zuchtbericht der thailändischen *Saturnia* (*Saturnia*) *pinratanai* (Lepidoptera: Saturniidae). – Nachrichten des Entomologischen Vereins Apollo, Frankfurt am Main, N.F. 15 (3): 305–314.

Lemaire, C. (1967): Description préliminaires d'Attacidae nouveaux d'Amerique du Sud (Lep.) (suite). – Bulletin de la Société entomologique de France (Paris); 71: 217–226.

– (1971): Révision du genre *Automeris* Hübner et des genres voisins. Biogéographie, éthologie, morphologie, taxonomie (Lep. Attacidae). – Mémoires du Muséum national d'Histoire naturelle (Nouvelle Série) Série A, Zoologie (Paris); LXVIII: [1]–232 pp. [+1 p. without pagination].

– (1973): Révision du genre *Lonomia* Walker [Lep. Attacidae]. – Annales de la Société entomologique de France. Revue internationale d'entomologie générale et appliquée (Paris) Nouvelle série, 8 (1972, published in 1973): 767–861.

– (1973): Révision du genre *Automeris* Hübner et des genres voisins. Biogéographie, éthologie, morphologie, taxonomie (Lep. Attacidae) (Suite). – Mémoires du Muséum national d'Histoire naturelle (Nouvelle Série) Série A, Zoologie (Paris); LXXIX: [233]–422 pp. [+2 pp. without pagination].

– (1974): Révision du genre *Automeris* Hübner et des genres voisins. Biogéographie, éthologie, morphologie, taxonomie (Lep. Attacidae) (Suite et fin). – Mémoires du Muséum national d'Histoire naturelle (Nouvelle Série) Série A, Zoologie (Paris); XCII: [423]–576 pp. [+2 pp. without pagination].

– (1978): Les Attacidae Americains. The Attacidae of America (= Saturniidae). – 238 pp. [+51 pp. without pagination]; Neuilly-sur-Seine (Éditions Lemaire).

– (1980): Les Attacidae Americains. The Attacidae of America (= Saturniidæ). ** Arsenurinae. – 199 pp. [+7 pp. without pagination]; Neuilly-sur-Seine (Editions Lemaire).

– (1980): Saturniidae. – In: Pearson, D. L. et al. (ed.): Preliminary floral and faunal survey, Tambopata reserved zone, Madre de Dios (Peru). II Insectes and spiders. – P. 24; Pennsylvania State University, University Park.

– (1982): Vingt-huit Saturniidae néotropicaux inédits. – Annales de la Société entomologique de France (Paris) (N.S.); 18 (1): 55–88.

– (1988): Les Saturniidae Americains. The Saturniidae of America. Los Saturniidae Americanos. (= Attacidae) *** Ceratocampinae. – 480 pp. [+66 pp. legends without pagination]; San José (Museo Nacional de Costa Rica).

– (1996): 117. Saturniidae. – In Heppner, J. B. (ed.) Atlas of Neotropical Lepidoptera; 1996); Volume 5B. Checklist: Part 4B. Drepanoidea – Bombycoidea – Sphingoidea. – Pp. 28–49, 61–62; Gainesville, Washington, Hamburg, Lima, Taipei, Tokyo.

– [with contributions by F. Bénéluz & N. Tangerini] (2002): The Saturniidae of America. Les Saturniidae Americains (= Attacidae). **** Hemileucinae. – Part A: [1]–688. Part B: [689]–1388; Part C: 143 pp. without pagination; Keltern (Goecke & Evers).

Lemaire, C. & M. J. Smith (1992): A new species of *Coloradia* from Sonora and Chihuahua, Mexico (Saturniidae: Hemileucinae). – Journal of the Lepidopterists' Society (Los Angeles); 46: 128–137.

Lemaire, C. & N. Venedictoff (1989): Catalogue and biogeography of the Lepidoptera of Ecuador. I – Saturniidae. With a description of a new species of *Meroleuca* Packard. – Bulletin of the Allyn Museum (Sarasota, FL); 129: [1]–60.

Lemaire, C. & K. L. Wolfe (1990): Three new species of *Paradirphia* (Saturniidae: Hemileucinae) from Mexico and Central America with notes on the immature stages. – The Journal of Research on the Lepidoptera (Beverly Hills, CA); 27 (3–4), 1988 (89) [published in 1990]: 197–212.

Lemaire, C., M. J. Smith & K. L. Wolfe (1992): A new *Automeris* from Arizona, including its life history and notes on the *Automeris colenon* complex (Lepidoptera: Saturniidae: Hemileucinae). – Tropical Lepidoptera (Gainesville, FL); 3 (2): 123–129.

Maes, J. M., P. Schmit, M. Laguerre & C. A. Conlan (2001): CD ROM Saturniidae/Sphingidae de Nicaragua. – Revista Nicaraguense de Entomologia. Publicación del Museo Entomológico. Associación Nicaragüense de, Entomología (Léon); N° 55/58. Diciembre 2001: 1–4 + CD ROM. [Saturniidae compiled by Maes, J.-M., Ch. Conlan & M. Laguerre; Sphingidae compiled by Maes, J.-M. & P. Schmit]

Meister, F. & A. Knorke (2004): *Hyperchiria schmiti* n. sp., eine neue Saturniide aus Argentinien (Lepidoptera: Saturniidae, Hemileucinae). – Bulletin des Lépidoptéristes Parisiens 13 (28): 65–70.

Meister, F. & S. Naumann (2006): *Leucanella atahualpa* n. sp. und *Leucanella yungasensis* n. sp., zwei neue Saturniiden aus dem Hochland von Peru und Südbolivien/Nordargentinien (Lepidoptera: Saturniidae, Hemileucinae). – Arthropoda 14 (1): 12–22.

Meister, F. & P. Schmit (2004): *Pseudodirphia knorkei* n. sp., eine neue Saturniide aus Argentinien (Lepidoptera: Saturniidae, Hemileucinae). – Bulletin des Lépidoptéristes Parisiens 13 (29): 114–120.

Meister, F. & B. Wenczel (2002): *Dirphia barinasensis* n. sp., eine neue Saturniide aus Venezuela (Lepidoptera: Saturniidae, Hemileucinae). – Arthropoda 10 (2): 13–25.

Mell, R. (1950): Aus der Biologie der chinesischen *Actias* Leach (*Argema* chapae sp. n., *A. sinensis f. virescens* f. n.) – Entomologische Zeitschrift, (Frankfurt am Main), 60 (6): 41–45, (7): 53–56.

Miller, J. C., D. H. Janzen & W. Hallwachs (2006): 100 Caterpillars – Portraits from the Tropical Forests of Costa Rica. – 3 + 264 pp.; Cambridge, Massachusetts – London, England (The Belknap Press of Harvard University Press)

Nässig, W. A. (1980 [»1979«]): Zur Zucht von *Actias sinensis* Walker (Attacidae). – Nachrichten des Entomologischen Vereins Apollo (Frankfurt am Main), A.F. 4 (3/4): 42–48.

– (1981): Ein Beitrag zur Kenntnis der Saturniidae und der Brahmaeidae des Iran und der Türkei, Teil 2: Saturniidae. – Nachrichten des Entomologischen Vereins Apollo (Frankfurt am Main), N.F. 2 (1): 1–39.

– (1983): Die Präimaginalstadien von *Attacus caesar* Maassen im Vergleich mit einigen verwandten Arten sowie deren Wehrdrüsen (Lep.: Saturniidae). – Nachrichten des Entomologischen Vereins Apollo (Frankfurt am Main), N.F. 3 (4): 129–152.

– (1986): Einleitung – Zur Systematik und Verbreitung von *Caligula thibeta*. – Pp. 12–15 in Nardelli, U.: Zur Kenntnis der Präimaginalstadien und zur Zucht *von Caligula thibeta* Westwood, 1853 (Lepidoptera, Saturniidae). – Nachrichten des Entomologischen Vereins Apollo (Frankfurt am Main), N.F. 7 (1): 11–23.

– (1989): Wehrorgane und Wehrmechanismen bei Saturniidenraupen (Lepidoptera, Saturniidae). – Verhandlungen des 1. Westdeutschen Entomologentages, Düsseldorf, 1988: 253–264.

– (1994): Zur Präimaginalmorphologie von *Saturnia (Rinaca) zuleika* und *S. (R.) thibeta* sowie Anmerkungen zu den Salassinae (Lepidoptera: Saturniidae). – Nachrichten des Entomologischen Vereins Apollo (Frankfurt am Main), N.F. 15 (3): 409–438.

Nässig, W. A. & R. G. Oberprieler (1990): Mimetischer oder aposematischer Habitus – ein Vergleich larvaler und imaginaler Strategien bei Saturniinen (Lepidoptera, Saturniidae). – [Poster 83. DZG-Tagung 1990 in Frankfurt.] Verhandlungen der Deutschen Zoologischen Gesellschaft 83: 655.

Nässig, W. A. & R. S. Peigler (1984): The life-history of *Actias maenas* (Saturniidae). – Journal of the Lepidopterists' Society 38 (2): 114–123.

Nässig, W. A. & G. Ragus (2001): Annotations on selected *Loepa* species and their preimaginal instars. 1. The larval morphology of *Loepa miranda* Atkinson in Moore, 1865 (Lepidoptera: Saturniidae, Saturniinae). – Nachrichten des Entomologischen Vereins Apollo, (Frankfurt am Main), N.F. 21 (4): 247–251

Nässig, W. A. & F. Taschner (1996): Beschreibung einer Altraupe von *Attacus aurantiacus* Rothschild, 1895 von den Kai-Inseln, Indonesien (Lepidoptera: Saturniidae). – Nachrichten des Entomologischen Vereins Apollo, (Frankfurt am Main), N.F. 17 (2): 153–159.

Nässig, W. A. & C. G. Treadaway (1998): The Saturniidae (Lepidoptera) of the Philippines. – Nachrichten des Entomologischen Vereins Apollo, (Frankfurt am Main) N. F. Supplement 17: 223–424.

Nässig, W. A., R. E. J. Lampe & S. Kager (1996): The Saturniidae of Sumatra (Lepidoptera). – Heterocera Sumatrana, Göttingen, 10: 3–110.

Nässig, W. A., R. E. J. Lampe & S. Kager (1996): Appendix I: The preimaginal instars of some Sumatran and South East Asian species of Saturniidae, including general notes on the genus *Antheraea* (Lepidoptera). – Heterocera Sumatrana, Göttingen, 10: 111–170.

Nässig, W. A., R. Brechlin & S. Naumann (1999): Notes on the *Cricula* Walker 1855 of Vietnam, with description of a new species (Insecta, Lepidoptera, Saturniidae). – Senckenbergiana biologica 78 (1/2): 183–192.

Naumann, S. (1986): Eine Zucht von *Saturnia mendocino* Behrens, 1876 (Lep.: Saturniidae). – Entomologische Zeitschrift 96 (19): 284–287.

– (1993): Short note on rearing the ZigZag Emperor, *Gonimbrasia t. tyrrhea*. – Newsletter Exotic Ent. Group, Summer 1993: 26.

- (1994): Bemerkungen zu einer rotbraunen Form von *Antheraea brunei* Allen & Holloway 1985 sowie zu den Präimaginalstadien der Art Lepidoptera: Saturniidae). – Nachrichten des entomologischen Vereins Apollo (Frankfurt/Main), N.F. 15 (3): 359–362.
- (1995): Die Saturniiden-Fauna von Sulawesi, Indonesien. – 145 S.; Dissertation, Freie Universität Berlin.
- (2006): Notes on *Actias dubernardi* (Oberthür, 1897), with description of the early instars (Lepidoptera: Saturniidae). – Nachrichten des Entomologischen Vereins Apollo, (Frankfurt am Main), N.F. 27 (1/2): 9–16.

Naumann, S. & R. Brechlin (2001): Zwei neue Arten der Gattung *Syntherata* Maassen, 1873 von den Molukken, Indonesien (Lepidoptera: Saturniidae). – Nachrichten des Entomologischen Vereins Apollo, Frankfurt am Main, N.F. 21 (4): 237–246.

Naumann, S. & U. Brosch [compilers] with contributions by Josè Antonio Böttger Cardenas, Diegi Bonilla, Ronald Brechlin, Ulrich Brosch, Josè Clavijo, Thibaud Decaens, Max Gaede†, Frank Meister, Carlos G. C. Mielke, Stefan Naumann, Richard S. Peigler, Luigi Racheli, Luz Dary Ramirez, David Robacker, Viktor Suter, Bernhard Wenczel, & Kirby L. Wolfe (2005): Contributions to the knowledge of Saturniidae (Lepidoptera) & entomohistory – Part I–III. – galathea. Berichte des Kreises Nürnberger Entomologen e. V. (Nürnberg); Supplement 15–17: Part I: 77 pp.

Naumann, S. & W. A. Nässig (2000): A rearing of *Antheraea (Antheraea)* sp. (probably *jana* (Stoll 1782)) from Bali, Indonesia (Lepidoptera: Saturniidae). – Nachrichten des Entomologischen Vereins Apollo, Frankfurt am Main, N.F. 21 (1): 25–30.

Naumann, S., U. Brosch & B. Wenczel (with contributions of R. Brechlin & F. Meister) (2005): Some new Saturniidae species from Peru (Lepidoptera). – galathea. Berichte des Kreises Nürnberger Entomologen e. V. (Nürnberg); Supplement 16, July 2005 [available: August 22, 2005]: 81–171.

Naumann, S., W. A. Nässig & S. Löffler (2008): Notes on the identity of *Loepa katinka diversiocellata* Bryk, 1944 and description of a new species, with notes on preimaginal morphology and some taxonomic remarks on other species (Lepidoptera: Saturniidae). – Nachrichten des Entomologischen Vereins Apollo, (Frankfurt am Main), N.F. 29 (3): 149–162.

Oberprieler, R. G. (1995): The emperor moths of Namibia. – 91 pp.; Hartbeespoort (Ekogilde).

Oberprieler, R. G. & W. A. Nässig (1994): Tarn – und Warntrachten – ein Vergleich larvaler und imaginaler Strategien bei Saturniinen (Lepidoptera: Saturniidae). – Nachrichten des Entomologischen Vereins Apollo (Frankfurt am Main), N.F. 15 (3): 267–303.

Packard, A. S. (1914): Monograph of the bombycine moths of North America, including their transformations and origin of the larval markings and armature, Teil 3 (ed. T. D. A. Cockerell): Families Ceratocampidae (exclusive of Ceratocampinae), Saturniidae, Hemileucidae, and Brahmaeidae. – Memoirs of the National Academy of Sciences 12 (1): ix + 1–516.

Paukstadt, L. H. & U. Paukstadt (1992): Beschreibung der Präimaginalstadien von *Attacus inopinatus* Jurriaanse & Lindemans, 1920 von Flores, Indonesien, sowie Angaben zur Biologie und Ökologie (Lepidoptera: Saturniidae). – Entomologische Zeitschrift (Essen), 102 (12): 213–220.
- (1993): Beschreibung der Präimaginalstadien von *Attacus dohertyi* Rothschild, 1895 n. stat., von Timor, Indonesien, sowie Angaben zur Biologie und Ökologie (Lepidoptera: Saturniidae). – Entomologische Zeitschrift (Essen), 103 (16): 281–293.
- (1993): Beschreibung der Präimaginalstadien von *Actias groenendaeli* Roepke (1954) n. stat., von Flores, Indonesien, sowie Angaben zur Biologie und Ökologie (Lepidoptera: Saturniidae). – Entomologische Zeitschrift (Essen), 103 (17): 305–314.
- (1993): Beschreibung der Präimaginalstadien von *Samia yayukae* Paukstadt, U., Peigler, R. S. & Paukstadt, L. H. (1993) von Timor, Indonesien, sowie Angaben zur Biologie und Ökologie (Lepidoptera: Saturniidae). – Entomologische Zeitschrift (Essen), 103 (20): 357–366.
- (1993): Beschreibung der Präimaginalstadien von *Cricula hayatia* Paukstadt, U. & Suhardjono, 1992 von Flores, Indonesien, sowie Angaben zur Biologie und Ökologie (Lepidoptera: Saturniidae). – Entomologische Zeitschrift (Essen), 103 (23): 429–438.
- (1996): Beschreibung der Präimaginalstadien von *Antheraea rosemariae* Holloway, Nässig & Naumann 1995 von Sulawesi, Indonesien, sowie Angaben zur Biologie und Ökologie (Lepidoptera: Saturniidae). – Entomologische Zeitschrift (Essen), 106 (12): 481–487.
- (1997): Beschreibung der Präimaginalstadien von *Antheraea cordifolia* Weymer, 1906 von Sulawesi, Indonesien, sowie Angaben zur Biologie und Ökologie (Lepidoptera: Saturniidae). – Entomologische Zeitschrift (Essen), 107 (4): 133–141.
- (1997): Beschreibung der Präimaginalstadien von *Attacus crameri* C. Felder, 1861 von Ambon, Indonesien (Lepidoptera: Saturniidae). – Entomologische Zeitschrift (Essen), 107 (10): 437–447.
- (1999): Beschreibung der Präimaginalstadien von *Antheraea (Antheraea) platessa* W. Rothschild, 1903 von Nord Borneo, sowie Angaben zur Biologie und Ökologie (Lepidoptera: Saturniidae). – Galathea, Berichte des Kreises Nürnberger Entomologen e. V. (Nürnberg), Suppl. 6: 13–22.
- (2000): Beschreibung der Präimaginalstadien von *Antheraea (Antheraea) pasteuri* Bouvier, 1928 [oder eines anderen nah verwandten Taxons] von Lombok, Indonesien (Lepidoptera: Saturniidae). – Galathea, Berichte des Kreises Nürnberger Entomologen e. V. (Nürnberg), Suppl. 7: 7–18.

– (2000): Beschreibung der Präimaginalstadien von *Actias isis* (Sonthonnax, 1897 [»1899«]) von Sulawesi, Indonesien (Lepidoptera: Saturniidae). – Galathea, Berichte des Kreises Nürnberger Entomologen e. V. (Nürnberg), Suppl. 7: 37–46.
– (2000): Beschreibung der Präimaginalstadien von *Antheraea (Antheraea) kelimutuensis* Paukstadt, Paukstadt & Suhardjono, 1997 von Flores, Indonesien (Lepidoptera: Saturniidae). – Galathea, Berichte des Kreises Nürnberger Entomologen e. V. (Nürnberg), Suppl. 8: 6–21.
– (2000): Beschreibung der Präimaginalstadien von *Antheraea (Antheraea) exspectata* Brechlin, 2000 von Sulawesi, Indonesien (Lepidoptera: Saturniidae). – Galathea, Berichte des Kreises Nürnberger Entomologen e. V. (Nürnberg), 16 (3): 109–124.
– (2000): Beschreibung der Präimaginalstadien von *Antheraea (Antheraea) raffrayi* Bouvier, 1928 von Bali, (Lepidoptera: Saturniidae). – Galathea, Berichte des Kreises Nürnberger Entomologen e. V. (Nürnberg), 16 (4): 129–138.
– (2001): Beschreibung der Präimaginalstadien von *Antheraea (Antheraeopsis) youngi* Watson, 1915 von Jawa, Indonesien (Lepidoptera: Saturniidae). – Galathea, Berichte des Kreises Nürnberger Entomologen e. V. (Nürnberg), Suppl. 10: 17–33.
– (2002): Die Präimaginalstadien von *Antheraea (Antheraea) larissa* (Westwood, 1847) von Jawa, Indonesien (Lepidoptera: Saturniidae). – Nachrichten des Entomologischen Vereins Apollo (Frankfurt am Main) 23 (1/2): 59–64.
– (2003): Die Präimaginalstadien von *Antheraea (Antheraea) imperator* Watson, 1913 von Jawa, Indonesien (Lepidoptera: Saturniidae). – Entomologische Zeitschrift (Stuttgart), 113 (1): 19–23.
Paukstadt, L. H., U. Paukstadt & S. Naumann (1996): Die Präimaginalstadien von *Antheraea rumphii rumphii* C. Felder, 1861 von Ambon, Indonesien, sowie taxonomische Bemerkungen (Lepidoptera: Saturniidae). – Entomologische Zeitschrift (Essen) 106 (5): 165–176.
Paukstadt, L. H., U. Paukstadt, S. Naumann & S. Kohll (1996): Die Präimaginalstadien von *Attacus erebus* Fruhstorfer, 1904 von Sulawesi, Indonesien, sowie Angaben zur Biologie und Ökologie (Lepidoptera: Saturniidae). – Entomologische Zeitschrift (Essen) 106 (8): 297–305.
Paukstadt, L. H., S. Naumann & U. Paukstadt (1996): Beschreibung der Präimaginalstadien von *Samia vandenberghi* (Watson, 1915) von Sulawesi, Indonesien, sowie Angaben zur Biologie und Ökologie (Lepidoptera: Saturniidae). – Entomologische Zeitschrift (Essen) 106 (9): 353–359.
Paukstadt, L. H., S. Naumann & U. Paukstadt (1996): Beschreibung der Präimaginalstadien von *Loepa minahassae* Mell, 1938 von Sulawesi, Indonesien, sowie Angaben zur Biologie und Ökologie (Lepidoptera: Saturniidae). – Entomologische Zeitschrift (Essen) 106 (10): 385–391.
Paukstadt, U. & L. H. Paukstadt (1985): Beschreibung und Abbildung der Präimaginalstadien von *Attacus atlas* Linnaeus, 1758 aus malaiischen Populationen (Lepidoptera: Saturniidae). – Entomologische Zeitschrift (Essen) 95 (9): 113–121.
– (1985): Beschreibung und Abbildung der Präimaginalstadien von *Attacus atlas* Linnaeus, 1758 aus javanischen Populationen (Lepidoptera: Saturniidae). – Entomologische Zeitschrift (Essen) 95 (23): 337–346.
– (1986): Beschreibung der Präimaginalstadien von *Attacus tabrobanis* Moore, 1882 aus singhalesischen Populationen, sowie Angaben zur Biologie und Ökologie (Lepidoptera: Saturniidae). – Nachrichten des Entomologischen Vereins Apollo (Frankfurt am Main) N. F., 7 (2/3): 71–90.
– (1989): Zuchtbeobachtungen zur Ökologie und Biologie von *Attacus* – und *Archeoattacus*-Raupen (Lepidoptera: Saturniidae). – Entomologische Zeitschrift (Essen) 99 (15): 209–215.
– (1990): Beitrag zur Kenntnis der Präimaginalstadien von *Cricula trifenestrata javana* (Watson, 1913) (Lepidoptera: Saturniidae). – Galathea, Berichte des Kreises Nürnberger Entomologen e. V. (Nürnberg), 6 (3): 90–100.
– (1995): Kurzbeschreibungen und Farbabbildungen der Raupenstadien einiger wilder Seidenspinner (Lepidoptera: Saturniidae) von den Kleinen Sunda-Inseln und aus Papua Neu Guinea). – Galathea, Berichte des Kreises Nürnberger Entomologen e. V. (Nürnberg), 11 (4): 155–164.
– (2000): The Life-History of *Actias maenas diana* Maassen in Maassen [& Weymer], 1872 from the Island of Bali, Indonesia (Lepidoptera: Saturniidae). – Galathea, Berichte des Kreises Nürnberger Entomologen e. V. (Nürnberg), Suppl. 8: 26–40.
– (2003): Das *Samia kikibudiamini*-Problem (Lepidoptera: Saturniidae). – Beiträge zur Kenntnis der wilden Seidenspinner (Wilhelmshaven), 1 (1): 44–64.
Paukstadt, U. & H. Probst (1990): Eine Zucht von *Coscinocera hercules* (Miskin) aus Papua-Neuguinea (Lepidoptera: Saturniidae). – Entomologische Zeitschrift (Essen), 100 (22): 413–426.
Paukstadt, U., R. S. Peigler & L. H. Paukstadt (1993): *Samia yayukae* n. sp., eine neue Saturniide (Lepidoptera) von Flores, Indonesien. – Entomologische Zeitschrift, Essen 103 (13): 229–235.
Paukstadt, U., R. S. Peigler & L. H. Paukstadt (1998): *Samia naumanni* n. sp., eine neue Saturniide von den Banggai- und Sula-Inseln, Indonesien (Lepidoptera: Saturniidae). – Entomologische Zeitschrift, Essen 108 (3): 114–121.
Paukstadt, U., L. H. Paukstadt & Y. R. Suhardjono (2002): *Samia kikibudiamini* n. sp., eine neue Saturniide von der Insel Alor, Indonesien (Lepidoptera: Saturniidae). – Galathea, Berichte des Kreises Nürnberger Entomologen e. V. (Nürnberg), 12: 3–11.
Paukstadt, U., Y. R. Suhardjono & L. H. Paukstadt (2003): Notes on the distribution of the genus *An-*

theraea Hübner, 1819 ("1816") and of some selected hosts of the larvae of this genus in Indonesia (Lepidoptera: Saturniidae). – Galathea, Berichte des Kreises Nürnberger Entomologen e. V. (Nürnberg), Supplement 14: 25–64

Peigler, R. S. (1989): A Revision of the Indo-Australian Genus *Attacus*. – XI + 167 pp.; Beverly Hills, California (The Lepidoptera Research Foundation).

Peigler, R. S. & S. Naumann (2003): A revision of the silkmoth genus *Samia*. – 230 pp.; University of the Incarnate Word San Antonio, Texas, USA.

Pinratana, A. & R. E. J. Lampe (1990): Saturniidae. – Moths of Thailand, Volume one. – v + 47 pp.; Bangkok (Brothers of St. Gabriel in Thailand/Bosco Offset).

Plontke, R. & R. Predel (1999): Einleitung der Metamorphose durch Ecdyson-Injektion bei weiblichen Puppen des Hybriden ♂ *Smerinthus ocellata* Linnaeus, 1758 × ♀ *Smerinthus planus* Walker, 1856 (Lep., Sphingidae). – Entomologische Nachrichten und Berichte, 43 (3): 1–8.

Prashanti, M., K. Veenakumari & S. Naumann (1998): *Samia fulva* Jordan, 1911 from the Andaman Islands, India (Indian Ocean) – preimaginal instars, host plants and taxonomical notes (Lepidoptera: Saturniidae). – Nachrichten des entomologischen Vereins Apollo (Frankfurt/Main), N.F. 19 (1): 51–63.

Rebel, H. (1925): Der Ailanthusspinner, ein heimisch gewordener Gross-Schmetterling, seine Lebensweise und Zucht, Rassen, Verbreitung und Einbürgerung, sowie dessen Bedeutung als Seidenspinner. – 20 pp.; Wien (Fritz Wagner).

Renner, F., R. Prange, R. Plontke & W. A. Nässig (2006): Die Zucht des Hybriden *Actias sinensis* (Walker, 1855) Männchen × *A. dubernardi* (Oberthür, 1897) Weibchen (Lepidoptera: Saturniidae). – Nachrichten des Entomologischen Vereins Apollo, Frankfurt am Main, N.F. 27 (1/2): 29–52.

Skell, F. (1942): Beitrag zur Kenntnis der Morphologie und Biologie von *Attacus atlas* L. – Mitt. Münch. Ent. Ges., 32: 110–140.

Stone, S. E. (1991): Foodplants of the world Saturniidae. – Memoirs of the Lepidopterists' Society, 4: xvi + 186 pp.

Travassos, L. (1946): Evolução de « *Copiopteryx semiramis* » (Cramer, 1775) (Lepidoptera, Adelocephalidae). Rev. Brasil. Biol., Rio de Janeiro, 6 (1): 1–6.

Voelschow, A. (1902): Die Zucht der Seidenspinner. – 3 + 82 pp.; Schwerin.

Wenczel, B. & S. Naumann (2005): New records for the genus *Rothschildia* Grote, 1897 from Peru and surrounding countries (Lepidoptera: Saturniidae). – Galathea, Berichte des Kreises Nürnberger Entomologen e. V. (Nürnberg), Supplement 16: 189–201.

Wolfe, K. L. (1993): The *Copaxa* of Mexico and their immature stages (Lepidoptera: Saturniidae). – Tropical Lepidoptera, Gainesville, 4 (Supp. 1): 1–26

– (2005): The Saturniidae. Wild Silkmoths. Giant Silkmoth. Emperor Moths. – The author (Escondido, CA); May 20, 2005: CD-R.

Wolfe, K. L. & F. Bénéluz (1997): *Copiopteryx jehovah* and its immature stages (Lepidoptera: Saturniidae: Arsenurinae). Tropical Lepidoptera (Gainesville, FL) 8 (2): 67–70.

Wolfe, K. L. & C. Lemaire (1993): A new *Copaxa* from Guatemala (Lepidoptera: Saturniidae: Hemileucinae). – Tropical Lepidoptera (Gainesville, FL); 4 (2): 147–149.

Wolfe, K. L. & A. Pescador (1994): *Caio richardsoni*: its immature stages and natural history (Lepidoptera: Saturniidae: Arsenurinae). – Tropical Lepidoptera (Gainesville, FL), 5 (1): 29–32.

Wolfe, K. L., C. Lemaire, S. A. Amarillo & C. A. Conlan (2003): A contribution to the systematics of the *Copaxa semioculata* species-group (Saturniidae), with notes on the early stages, and a description of *Copaxa lunula*, new species. – Journal of the Lepidopterists' Society, Los Angeles, 57 (1): 54–61.

Wolfe, K. L., D. Bonilla, L. D. Ramìrez & T. Decaens (2003): Rediscovery of *Copaxa sapatoza* and revealing of its immature stages (Lepidoptera : Saturniidae), Saturniinae). – Nachrichten des entomologischen Vereins Apollo (Frankfurt am Main) N.F. 24 (3): 143–146.

Wolfe, K. L., S. Naumann, U. Brosch, B. Wenczel & W. A. Nässig (2005): "*Copaxa bella* Wolfe, Naumann, Brosch, Wenczel & Nässig sp. n." [= contribution without title]. – In Wolfe, K. L. (2005): Revision of the *Copaxa semioculata* and *Copaxa medea* groups, with corrigenda of Wolfe et al. (2003a), descriptions of three new species, and notes on their early stages (Lepidoptera: Saturniidae). – Nachrichten des entomologischen Vereins Apollo N.F. (Frankfurt/Main); 26 (3): 121–136.

Wu, Y. & S. Naumann (2006): The preimaginal instars of *Actias chapae* (Mell, 1950) (Lepidoptera: Saturniidae). – Nachrichten des Entomologischen Vereins Apollo, Frankfurt am Main, N.F. 27 (1/2): 17–21.

Yen, S. H., W. A. Nässig, S. Naumann & R. Brechlin (2000): A new species of the *miranda*-group of the genus *Loepa* from Taiwan (Lepidoptera: Saturniidae). – Nachrichten des Entomologischen Vereins Apollo, Frankfurt am Main, N.F. 21 (3): 153–162.

Index of animals / Verzeichnis der Schmetterlinge

C

D

G

H

M

P

S

T

W

X

Y

Z

Index of plants / Verzeichnis der Pflanzen

D

E

F

Q

T

Rudolf E. J. Lampe

SATURNIIDAE
of the World

Their Life Stages from the
Eggs to the Adults

Pfauenspinner der Welt

Ihre Entwicklungsstadien
vom Ei zum Falter

368 pp., 336 colour plates
(2949 photographs),
9 coloured and 2 black-and-white
figures
30.3 × 21.5 cm. Hard cover

368 Seiten., 336 Farbtafeln
(2949 Abbildungen),
9 Farb- und 2 Schwarzweiß-
abbildungen

30.3 × 21.5 cm. Hard cover

ISBN 978-3-89937-084-3
68.00 Euro

There are about 1400 species world-wide in the moth family SATURNIIDAE. For over thirty years, Rudolf E. J. LAMPE has been studying more than three hundred of them. Furthermore, he has professionally photographed the stages of development from the egg, through the larval stages and pupation, to the adult moth. The result is a unique documentation of the metamorphosis of these fascinating animals. With 2949 illustrations, the author unfolds to us the unexpected variety and beauty of Saturniid larvae.

Etwa 1400 Arten der Schmetterlingsfamilie SATURNIIDAE sind weltweit bekannt. Rudolf E. J. LAMPE hat über einen Zeitraum von 30 Jahren mehr als 300 davon beobachtet und ihre Entwicklung vom Ei über die Larvenstadien, die Verpuppung bis hin zum Falter professionell fotografiert. Daraus ist eine einzigartige Dokumentation der Metamorphose dieser faszinierenden Tiere entstanden. Anhand von 2949 Abbildungen öffnet uns der Autor den Blick für die ungeahnte Vielfalt und Schönheit der Raupen.

Verlag Dr. Friedrich Pfeil
Wolfratshauser Straße 27 · 81379 München, Germany
phone: +49 89 742827-0 · fax: +49 89 7242772 · e-mail: info@pfeil-verlag.de
www.pfeil-verlag.de

Tafel 25 Saturniinae *Argema mittrei* Guérin-Meneville, 1847 Madagaskar

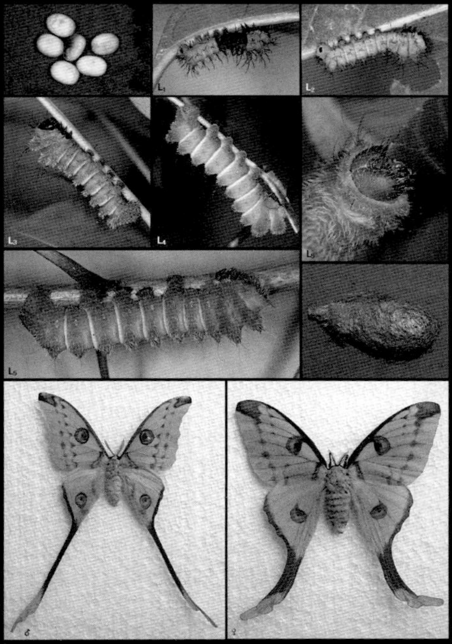

33

Tafel 136 Ceratocampinae *Eacles penelope* Cramer, 1775 Kolumbien

143

Tafel 326 Saturniinae *Rhodinia fugax* Butler, 1877 Japan

331

Tafel 40 Saturniinae *Melanocera menippe* Westwood, 1849 RSA

48